Microwave Circuit Design Using Programmable Calculators

THE ARTECH HOUSE MICROWAVE LIBRARY

Microwave Circuit Design Using Programmable Calculators

J.L. Allen, Ph.D.
Professor, Department of Electrical Engineering
University of South Florida, Tampa

M.W. Medley, Jr.
COMPACT Engineering, Inc.

Second Printing 1981

ARTECH HOUSE, INC.

Printed and bound in the United States of America.

Standard Book Number 1-58053-120-2

Library of Congress catalog card number: 80-17806

CONTENTS

CONTENTS

HP-41C COMPATIBILITY 109

EXAMPLES 111

APPENDIX A: TI-59 Program Listings

APPENDIX B: HP-67/97 Program Listings

FOREWORD

The network analysis programs presented in this book provide the user with a powerful but inexpensive and easy-to-use design and analysis tool. These programs provide the techniques which may be used for the steady-state analysis of a wide variety of complex networks, ranging from very low frequencies to within the millimeter wavelength regions, using only the personal programmable calculator. For the first time, it is possible to analyze highly complex circuits accurately and with relative ease, without resorting to large computing systems.

This book is the direct result of technological advances made over the past decade. Within this period, scientific calculators have progressed from simple machines with one data storage register and a few arithmetic functions, to machines which are programmable, contain more than 100 preprogrammed functions, have alphanumeric capability, and feature input/output ports for the attachment of peripheral devices. The programmable calculators are no larger than the early nonprogrammable scientific calculators and, in most cases, are less expensive. Furthermore, the computing power and ease of use of this class of calculators is far greater than that of the first large digital computers.

During this same time frame, network analysis techniques have been improved, while average circuit complexity increased significantly. A decade ago, very few transistors were furnished with S-parameter data or had the capability to operate at frequencies higher than a few GHz. Today, most RF transistors are supplied with this data and transistor amplifiers are routinely used through 18GHz. As the circuit complexity has increased, so has the need for more accurate analysis techniques. Stringent design requirements for today's "super components" have forced designers to use more accurate circuit models and faster design techniques. The simplifying assumptions which once enabled the designer to calculate with "slide rule accuracy", or using a simple calculator, are no longer acceptable. Many Computer Aided Design (CAD) programs have been developed as a result of these requirements. However, most of these rely on large computer systems and are expensive to use, but they do provide the necessary speed and accuracy.

The necessity of increased reliance on expensive, large-scale CAD programs has created a need for an alternative analysis approach. Fortunately, the evolution of the personal programmable calculator provides the means to achieve this goal. The network analysis programs described here for the HP-67, HP-97, HP-41C, and TI-59 calculators do represent a viable alternative to complete reliance on large computer systems. The package developed for the Hewlett-Packard series of Reverse Polish Notation calculators consists of over 5,000 program steps, while the one developed for the Texas Instrument Algebraic Operating System machine contains over 10,000 program steps, including various routines used with the compatible printer. These "pacs" were developed with the philosophy that they should require minimum user interaction, use standard program units, and operate similar to the large-scale CAD programs which they are meant to supplement or replace.

The programmable calculator user has the option of using any of five electrical parameter sets for the analysis. A given two-port network may be characterized to determine maximum available gain, matched source and load impedances, or stability of information. Two-ports may be created using the standard library of circuit elements, or interconnected in any of six ways. In most cases, the user need only load the desired program, enter the necessary data, and begin execution. Data storage and manipulation is automatic. Two-port analysis is accomplished by interconnecting the desired two-ports, in the proper order, at a single analysis frequency, and requires user interaction to initiate each calculation. A ladder analysis program is included which allows the user to program a topology consisting of lumped and distributed elements, and then analyze the network over a specified frequency band without any interaction.

This group of programs rivals the analysis capability of many large-scale CAD programs. A major advantage of these programs is their ease of use during the initial design phases in setting design goals and defining the network topology. A large-scale CAD program can then be used to analyze and optimize the complete circuit model, which has been previously defined by the calculator routines. A significant cost savings results. Alternatively, the user who does not have access to a large-scale CAD program can use these programs for the complete design, resulting in more accurate analysis and considerable time savings, both in the initial analysis and in avoiding empirical circuit modifications which might result from inaccurate, or nonexistent, analysis.

INTRODUCTION

The introduction of the scientific programmable calculator has made a significant impact on the computational techniques used by engineers and scientists. It is no longer necessary to rely on large computer systems, with their long turn-around times, for calculations of moderate complexity. Instead, the user who is willing to make a minimal investment can use one of these calculators. Although the execution times for a large computer system are much faster, immediate access and direct interaction can be a distinct advantage when using a programmable calculator. The basic philosophy of programming and user interaction for the programmable calculator, however, is quite different from the approach which is usually taken with a larger computer system.

When using a computer, there is a tendency to be rather inefficient when generating a program for a specified task. Similar computations are made at several different points in a program; variables are dimensioned and used to compute a single answer, and then not accessed for the remainder of the program. Format statements abound to an extent that the printout contains every program variable in a form suitable for framing, although only one variable is of temporary interest. This tendency only confirms the old adage that "every computer program will expand to meet the available memory". The programmable calculator user, on the other hand, must adopt a radically different philosophy, due to the limited computer resources available. The HP-41C and TI-59 calculators allow the user to allocate a portion of the total memory to program storage and the remainder to data storage, while the HP-65 and HP-67 have a fixed amount of memory for each. Although the programs can be "chained", each portion of a program must reside in memory with all the subroutines and data which it will access.

Limited program and data memory are not the only constraint imposed on the user; execution speed must be considered as well. Execution speed and the number of program steps are somewhat synonymous. In general, larger programs will take longer to run than programs with fewer lines of instructions. However, there are other factors which must be considered when programming a calculator. The basic calculator architecture and function algorithms will play an important role in the structure of an efficient program. For example, the label searching scheme of a particular machine has a definite bearing on the program structure. In the case of a calculator which starts at the top of program memory when searching for a label called as a subroutine or an unconditional branch, the most frequently used routines should be placed in this portion of memory. The HP-41C has several different types of labels, which means that the label type, as well as its location in memory, will influence execution speed. Most of the early programmable calculators had a very limited memory, which forced programmers to write subroutines for any series of program steps which could be used more than once and which would save overall program steps. Many hours, and even days, have been spent reducing a program by even a single program step. However, newer calculators have more merged instructions and program steps, which enables programmers to be more selective about which portions of a program should be accessed as a subroutine, and in the process, they can decrease the run time.

The various preprogrammed functions are actually microprograms which execute an algorithm for the desired function. Most of these functions are extremely slow in comparison with a typical program step because of the number of instructions which must be executed in the microprogram. Whenever possible, the programmer should take this into consideration and seek ways to either minimize the number of times a function is used, or use another function to obtain the desired results. A program which uses the COS of an angle several times, can be improved by computing the function once, storing the results, and then recalling whenever necessary. Another possibility is to mathematically manipulate an expression into another form. An expression which contains the SIN in the numerator and the COS in the denominator can be reduced to the single TAN function, rather than the two trigonometric functions. A complex number in polar form can be easily multiplied or divided by another complex number, but must be converted to rectangular form for addition and subtraction. Likewise, multiplication and division are not straightforward processes for complex numbers in rectangular form. Since either form can be retained in data memory with equal ease, and since the polar-to-rectangular and rectangular-to-polar conversions are two of the slower routines, the number of multiplications and divisions versus the number of additions and subtractions should be used to determine the complex form. All complex numbers are stored in polar form for this series of programs.

When programming a calculator, one point which is often overlooked is that the basic mathematical expressions which describe a particular problem are perhaps the single factor which has the most effect on the structure, size, and execution speed for a given program. An example published by one of the authors[1] serves to illustrate that minor changes in mathematical form can save several program steps. A portion of the equation which describes the effects of dispersion for a microstrip line is:

$$1 + G(f/D)^2$$

where $G = .6 + .009z,$

and $f/D = 20.32\pi hf/z$

Using the HP-65, the number of program steps required to enter the constants in the expression for "G" is 6, since the decimal point and each digit require one program step apiece. However, "G" can be multiplied by 100, thus

$$100G = 60 + .9z$$

The constants now require four program steps. The expression for f/D can be divided by 10, changing 20.32 to 2.032, so that when it is squared and multiplied by 100G, the desired result is obtained, in two less program steps.

Another set of equations describes the microstrip impedance as a function of the ratio of w/h:

$$z = \frac{60 \ln \{8h/w + w/4h\}}{\sqrt{\epsilon_{eff}}}$$

for $w/h \leqslant 1$, and

$$z = \frac{120\pi}{\sqrt{\epsilon_{eff}}\,\{w/h + 2.42 - .44\, h/w + (1 - h/w)^6\}}$$

for $w/h > 1$. The two equations have common coefficients, 60 and $\sqrt{\epsilon_{eff}}$, so that they can be written as

$$z = \frac{60P}{\sqrt{\epsilon_{eff}}},$$

and "P" will be calculated using the appropriate expression, depending upon which region w/h is in. "P" for $w/h \leqslant 1$ is first calculated without regard to the value of w/h, then 1 and w/h are entered in preparation for the calculation of the term $(1-h/w)^6$ for the case where $w/h > 1$. At this point a test is made to determine whether or not $w/h \leqslant 1$. If it is, a transfer is made to the point in the program where the multiplication by 60 and the division by $\sqrt{\epsilon_{eff}}$ is made. If the test is not true, the 1 and w/h are used to calculate this term along with the remaining terms of "P" for $w/h > 1$. The conclusion of this calculation ends at the common program step which results in the value of z.

[1]M.W. Medley, "Calculate Dispersive Microstrip Line Width in Minutes!," Microwaves, pp. 66-71, July 1978.

The techniques demonstrated above do, in fact, save several program steps. Notice that even the coefficients used in the test for w/h ≤ 1 are used, as entered; if the test fails, w/h is inverted, subtracted from 1, and the results raised to the sixth power. These manipulations are dictated both by the calculator structure and the mathematics of the problem. This program was written for the first programmable calculator, the HP-65. This calculator uses the Reverse Polish Notation (RPN) architecture, which requires the user to enter data into a four level "stack" before the operation is specified. The RPN allows data to be manipulated as above, and has the capacity to store data in the stack until needed. Actually, in the above program example, the term "60P" is carried in the stack for almost 50 program steps and used twice, without data store or recall statements. The HP-65 is limited to 100 program steps, and many of the label statements and arithmetic functions require two program steps. The problem for which this program was written cannot be solved directly — the solution must be obtained by iterative techniques. This requirement forces the programmer to fit the entire program within the allowed 100 program steps, so that an error function may be used to generate a new trial value for the variable, and the program loop used to generate a new calculation without operator intervention. The same program written for either an RPN calculator with more program memory, or an Algebraic Operating System (AOS) would be structured quite differently; emphasis could be placed on execution speed since there must be several iterations to determine the desired answer.

NETWORK ANALYSIS AND DESIGN USING THE HP—67/97 AND THE TI-59 PROGRAMMABLE CALCULATORS

The group of programs which are described in the following chapters were originally intended to be used in conjunction with a time-shared computer system for the design of microwave circuits, hence, the labels "MP" (Microwave Pac) for all programs. Computer-aided microwave design programs were being used on a time-shared basis for all phases of the design effort. This approach is time-consuming and expensive in the early design stage, where relatively simple calculations are required. For example, the initial stages of an amplifier design require the calculation of maximum available gain, or maximum stable gain if the device is not stable, at several frequencies across the band of interest. The next step would be to construct contours of constant gain on the load or source plane, along with the stability circles and noise contours, if noise is of interest, which then allows the designer to determine various matching networks. All of these calculations, which involve complex numbers, are of moderate complexity, but can be solved using a scientific calculator. However, due to the number of steps involved for each calculation, not to mention the frequency requirement, this approach is time-consuming and subject to many sources of error.

When Hewlett-Packard introduced the first personal scientific programmable calculator, the designer was provided with an attractive alternative to the complete reliance on computer-aided design (CAD) using large time-shared systems. For the first time, a designer could program complex calculations into a personal computing system and solve involved problems virtually error-free. In the process, the designer who is willing to make a minimal investment becomes more valuable to his employer due to increased efficiency, both in time and cost. Another, and perhaps more important, benefit is that programs, like text books, stay with the designer and become part of his experience and knowledge. The research and time necessary to solve difficult problems need only be done once; the solution can be programmed and used years later without additional time and effort.

The original purpose of this Microwave Pac was to calculate initial design parameters similar to those above. This soon expanded to include three-port mappings, since these could be done with the calculator almost as quickly as with a larger computer. Finally, as with most programs, the total analysis capability of this group of programs reached the point where it now rivals that of the CAD programs which it was meant to supplement. The first Microwave Pac was written for the HP-65, which was limited to 100 program steps and 9 data storage registers, or in certain cases 8, because some of the arithmetic functions used one of these registers. Since the analysis programs would be required to calculate and interconnect two-ports, the first programming problem which had to be solved was how to store 8 complex numbers, 16 numbers total (including real and imaginary or magnitude and angle) in 8 storage registers. This hurdle was overcome by devising a split register technique which could represent a complex number in polar form. This was realized by using the integer portion of the number as the magnitude, and the fraction as a positive angle between 0 and 360 degrees. Thus, the integer portion could be split from the entire number and divided by a constant to produce the magnitude, then the fractional portion could be multiplied by another constant to yield the angle. Some thought had to be given to the value of these constants, since they determine the range of values which can be represented. This technique works well and can represent the desired range of complex numbers; however, two additional subroutines must be included along with the complex arithmetic subroutines, which uses a substantial portion of the 100 program steps. Another problem which was encountered was storage for intermediate calculations. The order in which calculations were made had to be carefully thought out, and all data registers, including the stack, were used to their capacity. The resultant Microwave Pac for the HP-65 did, however, represent a vast improvement over previous methods, and was found to be a valuable asset during the several years it was in use.

The second generation of personal programmable calculators, which includes the HP-67/97 and TI-59, has provided the user with even more computing power at a significant cost reduction. The Microwave Pac designed to be used with these machines is much easier to use, and the execution time less than the original. It is no longer necessary to split registers using special subroutines, which consume program memory and increase run time. More programs can be contained on a single magnetic card, which reduces run time even further and offers more convenience to the user. As of this writing, the third generation calculator has been introduced by the Hewlett-Packard Company, the HP-41C. This calculator features alpha-numeric capability, variable program memory / data memory allocation (as with the TI-59), and memory expansion to over 2000 program steps (or bytes as they are called with this calculator). The possibilities are exciting: it may be possible to program a network and have the calculator analyze the entire circuit over several frequencies without user interaction. Network synthesis and optimization are also very real possibilities. Work is presently underway to develop this series of programs.

The network analysis programs contained in this book were intended to be used in conjunction with a larger CAD program, and should be used in this manner for maximum efficiency. In fact, the most cost-effective approach is to take advantage of recent developments for mini and microcomputers and use an analysis and optimization program based on one of these machines, which affords the user virtually unlimited execution time with no cost penalty. However, each user must determine the best point to switch from the calculator to a computer — although it is usually easier to analyze the maximum performance which may be obtained from a particular device or network, set design goals, and do the preliminary design using a calculator. A computer is then used to model the network, including parasitics, for complete analysis and optimization. The user who does not have access to a larger computing system can certainly use these programs for the complete design, and although this is somewhat time-consuming, it is certainly faster than the cut-and-try approach, or more accurate than simplifying assumptions.

The Microwave Pac is organized in such a fashion as to require minimum interaction and provide the user with a powerful, easy-to-use design tool. Circuit analysis using the programmable calculator is as easy to implement as most CAD programs. The network analysis and design programs described here for the HP-67/97 and TI-59 calculators include most of the features found in large CAD programs. All of the two-port interconnections, cascade, series, parallel, series-parallel, and parallel-series are included, as well as a special hybrid interconnection which is useful for amplifier or attenuator designs. A complete library of network elements is provided, which allows the user to easily analyze circuits containing lumped and distributed elements as well as special two-ports. Some of the standard elements are series and parallel combinations of resistors, inductors, capacitors, transformers, and transmission lines, including open and shorted lines connected in series and parallel. Additional two-ports which are provided include current and voltage dependent current and voltage sources, one-port networks, either measured or calculated, connected in series and parallel, and coupled transmission lines. One complete section of the Microwave Pac is dedicated to two-port characterization. An active or passive two-port may be characterized for maximum available gain, or maximum stable gain, matched source and load, and stability factor. This section can also be used to rapidly construct stability circles, contours of constant gain, and mismatch circles. The programs found in the mapping section may be used to map any parameter as a function of series or shunt feedback impedance, or one may easily calculate a mapping of the two-port input or output plane as a function of the entire load or source impedance plane, and device data measured in one configuration can be converted to another. The final program is useful for ladder analysis. A matching network may be analyzed for input impedance by inputting the element value and executing the program corresponding to the appropriate element; however, the topology may also be programmed and the network analyzed over frequency in the same manner as any CAD program.

Circuit analysis is implemented using two-port scattering parameter network theory. The scattering parameters are easy to measure from low VHF through the millimeter wavelengths and, unlike other parameters, provide direct insight into circuit performance. The reference impedance which is used for all calculations is 50 ohms, the normal RF measurement equipment reference; however, ideal transformers may be used to convert to any other desired real reference impedance, for example the 300 or 75 ohm reference associated with television transmission. A parameter conversion program which is used to convert between "S", "Y", "Z", "H", and "G" is included for low frequency analysis, where the S-parameters are not normally used, or where the user prefers some other parameter set. All internal calculations are made using S-parameters; however, this is not apparent to the user except for output data.

Most networks can be analyzed simply by storing the analysis frequency, loading the magnetic card which contains the desired circuit element, entering the element value, and executing the appropriate program to create the S-parameter matrix. The desired interconnection program is loaded and executed, which computes the new S-matrix as the specified connection of two two-ports, and overwrites the previous results. Another network element may now be calculated and interconnected in the same manner, without any user interaction other than loading and executing the desired program. The analysis frequency remains unchanged by all of the programs. At any given time, the calculator may contain a maximum of two two-ports, Two-Port A and Two-Port B. Two-Port A contains the results of the interconnection and is used as the input to the characterization and mapping programs; Two-Port B always contains the S-parameters of the calculated circuit element. The characterization programs do not modify the contents of Two-Port A, and may be used at any time to determine the two-port characteristics.

As an example of the use of Two-Ports A and B, consider the network analyzed below. Recall that all elements are stored in Two-Port B; therefore, the S-parameters for the device must be loaded in Two-Port A. A data entry program is provided for this purpose. The program which contains a series inductor

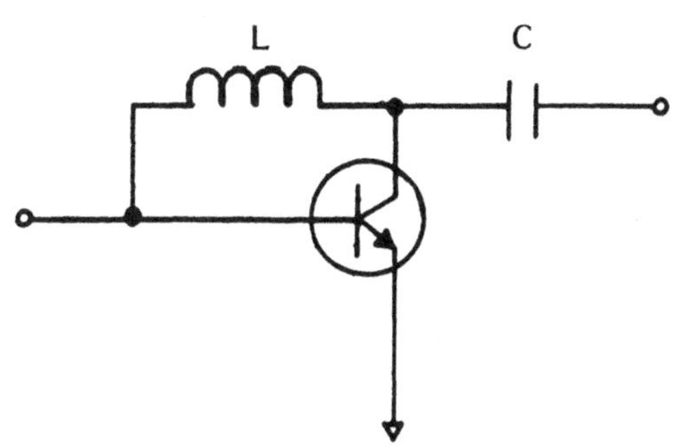

is loaded, the value of the inductor entered, and the program executed. At this point, Two-Port A contains the device parameters, while Two-Port B contains the s-parameters for the specified series inductor. The parallel interconnection is loaded and executed with the results shown. Two-Port A now

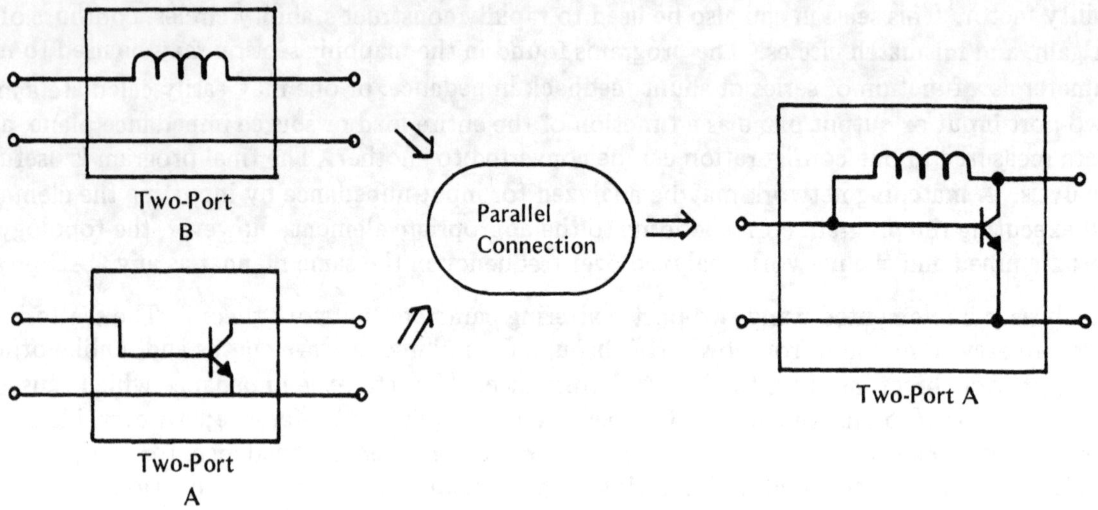

contains the parameters for the device and the inductor. The capacitor is now calculated in a similar fashion as Two-Port B, and the cascade interconnection is used to connect Two-Port A and B. After the final connection, Two-Port A contains the desired network. The magnitude of the four S-parameters, S_{11}, S_{12}, S_{21}, S_{22}, may now be recalled from Two-Port A to determine the input match, reverse isolation, gain, and output match, respectively. Or, the two-port characterization programs may be used to obtain desired information without changing the value of the resultant S-matrix.

Two-Port A or B may also be saved onto a data card, or loaded from a data card at any time. This feature is especially useful for device S-parameters which are often used, and is also necessary for cases where more than two two-ports are required. The complete network for the above example would contain an input matching network as well as the circuit shown. The input matching network can be analyzed by creating Two-Port B for the desired element and interconnecting to form the resultant input network as Two-Port A. Note that the first element must be calculated, placed in Two-Port A, then the second element calculated as Two-Port B before the first interconnection. Another program is provided to exchange the contents of A and B. The input matching network may be saved on a data card while the parallel combination is calculated, the resultant Two-Port A placed in Two-Port B, and the input network reloaded as A and cascaded. The parallel combination may also be calculated first, saved as Two-Port B, and loaded following the last interconnection of the input matching network.

The details of the actual location of Two-Port A and B, as well as the specific data entry method and method of program execution, can be found in the section applicable to the desired calculator and program. The user may easily create special two-port elements for a particular application. The two-port S-matrix must be calculated without modification of the data registers which contain Two-Port A and the analysis frequency. The results should be stored in Two-Port B in order to be compatible with the remaining programs. The complex arithmetic subroutines which are contained in most of the programs may be used for these calculations. The subroutines for the TI-59 are labeled "+", "—", and "x". These routines are called with the register number for the magnitude of the first complex number in "t", and the second in "x". The results are returned as magnitude and angle in "x" and "t", respectively. Note that data registers 1 through 4 are used by these subroutines, and that the angle of the complex number is in the next highest numbered data register. The HP-67/97 subroutines can be used by loading the two complex numbers in the stack before a call to the desired subroutine. The first complex number should be placed, magnitude and angle, in "x" and "y", and the second in "z" and "t". The results are returned as magnitude and angle in "x" and "y", respectively. These routines are not always labeled the same; however, the complex addition is label 8 (program step 89), and multiplication is label 9 (step 102), in program MP-21, cascade connection.

The programs are divided into seven broad categories, which makes it easier to find a particular program. A technique which has been found to work well for the organization of a large number of programs is to use a 3 x 5 inch card file. A pocket is made from clear mylar to allow the user to see a particular card. Each 3 x 5 card is numbered, and all information pertaining to the use of the program is printed on the card. Colored dividers are used to separate the various program segments, MP-00, MP-10, etc. Programs are numbered within the broader category — for example, MP-23 (Hybrid connection) is under the MP-20 section (Interconnection and Parameter Conversions). A decimal number is used when more than one magnetic card is required for a particular program. For example, MP-54.1 is the first card, and MP-54.2 is the second card of the Three-Port Mapping Program.

The next two chapters give detailed instructions for the purpose and method of execution for each program, as well as numerical examples which may be used for program verification. Calculator keystrokes are indicated by the function printed on the key (or above and below if a shifted function). For example, [R/S] would require the user to press this key, which would start or continue program execution. Data registers are indicated by "Reg (00)". The final chapter contains examples which will serve to clarify the concepts involved in the use of these programs for network analysis and design.

PROGRAM UNITS

Frequency	GHz
Resistance	Ohms
Capacitance	Picofarads
Inductance	Nanohenrys
Length	Inches
Reference Impedance	50 Ohms

TI-59

INTRODUCTION

The network analysis programs for the TI-59 are structured to use 60 registers for program memory and 60 registers for data storage, which is the initial partition each time the calculator is turned on. Side one of each program card is always loaded into Bank 1, program memory locations 000 through 239; and side two is loaded into Bank 2, program memory locations 240 through 479. In the case of a program which requires more than 480 program steps, card 1 is loaded and executed; then the user is prompted to load card 2, which over-writes the first portion of the program without modification of the data registers.

Data storage registers (00) through (29), Bank 4, always contain the current frequency, Two-Port A and Two-Port B, as shown below:

Reg (00) Frequency

TWO-PORT A		TWO-PORT B	
Reg (14)	$\lvert S_{11} \rvert$	Reg (22)	$\lvert S_{11} \rvert$
Reg (15)	Θ_{11}	Reg (23)	Θ_{11}
Reg (16)	$\lvert S_{12} \rvert$	Reg (24)	$\lvert S_{12} \rvert$
Reg (17)	Θ_{12}	Reg (25)	Θ_{12}
Reg (18)	$\lvert S_{21} \rvert$	Reg (26)	$\lvert S_{21} \rvert$
Reg (19)	Θ_{21}	Reg (27)	Θ_{21}
Reg (20)	$\lvert S_{22} \rvert$	Reg (28)	$\lvert S_{22} \rvert$
Reg (21)	Θ_{22}	Reg (29)	Θ_{22} .

The remaining data registers of Bank 4 are used as scratch data memory; that is, they contain intermediate program calculations. Several of the more complex programs use data registers in Bank 3 for temporary data storage; however, Bank 3 is primarily used to save and load two-port data from data cards without modification of the current values of either Two-Port A or B. In the course of normal program execution, the S-parameters for any element are stored as Two-Port B, then interconnected with Two-Port A and the results placed in Two-Port A. This approach allows a complex network to be analyzed with minimum user interaction.

Complete program listings for all TI-59 network analysis programs can be found in Appendix A. A suggested card labeling scheme is also shown with each program. This labeling should be sufficient to prompt a user who is moderately familiar with these programs. The complex subroutines which are used by most of the programs are located in program memory locations 000 through 084. It is suggested that these steps be programmed on a separate card and loaded into program memory before the remaining program steps are entered. To verify the existence of this set of subroutines, check the location of three labels. LBL "+" is at 000 and 001, LBL "–" at 004 and 005, LBL "x" at 052 and 053, while the RTN is located at memory location 084. The user should also note the cards MP-54.2 and MP-55.2 are identical, which allows programs MP-54 and MP-55 to stand alone.

The remainder of this section explains in detail the description and use of each program, as well as a numerical example of each function — which is intended to both familiarize the user with the program and to verify correct programming. The final section contains an example of the analysis of a complex network which will demonstrate the power of this set of programs.

MP-00 MISCELLANEOUS

These three programs provide the user with an easy way to enter or recall data, as well as some miscellaneous relations which are often used for circuit analysis and design.

MP-01 Data Input/Output

This program is used to enter or recall two-port data, and to interchange Two-Port A and Two-Port B, as well as to record and load data from a data card. MP-01 expects normalized parameter input. The parameters can be any which are used by the programs; however, MP-02 should be used to enter or recall non-normalized data.

ENTER TWO-PORT A	: Run [A] at "11", [CLR], enter $\lvert S_{11} \rvert$, [x⇌t], enter Θ_{11}, [R/S] at "12", [CLR], enter $\lvert S_{12} \rvert$, [x⇌t], enter Θ_{12}, [R/S] at "21", [CLR], enter $\lvert S_{21} \rvert$, [x⇌t], enter Θ_{21}, [R/S] at "22", [CLR], enter $\lvert S_{22} \rvert$, [x⇌t], enter Θ_{22}, [R/S] at "99" Program Stop, data in Reg (14) thru Reg (21) .
ENTER TWO-PORT B	: Run [B] at "11", [CLR], enter $\lvert S_{11} \rvert$, [x⇌t], enter Θ_{11}, [R/S] at "12", [CLR], enter $\lvert S_{12} \rvert$, [x⇌t], enter Θ_{12}, [R/S] at "21", [CLR], enter $\lvert S_{21} \rvert$, [x⇌t], enter Θ_{21}, [R/S] at "22", [CLR], enter $\lvert S_{22} \rvert$, [x⇌t], enter Θ_{22}, [R/S] at "99" Program Stop, data in Reg (22) thru Reg (29) .
INTERCHANGE A & B	: Run [C] at "99" Program Stop, Reg (14) thru Reg (21) interchanged with Reg (22) thru Reg (29) .
LOAD B	: Run [D] load data card which was saved using MP-01 [E] at "99", data in Reg (22) thru Reg (29) .
SAVE A	: Run [E] load data card which will contain Two-Port A at "99", Two-Port A stored on data card, Reg (14) thru Reg (21) unchanged.

LIST A	: Run [A'] at stop, $\|S_{11}\|$ in display, [x⇌t] , Θ_{11} in display [R/S] , at stop, $\|S_{12}\|$ in display, [x⇌t] , Θ_{12} in display [R/S] , at stop, $\|S_{21}\|$ in display, [x⇌t] , Θ_{21} in display [R/S] , at stop, $\|S_{22}\|$ in display, [x⇌t] , Θ_{22} in display [R/S] , at "99", Program Stop.
PRT A	: Run [E'] set print flag run [A'] program requires no other user interaction.
LIST B	: Run [B'] same execution as LIST A above.
PRT B	: Run [E'] set print flag run [B'] program requires no other user interaction.

Program verification can be accomplished by entering a known set of numbers using the appropriate label, then recalling this data from the corresponding registers.

Enter the following using [A] : $A_{11} = 1\angle 2$ $A_{12} = 3\angle 4$
$A_{21} = 5\angle 6$ $A_{22} = 7\angle 8$.

Note that the convention, $A_{11} = 1\angle 2$, is used to indicate the complex number associated with S_{11} of Two-Port A. The "1" is the magnitude of S_{11}, $|S_{11}|$; while the "2" is the angle, or Θ_{11}.

Enter the following using [B] : $B_{11} = 9\angle 10$ $B_{12} = 11\angle 12$
$B_{21} = 13\angle 14$ $B_{22} = 15\angle 16$.

Recall registers (14) through (21) , which correspond to Two-Port A. The results should be the numbers 1 through 8, corresponding to registers (14) through (21) , respectively. Registers (22) through (29) , Two-Port B, should correspond to the numbers 9 through 16. Use [A'] and [B'] to list Two-Ports A and B, the results should be the same as the data entered. If a printer is available, [E'] [A'] and [E'] [B'] can be used to print Two-Ports A and B.

Run [C] , interchange Two-Ports A and B, then recall the data as above. The results should be interchanged, that is the numbers 1 through 8 should now correspond to Two-Port B, while the numbers 9 through 16 should correspond to Two-Port A. Run [C] again to restore Two-Port A and B.

Run [E] to save Two-Port A on a data card. Clear the data memory using the function [2nd] [CMs] . Run [D] to load the saved data into Two-Port B. Recall Two-Port B as above and ensure that the numbers 1 through 8 are loaded into the corresponding registers. Should a problem exist with these two functions, again clear the data memory and load the data card into Bank 3 using keyboard commands. The numbers 1 through 8 should be found in registers (30) through (37) , which indicates that the SAVE A, [E] , function is correct.

MP-02 Parameter Normalization

This program is used to enter and recall non-normalized Z, Y, H and G parameters into Two-Port A. All network analysis is performed using S-parameters; therefore, the MP-25, Parameter Conversions, program should be used along with MP-02 to convert to or from S-parameters.

ENTER Z-PARAMETERS : Run [A]
at "11", [CLR], enter $|Z_{11}|$, [x⇌t], enter Θ_{11}, [R/S]
at "12", [CLR], enter $|Z_{12}|$, [x⇌t], enter Θ_{12}, [R/S]
at "21", [CLR], enter $|Z_{21}|$, [x⇌t], enter Θ_{21}, [R/S]
at "22", [CLR], enter $|Z_{22}|$, [x⇌t], enter Θ_{22}, [R/S]
at "99", Program Stop, data in Reg (14) thru Reg (21)

RECALL Z-PARAMETERS : Run [A′]
at "11", [R/S] at stop, $|Z_{11}|$ in display, [x⇌t], Θ_{11} in display, [R/S]
at "12", [R/S] at stop, $|Z_{12}|$ in display, [x⇌t], Θ_{12} in display, [R/S]
at "21", [R/S] at stop, $|Z_{21}|$ in display, [x⇌t], Θ_{21} in display, [R/S]
at "22", [R/S] at stop, $|Z_{22}|$ in display, [x⇌t], Θ_{22} in display
[R/S] at "99" Program Stop.

PRINT Z-PARAMETERS : Run [E′], then [A′].

ENTER Y-PARAMETERS : Run [B]
(in mmho) program entry same as LBL [A] above.

RECALL Y-PARAMETERS : Run [B′]
(in mmho) program output same as [A′] above.

PRINT Y-PARAMETERS : Run [E′], then [B′].

ENTER H-PARAMETERS : Run [C]
program entry same as LBL [A] above.

RECALL H-PARAMETERS : Run [C′]
program output same as [A′] above.

PRINT H-PARAMETERS : Run [E′], then [C′].

ENTER G-PARAMETERS : Run [D]
program entry same as LBL [A] above.

RECALL G-PARAMETERS : Run [D′]
program output same as [A′] above.

PRINT G-PARAMETERS: Run [E′], then [D′].

Program verification is accomplished by entering a particular set of values for each of the parameter sets. In each case the normalized parameters are loaded into Two-Port A. The values shown below will load the numbers 1 through 8 into registers (14) through (21), respectively. Data memory should be cleared each time to ensure that the new parameters are actually stored in data memory. Recall of the parameters should correspond to the non-normalized values entered. The following parameters should be entered:

$Z_{11} = 50\angle 2$ $\quad$ $Z_{12} = 150\angle 4$

$Z_{21} = 250\angle 6$ $\quad$ $Z_{22} = 350\angle 8$

$Y_{11} = 20\angle 2$ $\quad$ $Y_{12} = 60\angle 4$

$Y_{21} = 100\angle 6$ $\quad$ $Y_{22} = 140\angle 8$

$H_{11} = 50\angle 2$ $\quad$ $H_{12} = 3\angle 4$

$H_{21} = 5\angle 6$ $\quad$ $H_{22} = 0.14\angle 8$

$G_{11} = 0.02\angle 2$ $\quad$ $G_{12} = 3\angle 4$

$G_{21} = 5\angle 6$ $\quad$ $G_{22} = 350\angle 8$

MP-03 MISCELLANEOUS RELATIONS

This program contains often-used relations between reflection coefficient and impedance, admittance, VSWR, return loss, and reflection loss. Equivalences between series and parallel loads are also included.

REFLECTION COEFFICIENT ⇒ IMPEDANCE, Γ⇒Z

Enter |Γ|, [x⇌t], enter Θ run [A]
at Program Stop, |Z| in display, [x⇌t], Θ in display.

IMPEDANCE ⇒ REFLECTION COEFFICIENT, Z⇒Γ

Enter |Z|, [x⇌t], enter Θ run [A′]
at Program Stop, |Γ| in display, [x⇌t], Θ in display.

REFLECTION COEFFICIENT ⇒ ADMITTANCE, Γ⇒Y (mmho)

Enter |Γ|, [x⇌t], enter Θ run [B]
at Program Stop, |Y| mmho in display, [x⇌t], Θ in display.

ADMITTANCE ⇒ REFLECTION COEFFICIENT, Ymmho ⇒ Γ

Enter |Y| mmho, [x⇌t], enter Θ run [B′]
at Program Stop, |Γ| in display, [x⇌t], Θ in display.

RETURN LOSS ⇌ REFLECTION COEFFICIENT, RTN LOSS ⇌ Γ

Enter return loss in dB run [C]
at Program Stop, |Γ| in display

or enter |Γ| run [C]
at Program Stop, return loss (dB) in display.

REFLECTION LOSS ⇌ REFLECTION COEFFICIENT, RFL LOSS ⇌ Γ

Enter reflection loss in dB run [C′]
at Program Stop, |Γ| in display

or enter |Γ| run [C′]
at Program Stop, reflection loss (dB) in display.

VSWR ⇌ REFLECTION COEFFICIENT, VSWR ⇌ Γ

Enter VSWR, run [D]
at Program Stop, |Γ| in display

or enter |Γ|, run [D]
at Program Stop, VSWR in display.

MILLIWATT ⇒ dBm, MW ⇒ dBm

Enter MW, run [D′]
at Program Stop, dBm in display.

SERIES ⇒ SHUNT EQUIVALENT LOAD Rs±jXs ⇒ Rp±jXp

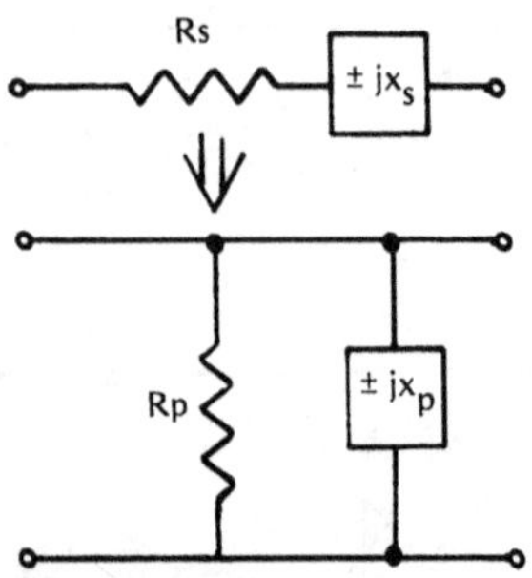

Enter Rs, [x⇌t], enter Xs run [E]
at Program Stop, Rp in display, [x⇌t], Xp in display.

SHUNT ⇒ SERIES EQUIVALENT LOAD Rp±jXp ⇒ Rs±jXs

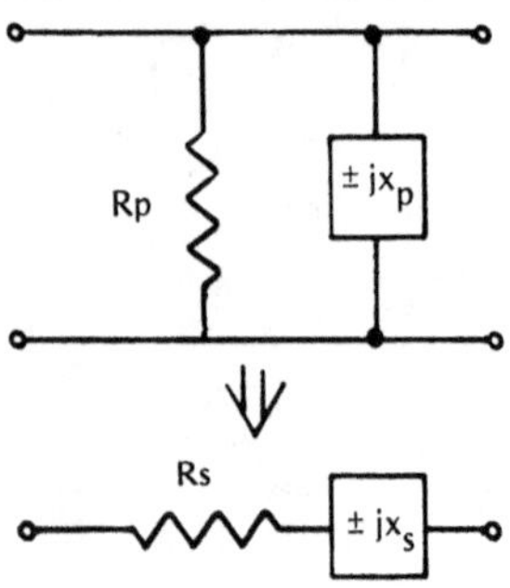

Enter Rp, [x⇌t], enter Xp run [E′]
at Program Stop, Rs in display, [x⇌t], Xs in display.

The following can be used for program verification.

$\Gamma = 0.5\angle 45$, $Z = 94.93362329\angle 43.31385666$
$= 69.07435698 + j65.1239283$ ([2nd] [p→R])

$\Gamma = 0.5\angle 45$, $Y = 10.53367569\angle -43.31385666$ mmho
$= 7.664374852 - j7.226041909$ mmho ([2nd] [p→R])

$|\Gamma| = 0.5$, RTN Loss = −6.020599913 dB

$|\Gamma| = 0.5$, RFL Loss = −1.249387366 dB

$|\Gamma| = 0.5$, VSWR = 3:1

100 MW, dBm = 20

Rs + jXs = 69.07435698 + j65.1239283 ohms

Rp + jXp = 130.4737854 + j138.3883476 ohms

MP-10 TWO-PORT ANALYSIS

The following four programs are used to characterize a two-port network whose S-parameters are stored in Two-Port A. These programs provide the user with valuable design information such as stability, maximum available gain, simultaneous matched load and source, S'_{11}, S'_{22}, and transducer gain. The frequency, in GHz, should be stored in register (00) for each of these programs. All of the analysis programs analyze Two-Port A; Two-Port B is not changed by any of these programs.

MP-11 Two-Port Characterization

This program analyzes the two-port to determine K, the stability factor; GA max, the maximum available gain; and Γms and Γml, the simultaneous matched source and load. If the stability factor is less than 1, the maximum available gain is not defined, and a simultaneous matched source and load do not exist. Under these conditions, maximum stable gain, MSG, must be used as the maximum design gain.

Execute [A] , ANALYZE, before any other labels are executed.

at Program Stop, K will be in the display.

If the stability factor, K, is less than 1, calculate maximum stable gain, MSG.

MSG: Run [A′]

at Program Stop, MSG in dB will be in the display.

If the stability factor, K, is greater than 1, other parameters may be calculated in any order.

MAXIMUM AVAILABLE GAIN, GA max: Run [B]

at Program Stop, GA max in dB will be in the display.

MATCHED LOAD, Γml: Run [C]

at Program Stop, |Γml| is displayed, [x⇌t] , Θ displayed.

MATCHED SOURCE, Γms: Run [D]

at Program Stop, |Γms| is displayed, [x⇌t] , Θ displayed.

Stability factor, K, may be recalled at any time by running [E].

Any parameter may be printed by running [D′] before the desired label. The frequency may be printed using label [E′].

As an example, enter the following S-parameters using MP-01. These are the S-parameters for a 1 micron gate-length GaAs FET at 2.0 GHz.

```
TWO-PORT  A
      0.979      S11
       -13.
       0.02      S12
        89.
      1.629      S21
       165.
      0.824      S22
        -2.
```

Analyze the two-port by running label [A]. The stability factor is less than 1, which indicates that the device is potentially unstable and the maximum stable gain must be used. Program output is shown on the right. This data is obtained using [E′] for frequency, [D′] [E] for stability factor, and [D′] [A′] for MSG.

```
          2.     FREQ
 .2351176572        K
 19.10891089      MSG
```

An example of a stable two-port may be obtained by reducing the magnitude of the reverse transmission coefficient, S_{12}, of the previous example, which will tend to stabilize the network. Store 0.001 in Reg (16), which is the magnitude of S_{12}, as shown by the listing to the right.

```
TWO-PORT  A
      0.979      S11
       -13.
      0.001      S12
        89.
      1.629      S21
       165.
      0.824      S22
        -2.
```

Run label [A] to analyze the two-port. In this case, the stability factor is greater than 1, so that the maximum available gain, along with the matched source and load, all exist and may be calculated. The listing on the right-hand side of this page shows the results of these calculations.

```
          2.     FREQ
 4.109870701        K
 23.03636404       GA
 .9795498956       MS
 13.24468893
 .8282336333       ML
  4.66813005
```

MP-12 Stability

This program determines stability information necessary for the design of networks using two-ports which are potentially unstable. The program computes stability circles on the input and output planes and defines the region of stability. This program should be used in conjunction with MP-11, even though the stability factor is greater than 1, since K>1 is a necessary but not sufficient condition for absolute stability.

Execute [A], before any other subroutines are executed.

At Program Stop, K will be in the display.

TO DETERMINE A STABILITY CIRCLE ON THE INPUT PLANE:

Run [B]

at Program Stop, magnitude of the stability circle center will be in the display, [x⇌t], the angle of stability circle center will be in the display.

[R/S] at Program Stop, radius of stability circle will be in the display, [x⇌t], code which describes the stable region will be in the display.

TO DETERMINE A STABILITY CIRCLE ON THE OUTPUT PLANE:

Run [C]

program outputs are as above.

The code which describes the stable region is as follows:

"0" stable region is outside the circle

"1" stable region is inside the circle

"10" absolute stability.

The data may be printed by using [D] before [B] or [C]. The printed output is "IN" or "OUT", which indicates input or output plane, magnitude of stability circle center, angle of stability circle center, radius of stability circle, and stability code.

The center and radius of the stability circle are with respect to the coordinate system of the input or output plane. Thus, with a standard Smith Chart, the center of the stability circle would be determined in the same manner as a reflection coefficient, and the radius would be determined from the magnitude scale of the Smith Chart. In those cases where the stability circle is too large for a standard Smith Chart, program MP-54 may be used to construct a Smith Chart with a more convenient scale.

The two examples from the previous program will be used for an example of the stability program.

TWO-PORT A		IN	INPUT PLANE
0.979	S11	1.030277982	MAGNITUDE OF CENTER
-13.		17.88885019	ANGLE OF CENTER
0.02	S12	0.106573772	RADIUS OF CIRCLE
89.		0.	STABLE OUTSIDE
1.629	S21		
165.		OUT	OUTPUT PLANE
0.824	S22	1.767947556	MAGNITUDE OF CENTER
-2.		45.4303725	ANGLE OF CENTER
		1.241675786	RADIUS OF CIRCLE
		0.	STABLE OUTSIDE

```
TWO-PORT A                     IN        INPUT PLANE
  0.979      S11    1.021544796          MAGNITUDE OF CENTER
   -13.             13.24468803          ANGLE OF CENTER
  0.001      S12    .0052952527          RADIUS OF CIRCLE
    89.                     10.          ABSOLUTE STABILITY
  1.629      S21
   165.                     OUT          OUTPUT PLANE
  0.824      S22    1.216012879          MAGNITUDE OF CENTER
    -2.              4.66812005          ANGLE OF CENTER
                    .0578294442          RADIUS OF CIRCLE
                            10.          ABSOLUTE STABILITY
```

MP-13 Constant Gain, S'_{11}, and S'_{22}

This program computes the contours of constant gain on either the input or the output plane. These calculations are made with the assumption that the other port of the network is conjugately matched. The S'_{11} and S'_{22} programs can be used to determine the input or output reflection coefficient for a specific load or source termination.

INPUT PLANE CONSTANT GAIN : Run [A]

at "21", [CLR], enter desired gain in dB, [R/S]

at Program Stop, magnitude of center in display [x⇌t], angle of center in display, [R/S]

at Program Stop, radius of circle in display [x⇌t], gain (dB) in display [R/S]

at "21", [CLR], enter another gain; repeat above

or at "21", [RST] to stop.

[2nd] [St flg] 2 to print results, "I" indicates input plane.

OUTPUT PLANE CONSTANT GAIN : Run [B]

program execution is the same as [A] above.

[2nd] [St flg] 2 to print results, "O" indicates output plane.

CALCULATE S'_{11} :

Enter $|\Gamma l|$, [x⇌t], Θl run [C]

at Program Stop, $|S'_{11}|$ in display, [x⇌t], Θ'_{11} in display.

[2nd] [St flg] 2 to print calculations, "11" is S'_{11}.

CALCULATE S'_{22} :

Enter $|\Gamma s|$, [x⇌t], Θs run [D]

at Program Stop, $|S'_{22}|$ in display, [x⇌t], Θ'_{22} in display.

[2nd] [St flg] 2 to print calculations, "22" is S'_{22}.

As an example, enter the FET S-parameters used in the previous examples and determine the 10 dB constant gain contours on the input and output planes.

TWO-PORT A		10.	I	10 dB GAIN, INPUT PLANE
0.574	S11	.5515278696		MAGNITUDE OF CENTER
-23.		17.88885019		ANGLE OF CENTER
0.02	S12	.4546610125		RADIUS OF CIRCLE
69.				
1.629	S21		O	10 dB GAIN, OUTPUT PLANE
165.		10.		MAGNITUDE OF CENTER
0.024	S22	.1590818037		ANGLE OF CENTER
-2.		45.4303725		RADIUS OF CIRCLE
		0.890396447		

S'_{22} and S'_{11} for respective source and load reflection coefficients which fall on the 10 dB constant gain contours, will now be calculated. The values of these reflection coefficients are determined from the radius of the constant gain contour subtracted from its center, located at the angle of the center. The reader should note that the constant gain contours should be constructed on a Smith Chart to obtain full advantage of these graphic aids. Also, there may be an advantage to using one impedance point over another. For example, a point may be found which requires a single matching element, or one region may result in lower noise figures, VSWR, or better stability. The source and load mappings of MP-55 also can be used in conjunction with this program to provide more useful information. MP-55 provides a graphic representation of S'_{11} and S'_{22} for the entire load and source planes. The user may then investigate all regions of the gain contours, and immediately determine the corresponding load or source match.

.9720349514	11	S'_{11}	MAGNITUDE	INPUT DATA
-12.15883069		S'_{11}	ANGLE	
.7313151663		Γl	MAGNITUDE	
-134.5696275		Γl	ANGLE	

.8242747105	22	S'_{22}	MAGNITUDE	INPUT DATA
-2.241537498		S'_{22}	ANGLE	
.0968668571		Γs	MAGNITUDE	
17.88885019		Γs	ANGLE	

MP-14 Transducer Gain and Mismatch

This program computes the transducer gain of a two-port with arbitrary source and load terminations, along with mismatch loss. Mismatch circles can be constructed for a given reflection coefficient to determine the contours of constant mismatch, or the mismatch in dB can be calculated for two reflection coefficients.

> *TRANSDUCER GAIN, GT* : Run [A]
>
> at "11", [CLR] , enter $|\Gamma s|$, [x⇌t] , enter Θs, [R/S]
>
> at "22", [CLR] , enter $|\Gamma l|$, [x⇌t] , enter Θl, [R/S]
>
> at Program Stop, GT (dB) is in the display.

[2nd] [St flg] 1 to print results. Output is GT in dB,

Γs, magnitude and angle, Γl, magnitude and angle.

MISMATCH CIRCLES : Run [B]

at "11", [CLR], enter |Γ|, [x⇌t], Θ, [R/S]

at "21", [CLR], enter mismatch (positive dB), [R/S]

at Program Stop, magnitude of center in display, [x⇌t], angle of center in display [R/S]

at Program Stop, radius of circle in display, [x⇌t], mismatch (positive dB) in display

[RST] to stop program

[R/S] to branch to "21" for another circle.

[2nd] [St flg] 1 to print results.

MISMATCH LOSS : Run [C]

at "1", [CLR], enter $|\Gamma_1|$, [x⇌t], Θ_1, [R/S]

at "2", [CLR], enter $|\Gamma_2|$, [x⇌t], Θ_2, [R/S]

at Program Stop, mismatch (dB) in display.

[2nd] [St flg] 1 to print results.

The transducer gain portion of this program is very useful for checking previous calculations, as well as gain provided by a particular matching network. As a first example, the maximum available gain for the matched source and load reflection coefficients from the stable example of MP-11 will be checked. Store 0.001 into register (16) and run the transducer gain for the calculated Γms and Γml.

```
TWO-PORT A
   0.979      S11              23.03638404      GT
    -13.                       .9795498956      ΓS
   0.001      S12              13.24468893
     89.                       .8282336333      ΓL
   1.629      S21               4.66813005
    165.
   0.824      S22
     -2.
```

Thus, the transducer gain using the matched load and source does produce the maximum available gain. The constant gain, S'_{11}, and S'_{22} calculations of the previous program can also be easily checked. Store 0.02 into register (16) which restores Two-Port A. Now, calculate the transducer gain for the examples of MP-13. Note that source and load reflection coefficients are the complex conjugates of S'_{11} and S'_{22}, respectively. Recall that the constant gain contours assume a complex conjugate match.

```
TWO-PORT A                       10.        GT
  0.979        S11      .9720349314         ΓS
  -10.                  12.15883069
  0.02         S12      .7313151663         ΓL
  89.                   -134.5696275
  1.629        S21
  165.
  0.624        S22               10.        GT
  -2.                   .0968668571         ΓS
                        17.88885019
                        .8242747105         ΓL
                        2.241537498
```

10 dB TRANSDUCER GAIN
S'^*_{11} FROM FIRST EXAMPLE
Γl FROM FIRST EXAMPLE

10 dB TRANSDUCER GAIN
Γs FROM SECOND EXAMPLE
S'^*_{22} FROM SECOND EXAMPLE

Thus the constant gain calculations using S'_{11} and S'_{22} are correct. The second example above resulted from constructing a constant 10 dB gain contour on the input plane, assuming a value of source reflection coefficient which falls on this contour, calculating S'_{22} corresponding to this Γs, and using S'^*_{22} for Γl. This procedure works well, as the example demonstrates; however, the designer is constrained to matching to a single value on the output plane. A different Γs will require another Γl, so that the designer can try various values of Γs, or use MP-55 to map the entire plane, until a desired value is obtained. The designer has another option: the 12 dB gain contour may be constructed on the input (or output) plane, then a 2 dB mismatch contour may be constructed about Γl (Γs), This technique allows the designer to select any Γs on the input plane contour, and any Γl on the output plane mismatch contour, which affords much greater flexibility in the selection of matching networks. As an example, a 2 dB mismatch contour will be calculated for S'_{22} of the last example above.

```
          2.        G
 .6941261225      CTR
 2.241537498
 0.2599137218     RAD
 .8242747105        Γ
-2.241537498
```

2 dB MISMATCH
CENTER OF MISMATCH CIRCLE
ANGLE OF MISMATCH CIRCLE
RADIUS OF MISMATCH CIRCLE
S'_{22}

The transducer gain can now be calculated for Γs of the previous example, and a Γl which falls on the mismatch contour. Use the magnitude of center minus the radius, and the angle of the center.

```
          8.       GT
 .0968668571       ΓS
 17.88885019
 .4342124045       ΓL
 2.241537498
```

Which results in the expected 2 dB gain reduction.

As an example of the mismatch loss calculation, use S'_{22} and Γl from the above, which results in the expected 2 dB mismatch.

```
         -2.       GT
 .8242747105       ΓS
-2.241537498
 .4342124045       ΓL
 2.241537498
```

MISMATCH LOSS IN dB
MAGNITUDE OF Γ_1
ANGLE OF Γ_1
MAGNITUDE OF Γ_2
ANGLE OF Γ_2

MP-20 INTERCONNECTIONS AND PARAMETER CONVERSIONS

This group of six programs is used to interconnect two-ports in cascade, series, parallel, series-parallel, and parallel-series. The Hybrid connection is also included for amplifier design. One-port networks connected in series or parallel and the various parameter set conversions are also included in this section. The examples for the interconnection of these programs will use the same Two-Ports A and B. The reader may find it convenient to save Two-Ports A and B from Bank 4 on a data card, which can then be loaded for each example.

MP-21 Cascade Connection

RUN [A]

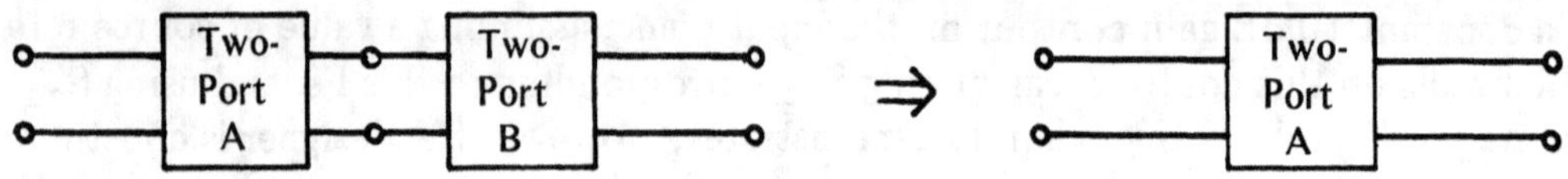

Two-Port A in Reg (14) thru Reg (21)

Two-Port B in Reg (22) thru Reg (29)

resulting Two-Port A in Reg (14) thru Reg (21) .

EXAMPLE:

```
TWO-PORT A                     TWO-PORT B
   0.979       S11                0.5        S11
   -13.                           45.
   0.02        S12                0.1        S12
   89.                           -30.
   1.629       S21                1.1        S21
   165.                           60.
   0.324       S22                0.7        S22
   -2.                           -70.
```

```
    TWO-PORT A
 .9984724171         S11
-13.54594423
 .0026558094         S12
 80.90810764
 2.379472402         S21
 246.9081076
 .6484351104         S22
-60.74094112
```

RESULTS OF CASCADE OF A AND B ABOVE

MP-22 Series and Parallel Connection

SERIES CONNECTION, SER : Run [A]

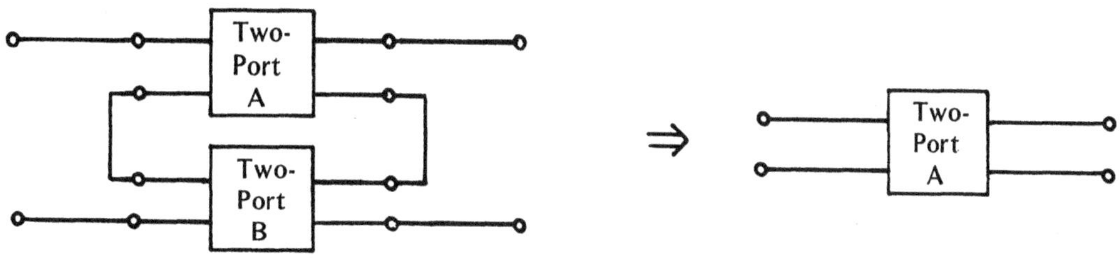

Two-Port A in Reg (14) thru Reg (21)

Two-Port B in Reg (22) thru Reg (29)

resulting Two-Port A in Reg (14) thru Reg (21) .

EXAMPLE:

```
TWO-PORT A
     0.979      S11
      -13.
      0.02      S12
       89.
     1.629      S21
      165.
     0.824      S22
       -2.
```

```
TWO-PORT B
       0.5      S11
       45.
       0.1      S12
      -30.
       1.1      S21
       60.
       0.7      S22
      -70.
```

```
        TWO-PORT A
  .9506040427      S11
 -13.03411216
  .0250417181      S12
  76.11194466
  1.499264648      S21
  166.4502022
  .9012212565      S22
 -5.825530209
```

RESULTS OF SERIES CONNECTION OF TWO-PORTS A AND B ABOVE

PARALLEL CONNECTION, PAR : Run [B]

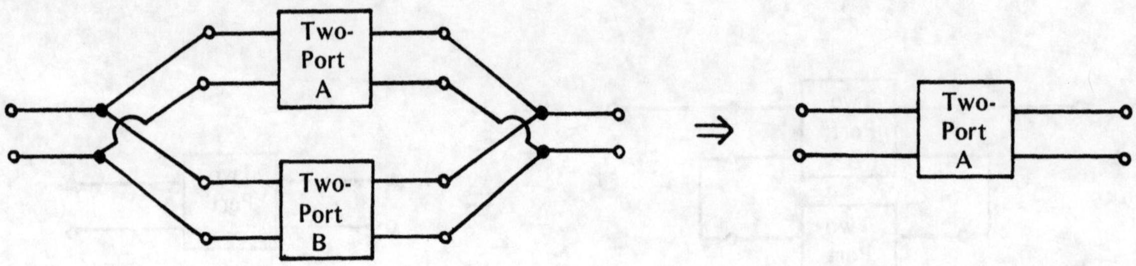

Two-Port A in Reg (14) thru Reg (21)
Two-Port B in Reg (22) thru Reg (29)
resulting Two-Port A in Reg (14) thru Reg (21)

EXAMPLE:

```
TWO-PORT A                  TWO-PORT B
 0.979      S11                0.5       S11
 -13.                          45.
 0.02       S12                0.1       S12
 89.                          -30.
 1.629      S21                1.1       S21
 165.                          60.
 0.824      S22                0.7       S22
 -2.                          -70.
```

```
TWO-PORT A
 .4268155617     S11
 40.57866628
 .0885367006     S12
-25.44936691
 1.204625677     S21
 98.86158428
 .5640421209     S22
-74.19218996
```

RESULTS OF PARALLEL CONNECTION OF TWO-PORTS A AND B ABOVE.

MP-23 Hybrid Connection

HYBRID CONNECTION, HYB :

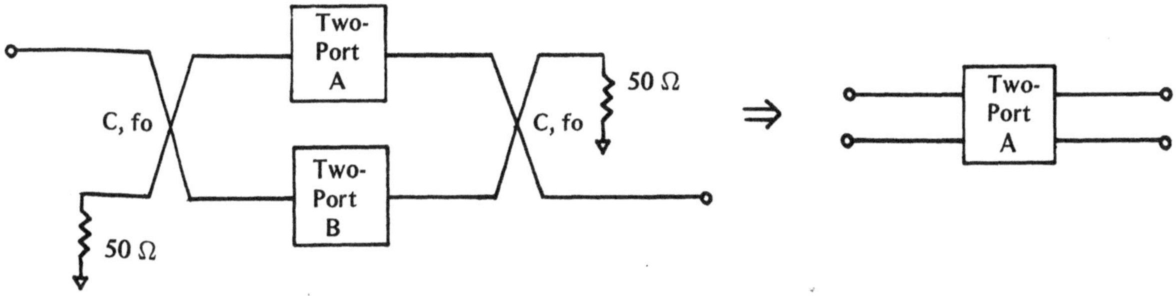

Frequency (GHz) in Reg (00)

enter coupling coefficient (+dB), [x⇌t], enter coupler center frequency

run [A]

Two-Port A in Reg (14) thru Reg (21)

Two-Port B in Reg (22) thru Reg (29)

resulting Two-Port A in Reg (14) thru Reg (21) .

NOTE: Coupler center frequency cannot be the exact analysis frequency (Reg (00)); offset the coupler frequency slightly.

EXAMPLE: Store analysis frequency in Reg (00) , 2.00 GHz. Coupling coefficient is −3.0 dB, center frequency is 2.000001 GHz.

TWO-PORT A	
0.979	S11
-13.	
0.02	S12
89.	
1.629	S21
165.	
0.824	S22
-2.	

TWO-PORT B	
0.5	S11
45.	
0.1	S12
-30.	
1.1	S21
60.	
0.7	S22
-70.	

TWO-PORT A	
.4162403508	S11
-43.53998956	
.0459910693	S12
-109.0371666	
.8567480307	S21
36.67762755	
.4289351427	S22
-132.6805878	

RESULTS OF HYBRID CONNECTION OF TWO-PORTS A AND B ABOVE

MP-24 Series-Parallel and Parallel-Series Interconnection

SERIES · PARALLEL CONNECTION, SER · PAR : Run [A]

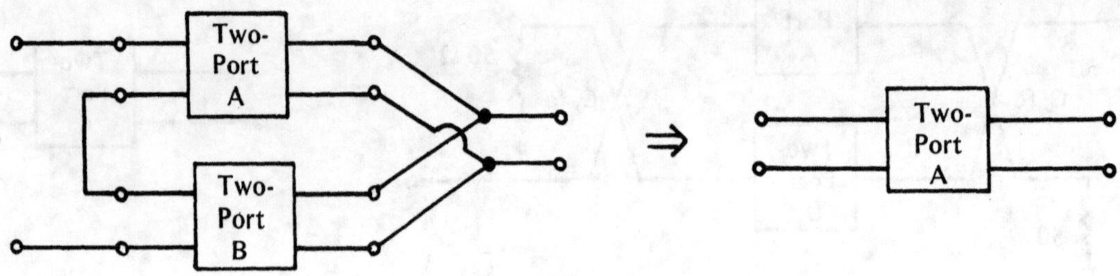

Two-Port A in Reg (14) thru Reg (21)
Two-Port B in Reg (22) thru Reg (29)

resulting Two-Port A in Reg (14) thru Reg (21) .

EXAMPLE:

```
TWO-PORT A
     0.979      S11
      -13.
      0.02      S12
       69.
     1.629      S21
      165.
     0.824      S22
       -2.
```

```
TWO-PORT B
     0.5        S11
     45.
     0.1        S12
    -30.
     1.1        S21
     60.
     0.7        S22
    -70.
```

```
   TWO-PORT A
.9046394276      S11
-15.5001811
.0509235269      S12
59.60071887
1.820598139      S21
134.1812066
.7913984772      S22
-58.9906427
```

RESULTS OF SERIES · PARALLEL CONNECTION OF TWO-PORTS A AND B ABOVE

PARALLEL · SERIES CONNECTION, PAR · SER : Run [B]

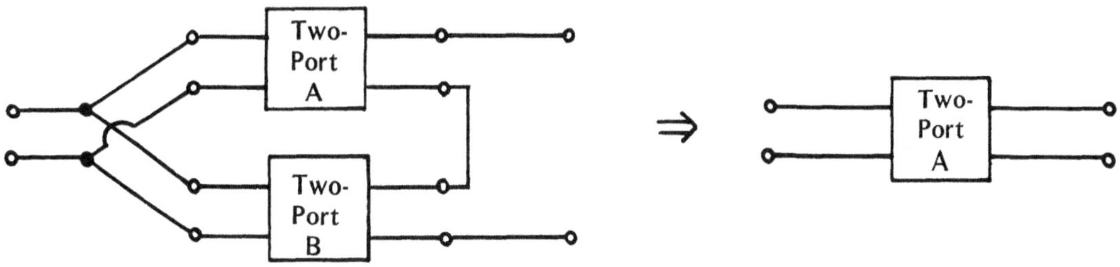

Two-Port A in Reg (14) thru Reg (21)
Two-Port B in Reg (22) thru Reg (29)
resulting Two-Port A in Reg (14) thru Reg (21) .

EXAMPLE:

```
  TWO-PORT A
     0.979      S11
      -13.
      0.02      S12
       89.
     1.629      S21
      165.
     0.824      S22
       -2.
```

```
  TWO-PORT B
       0.5      S11
       45.
       0.1      S12
      -30.
       1.1      S21
       60.
       0.7      S22
      -70.
```

```
     TWO-PORT A
 .5203576922      S11
 9.601709862
 0.002419354      S12
-39.25069507
 1.023637195      S21
 176.7767929
 .8356954757      S22
-1.794891369
```

RESULTS OF PARALLEL · SERIES CONNECTION OF TWO-PORTS A AND B ABOVE

MP-25 Parameter Conversions

S ⇒ Y CONVERSION : Run [A]

normalized S in Two-Port A at start
normalized Y in Two-Port A at stop.

Y ⇒ S CONVERSION : Run [A']

normalized Y in Two-Port A at start
normalized S in Two-Port A at stop.

S ⇒ Z CONVERSION : Run [B]

normalized S in Two-Port A at start
normalized Z in Two-Port A at stop.

Z ⇒ S CONVERSION : Run [B']

normalized Z in Two-Port A at start
normalized S in Two-Port A at stop.

S⇒H CONVERSION : Run [C]

normalized S in Two-Port A at start
normalized H in Two-Port A at stop.

H⇒S CONVERSION : Run [C']

normalized H in Two-Port A at start
normalized S in Two-Port A at stop.

S⇒G CONVERSION : Run [D]

normalized S in Two-Port A at start
normalized G in Two-Port A at stop.

G⇒S CONVERSION : Run [D']

normalized G in Two-Port A at start
normalized S in Two-Port A at stop.

Z ⇋ Y *CONVERSION* : Run [E]

normalized Z(Y) in Two-Port A at start
normalized Y(Z) in Two-Port A at stop.

G ⇋ H *CONVERSION* : Run [E']

normalized G(H) in Two-Port A at start
normalized H(G) in Two-Port A at stop.

EXAMPLES: The Two-Port A S-parameters which have been used for the previous examples are used to convert to the various parameter sets. The S-parameters may be converted to any parameter set, checked, then converted back to the S-parameter set. This double conversion will cause a small error due to the numerical accuracy of the calculator. MP-02 may be used to display or print the non-normalized values of the various parameters; MP-01 was used for this example.

NORMALIZED S-PARAMETERS

```
TWO-PORT A
   0.979      S11
    -13.
   0.02       S12
     89.
   1.629      S21
    165.
   0.824      S22
     -2.
```

NORMALIZED Y-PARAMETERS

```
   TWO-PORT A
 .1052110178      S11
 84.37823549
 .0111389365      S12
-84.17967051
 .9072663768      S21
-8.179670508
 0.095379241      S22
 4.477793541
```

NORMALIZED Z-PARAMETERS

```
     TWO-PORT A
 4.735860406      S11
-83.76840726
 .5530807337      S12
 7.574128637
 45.04842576      S21
 83.57412864
 5.224034357      S22
-3.867965362
```

NORMALIZED H-PARAMETERS

```
    TWO-PORT A
 9.504707979      S11
-84.37823549
 .1058723385      S12
   11.442094
  8.62330197      S21
  267.442094
 .1914229371      S22
 3.867965362
```

NORMALIZED G-PARAMETERS

```
     TWO-PORT A
 .2111548725      S11
 83.76840726
 .1167856918      S12
-88.65746411
 9.512194597      S21
 167.3425359
 10.48445621      S22
  -4.4777936
```

MP-26 One Port Used in Series or Parallel

This program will convert a one-port to a two-port network, with the one-port connected in either series or parallel. This program is useful for analyzing networks with parallel or series branches, such as amplifier bias networks, power dividers, or switches. When using a cascade of two-ports as the network branch, the user should be aware that the input to the two-port is calculated with a 50 ohm termination when using S-parameters. Therefore, a short or an open should be cascaded with the network to obtain the desired input.

USE SERIES : Run [A]

Γin ⇒

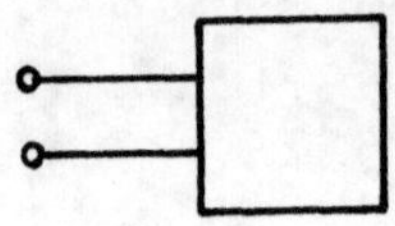

⇒

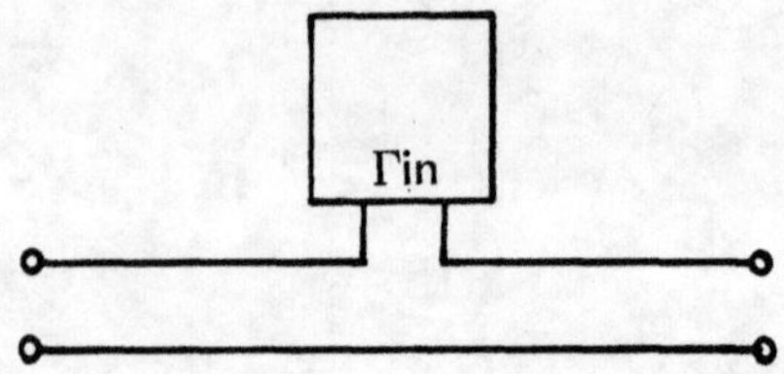

Γin in Reg (14) and Reg (15) at start
Two-Port B in Reg (22) thru Reg (29) at stop.

USE PARALLEL : Run [B]

Γin ⇒

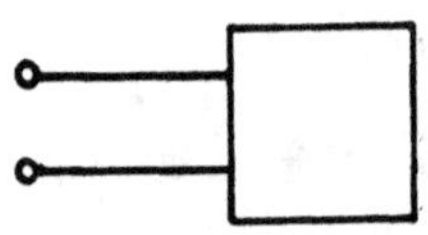

⇒

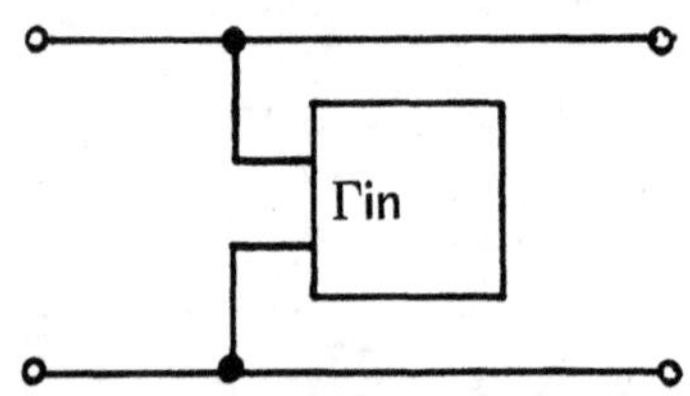

Γin in Reg (14) and Reg (15) at start
Two-Port B in Reg (22) thru Reg (29) at stop.

ENTER Z∠Θ : Zin ⇒ Γin
enter |Z|, [x⇌t], enter Θ, run [C]
Γin in Reg (14) and Reg (15) at stop

Zin ⇒

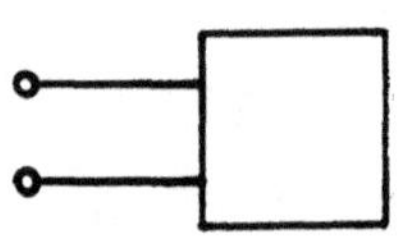

⇒ Γin

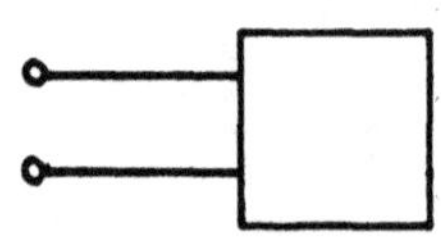

run either (A) or (B) above.

TERMINATE SHORT : Run [D]

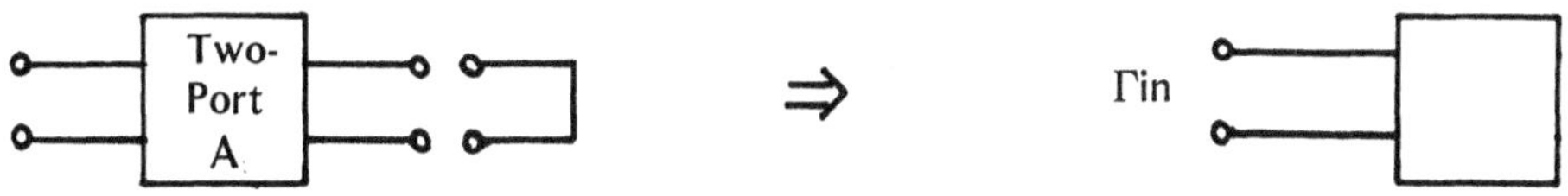

Two-Port A in Reg (14) thru Reg (21) at start
Γin (S'_{11} with short) in Reg (14) and Reg (15) at stop
run either [A] or [B] above.

TERMINATE OPEN : Run [E]

Two-Port A in Reg (14) thru Reg (21) at start
Γin (S'_{11} with open) in Reg (14) and Reg (15) at stop
run either [A] or [B] above.

EXAMPLES:

Store Γ = 0.5 in Reg (14)

Θ = 45 in Reg (15)

connect Γin = 0.5 $\angle$45 in series:
Run [A]

Store Γ = 0.5 in Reg (14)

Θ = 45 in Reg (15)

Connect Γin = 0.5$\angle$45 in parallel:
Run [B]

Γ = 0.5$\angle$45 IN SERIES

```
        TWO-PORT B
 0.523965523          S11
 22.24820632
 .5519282893          S12
-21.06564704
 .5519282893          S21
-21.06564704
 0.523965523          S22
 22.24820632
```

Γ = 0.5$\angle$45 IN PARALLEL

```
        TWO-PORT B
-0.218500161          S11
-34.69331201
 0.829720479          S12
 8.620544645
 0.829720479          S21
 8.620544645
-0.218500161          S22
-34.69331201
```

Convert Z = 20 ∠30 to Γin:
enter 20, [x⇌t], enter 30, run [C]
at Program Stop, Γin in Reg (14) and Reg (15)

Reg (14)	0.502140578
Reg (15)	-205.4633451

Terminate Two-Port A with a short and open. The two-port parameters of the previous examples will be used, as shown to the right.

```
TWO-PORT A
  0.979       S11
   -13.
   0.02       S12
    89.
  1.629       S21
   165.
  0.824       S22
    -2.
```

Run [D] to determine Γin of this two-port with a short termination.

Reg (14)	.9798161801
Reg (15)	-11.9559939

Reload the S-parameters and run [E] to determine Γin with an open termination.

Reg (14)	0.957037934
Reg (15)	-23.72097227

MP-30 Lumped and Distributed Elements

This group of three programs provides the user with a wide range of lumped and distributed elements. The two-port S-parameters for each element are calculated at the frequency specified by register (00), and are stored as Two-Port B, which can then be interconnected with Two-Port A.

MP-31 Series Resistor, Inductor, and Capacitor

Frequency (GHz) must be in Reg (00).

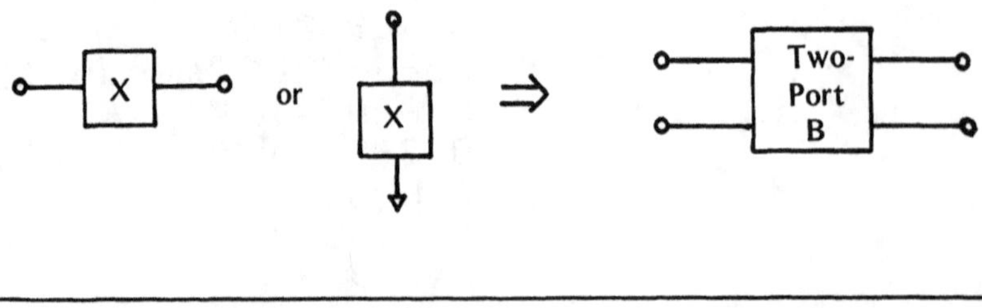

Reg (22) thru Reg (29)

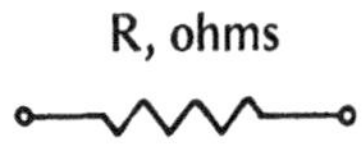	Enter R, run [A].	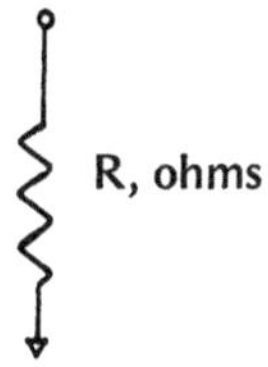	Enter R, run [E′], run [A].
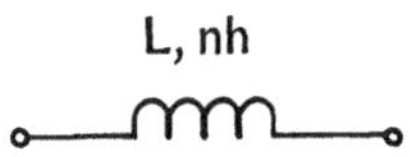	Enter L, run [B].	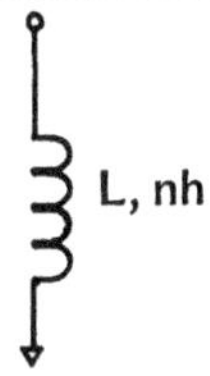	Enter L, run [E′], run [B].
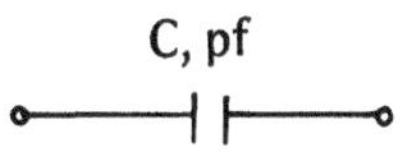	Enter C, run [C].	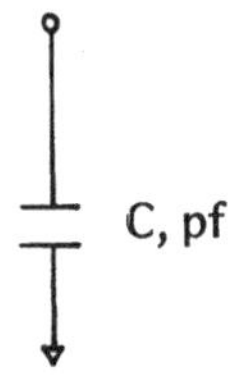	Enter C, run [E′], run [C].
	Enter R, [x⇌t], enter L, run [D].		Enter R, [x⇌t], enter L run [E′], run [D].
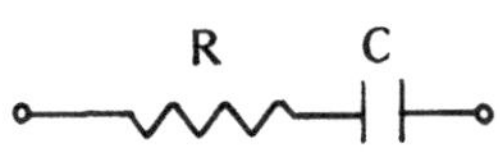	Enter R, [x⇌t], enter C, run [E].		Enter R, [x⇌t], enter C run [E′], run [E].
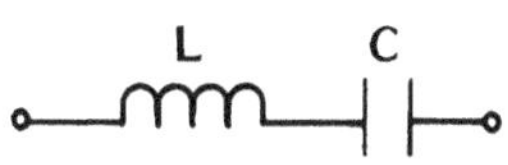	Enter L, [x⇌t], enter C, run [A′].		Enter L, [x⇌t], enter C run [E′], run [A′].
	Enter R, [x⇌t], enter L, run [B′] at stop, enter C, [R/S].		Enter R, [x⇌t], enter L run [E′], run [B′] at stop, enter C, [R/S].

The following examples are various combinations of a 75 ohm resistor, 5 nanohenry inductor, and a 1 picofarad capacitor at 2.0 GHz. Store 2 in register (00), then use the above table and the schematic representation for the element to verify the program.

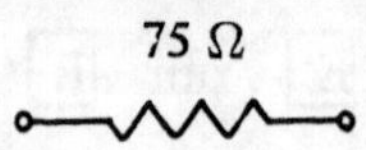


```
      TWO-PORT B
.4285714286        S11
  -3.4738692-49
.5714285714        S12
  2.6054019-49
.5714285714        S21
  2.6054019-49
.4285714286        S22
  -3.4738692-49
```

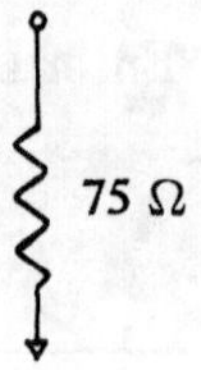


```
      TWO-PORT B
        0.25       S11
        180.
        0.75       S12
.0000000001
        0.75       S21
.0000000001
        0.25       S22
        180.
```

```
      TWO-PORT B
.5320180445        S11
57.85809236
0.846733016        S12
-32.14190764
0.846733016        S21
-32.14190764
.5320180445        S22
57.85809236
```

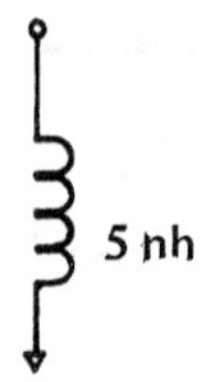


```
      TWO-PORT B
.3696978476        S11
 111.696984
.9291520336        S12
21.69698397
.9291520336        S21
21.69698397
.3696978476        S22
 111.696984
```

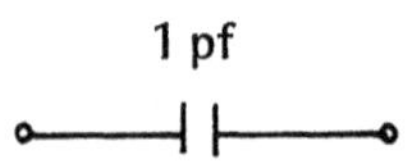


```
      TWO-PORT B
.6226769923        S11
-51.48811275
.7824789858        S12
38.51188725
.7824789858        S21
38.51188725
.6226769923        S22
-51.48811275
```

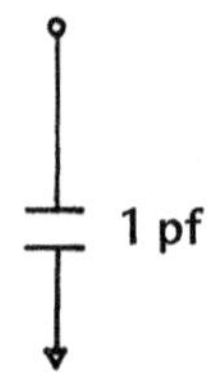


```
      TWO-PORT B
.2997168036        S11
252.5594055
.9540282164        S12
-17.44059449
.9540282164        S21
-17.44059449
.2997168036        S22
252.5594055
```

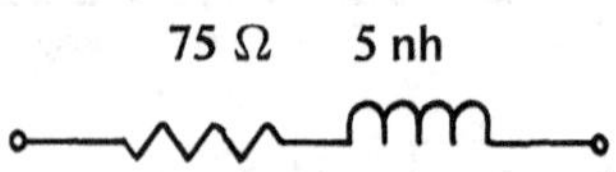


```
      TWO-PORT B
.5262024924        S11
20.20473696
.5378144427        S12
-19.75012556
.5378144427        S21
-19.75012556
.5262024924        S22
20.20473696
```

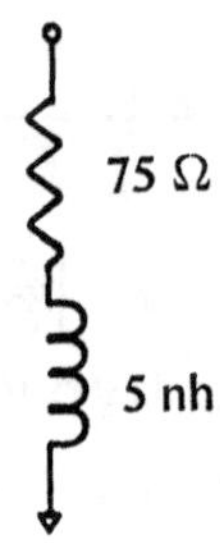


```
      TWO-PORT B
0.211683254        S11
147.8580924
.8284512055        S12
 7.81295488
.8284512055        S21
 7.81295488
0.211683254        S22
147.8580924
```

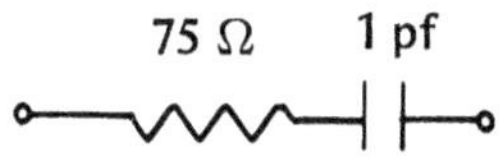


```
   TWO-PORT B
 .5688133894     S11
-22.24355095
 .5201735508     S12
 24.45264174
 .5201735508     S21
 24.45264174
 .5688133894     S22
-22.24355095
```

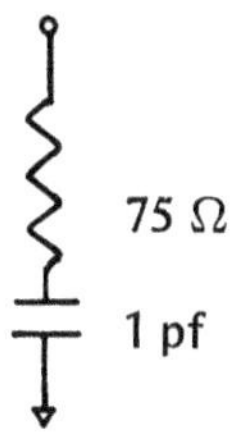


```
   TWO-PORT B
 .1956197465     S11
 218.5118873
 .8556461906     S12
-8.184305439
 .8556461906     S21
-8.184305439
 .1956197465     S22
 218.5118873
```

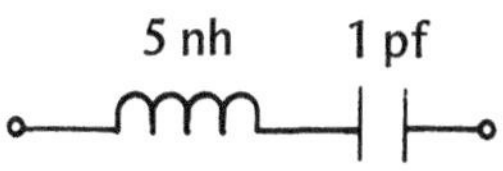


```
   TWO-PORT B
 .1651565694     S11
-80.49366998
 .9862673611     S12
 9.506330022
 .9862673611     S21
 9.506330022
 .1651565694     S22
-80.49366998
```

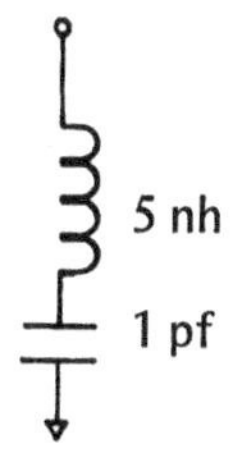


```
   TWO-PORT B
 .8308373064     S11
 213.8151541
 .5565153619     S12
-56.18484592
 .5565153819     S21
-56.18484592
 .8308373064     S22
 213.8151541
```

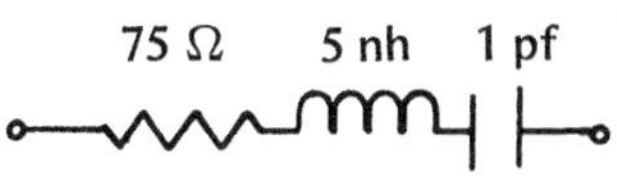


```
   TWO-PORT B
 .4371273132     S11
-7.120323517
 .5688302777     S12
 5.46594776
 .5688302777     S21
 5.46594776
 .4371273132     S22
-7.120323517
```

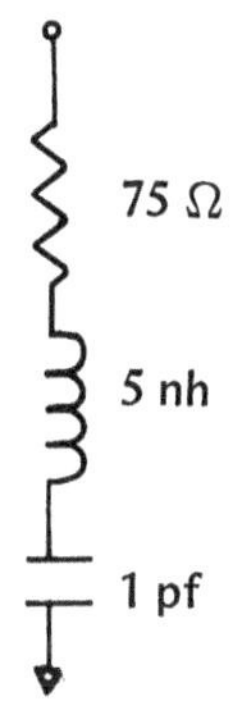


```
   TWO-PORT B
 .2465668403     S11
   189.50633
 .7579139482     S12
-3.079941255
 .7579139482     S21
-3.079941255
 .2465668403     S22
   189.50633
```

MP-32 Parallel Resistor, Industor, and Capacitor

Frequency (GHZ) must be in Reg (00).

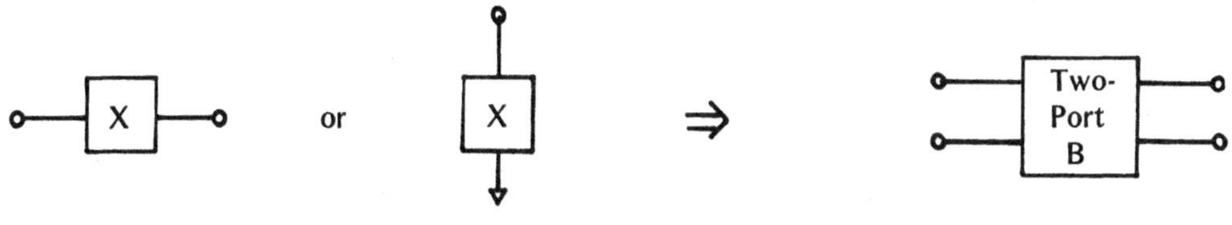

Reg (22) thru
Reg (29)

Enter R, run [A].

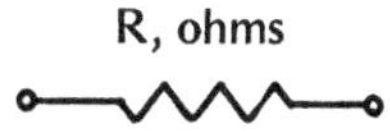

Enter R., run [E′], run [A].

Circuit	Procedure	Circuit	Procedure
	Enter L, run [B].	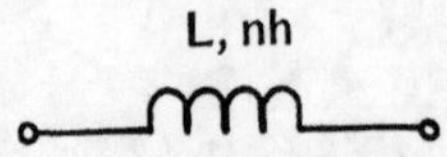	Enter L, run [E′], run B.
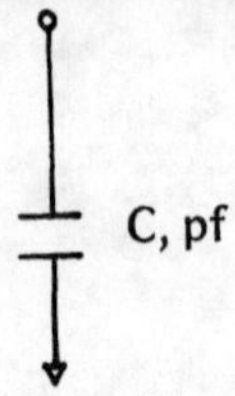	Enter C, run [C].	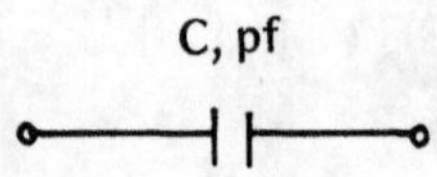	Enter C, run [E′], run C.
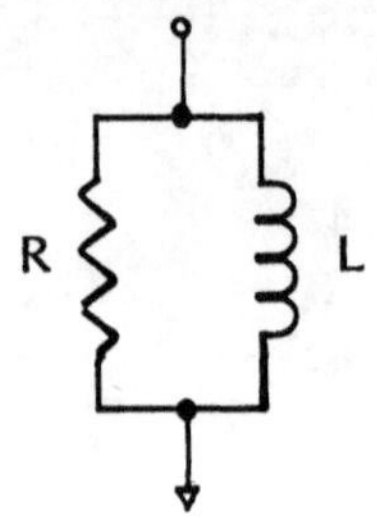	Enter R, [x⇌t], enter L run [D].	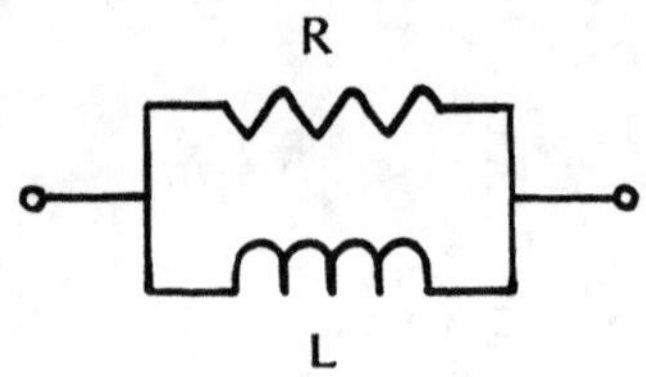	Enter R, [x⇌t], enter L run [E′], run [D].
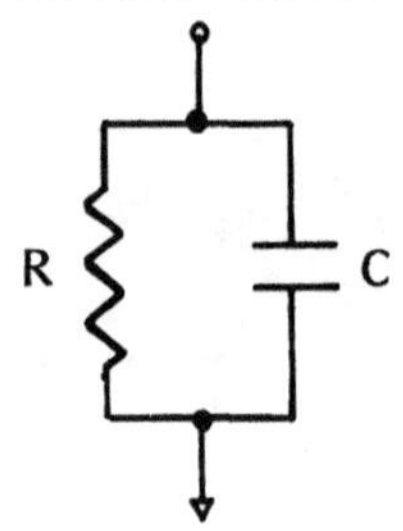	Enter R, [x⇌t], enter C run [E].	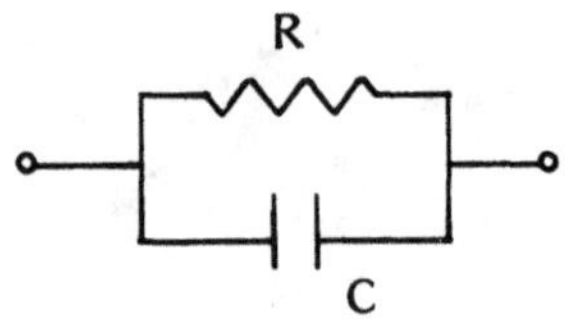	Enter R, [x⇌t], enter C run [E′], run [E].
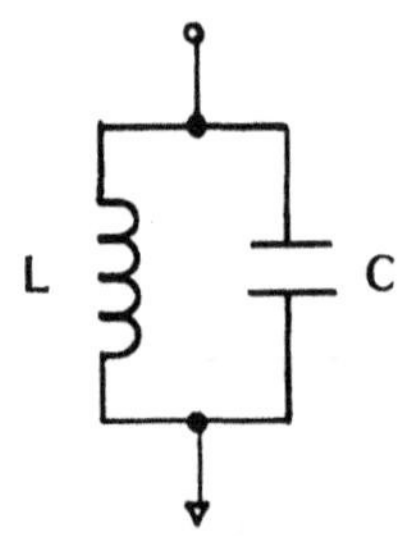	Enter L, [x⇌t], enter C run [A′].	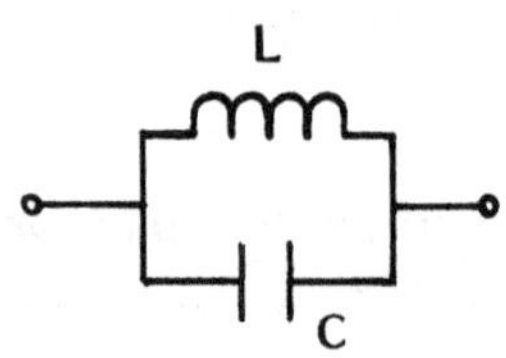	Enter L, [x⇌t], enter C run [E′], run [A′].
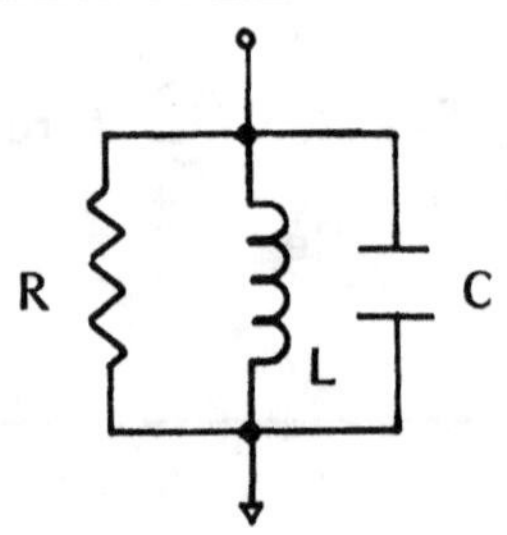	Enter R, [x⇌t], enter L run [B′] at stop, enter C, [R/S].	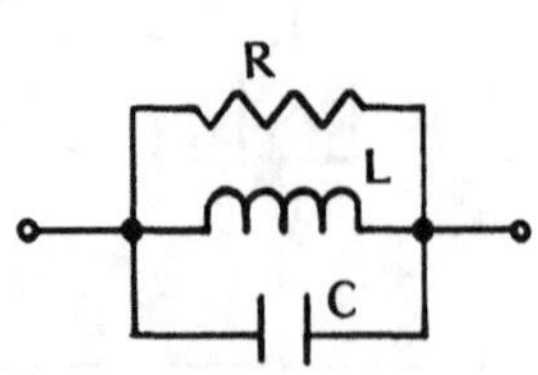	Enter R, [x⇌t], enter L run [E′], run [B′] at stop, enter C, [R/S].

Store 2 in Reg (00) (analysis frequency is 2.0 GHz). Using the above table, enter the appropriate elements to calculate Two-Port B.

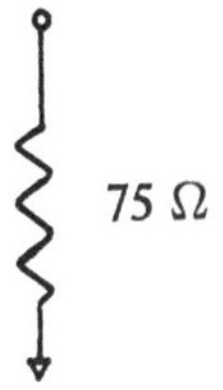

TWO-PORT B	
0.25	S11
180.	
0.75	S12
8.5489749-49	
0.75	S21
8.5489749-49	
0.25	S22
180.	

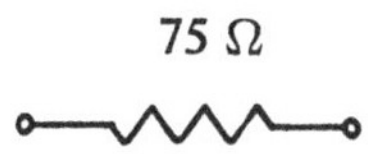

TWO-PORT B	
.4285714286	S11
1.9540514-48	
.5714285714	S12
-1.4655385-48	
.5714285714	S21
-1.4655385-48	
.4285714286	S22
1.9540514-48	

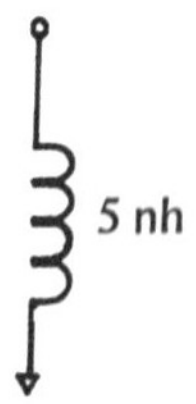

TWO-PORT B	
.3696978476	S11
111.696984	
.9291520336	S12
21.69698397	
.9291520336	S21
21.69698397	
.3696978476	S22
111.696984	

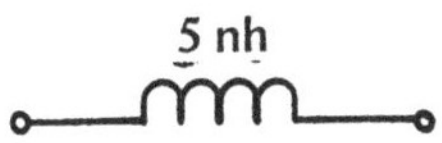

TWO-PORT B	
.5320180445	S11
57.85809236	
0.8467330116	S12
-32.14190764	
0.8467330116	S21
-32.14190764	
.5320180445	S22
57.85809236	

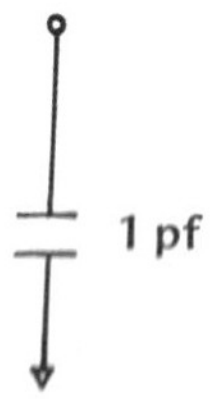

TWO-PORT B	
.2997168036	S11
252.5594055	
.9540282164	S12
-17.44059449	
.9540282164	S21
-17.44059449	
.2997168036	S22
252.5594055	

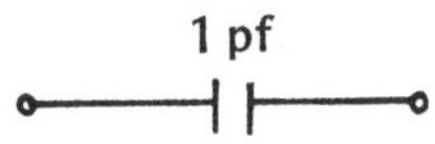

TWO-PORT B	
.6226769923	S11
-51.48811275	
.7824789858	S12
38.51188725	
.7824789858	S21
38.51188725	
.6226769923	S22
-51.48811275	

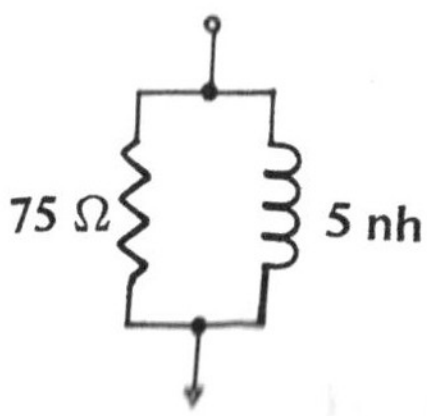

TWO-PORT B	
.3730407774	S11
146.5707823	
.7186823687	S12
16.61591978	
.7186823687	S21
16.61591978	
.3730407774	S22
146.5707823	

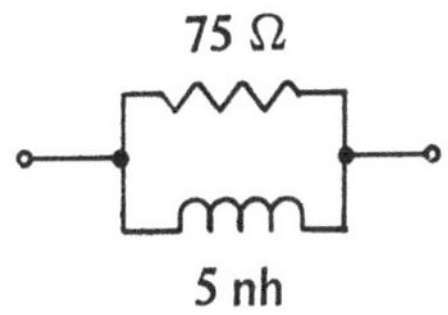

TWO-PORT B	
.3540522024	S11
34.29760946	
.7351003146	S12
-15.74752803	
.7351003146	S21
-15.74752803	
.3540522024	S22
34.29760946	

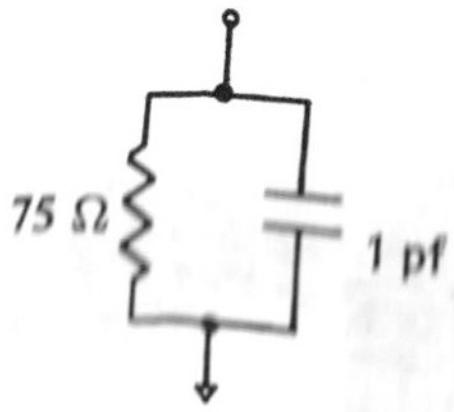

TWO-PORT B	
.3343789213	S11
210.0456264	
.7300098918	S12
-13.25818091	
.7300098918	S21
-13.25818091	
.3343789213	S22
210.0456264	

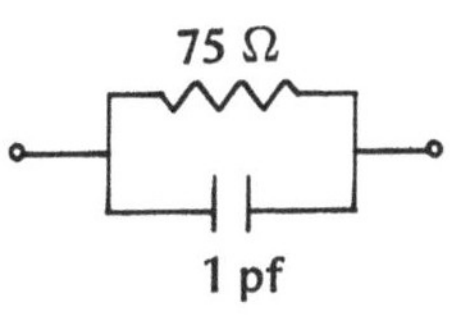

TWO-PORT B	
.3773294458	S11
-28.30507305	
.6913386489	S12
14.99873425	
.6913386489	S21
14.99873425	
.3773294458	S22
-28.30507305	

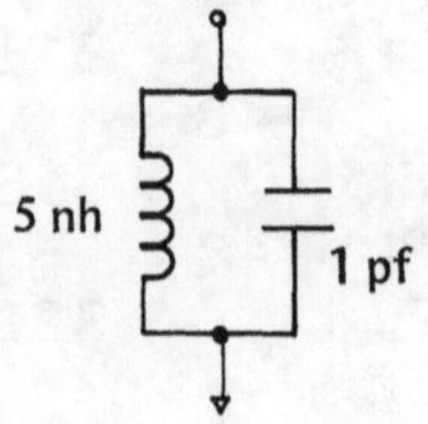


```
     TWO-PORT B
 0834361431        S11
   94.786103
.9965131259        S12
4.786102996
.9965131259        S21
4.786102996
.0834361431        S22
   94.786103
```

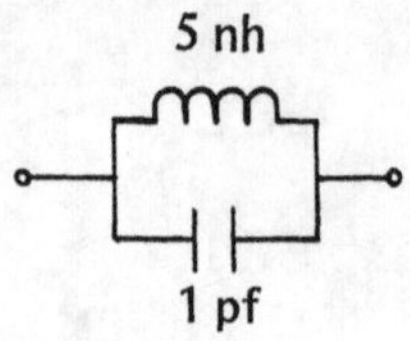


```
     TWO-PORT B
.9482331526        S11
18.51633511
0.317575012        S12
-71.48366409
0.317575012        S21
-71.48366409
.9482331526        S22
18.51633511
```

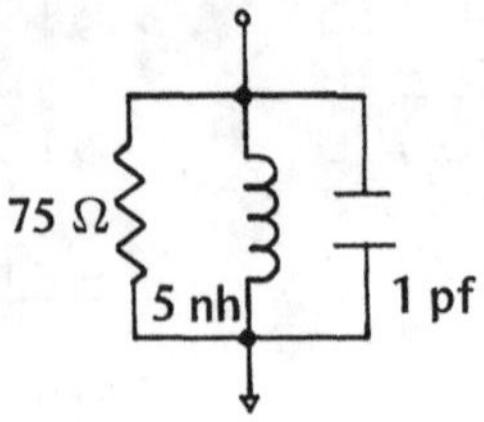


```
     TWO-PORT B
.2572593367        S11
169.4931433
.7485256042        S12
3.593231577
.7485256042        S21
3.593231577
.2572593367        S22
169.4931433
```

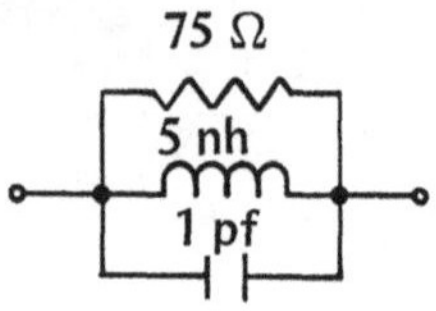


```
     TWO-PORT B
.4242237835        S11
8.168097034
.5832026509        S12
-5.931991281
.5832026509        S21
-5.931991281
.4242237835        S22
8.168097034
```

MP-33 Transmission Line Elements

Frequency (GHz) must be in Reg (00). Characteristic impedance is in ohms, length is inches in air dielectric.

Transmission Line element is stored as Two-Port B, Reg (22) thru Reg (29).

TRANSMISSION LINE, TRL

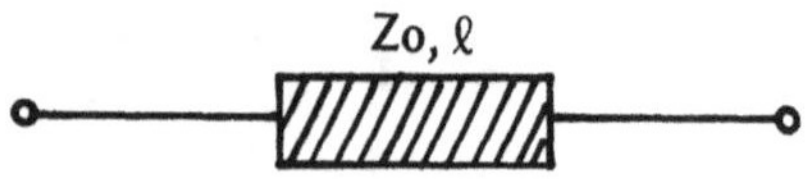

Enter characteristic impedance (Zo), [x⇌t]
enter length (inches), run [A].

OPEN-CIRCUIT TRANSMISSION LINE CONNECTED IN PARALLEL, OTRL

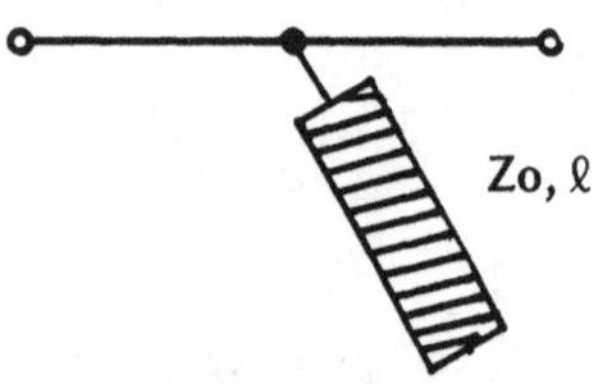

Enter characteristic impedance (Zo), [x⇌t]
enter length (inches), run [B].

OPEN-CIRCUIT TRANSMISSION LINE CONNECTED IN SERIES, OTRL

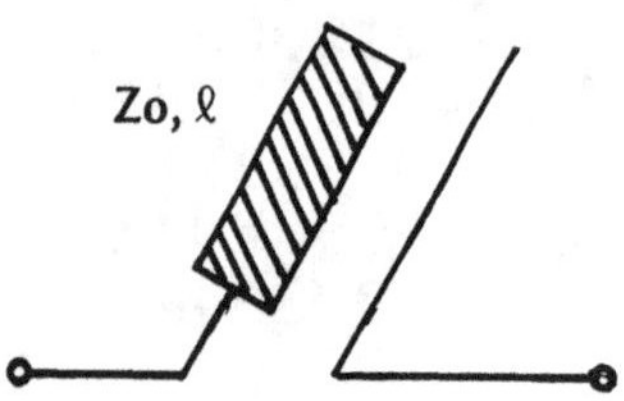

Enter characteristic impedance (Zo), [x⇌t]
enter length (inches), run [E′], run [B].

SHORT-CIRCUIT TRANSMISSION LINE CONNECTED IN PARALLEL, STRL

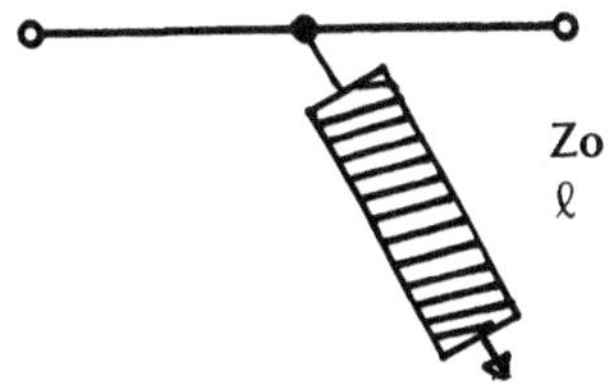

Enter characteristic impedance (Zo), [x⇌t] enter length (inches), run [C] .

SHORT-CIRCUIT TRANSMISSION LINE CONNECTED IN SERIES, STRL

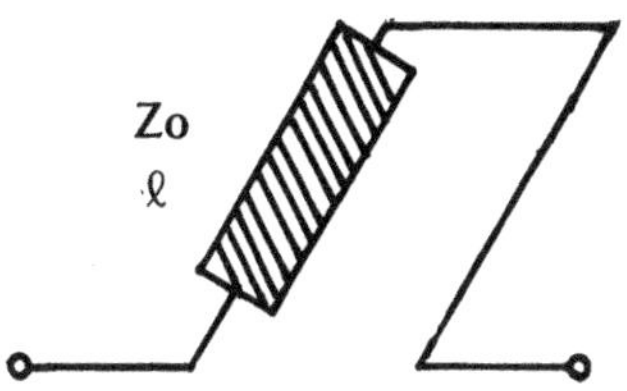

Enter characteristic impedance (Zo), [x⇌t] enter length (inches), run [E′] , run [C] .

EXAMPLES: The following examples are for a 60 ohm transmission line, 0.5 inches long. The analysis frequency is 2.0 GHz.

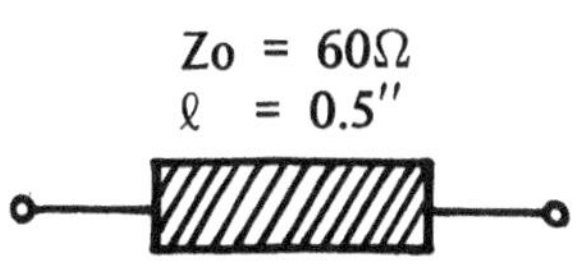


```
      TWO-PORT  B
 .0925940483        S11
59.10435125
 .9957039431        S12
-30.89564875
 .9957039431        S21
-30.89564875
 .0925940483        S22
59.10435125
```

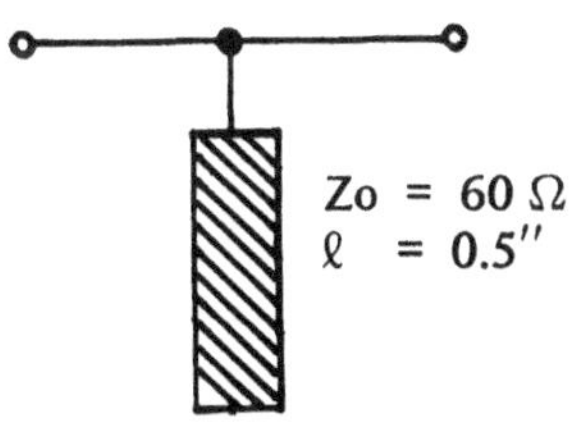


```
      TWO-PORT  B
 .2381817096        S11
256.2207518
 0.971220611        S12
-13.7792482
 0.971220611        S21
-13.7792482
 .2381817096        S22
256.2207518
```

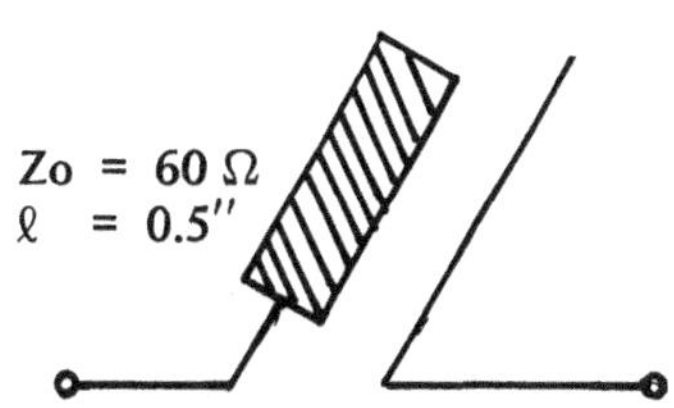


```
      TWO-PORT  B
 0.713870802        S11
-44.4492644
 .7002774294        S12
45.55073356
 .7002774294        S21
45.55073356
 0.713870802        S22
-44.4492644
```

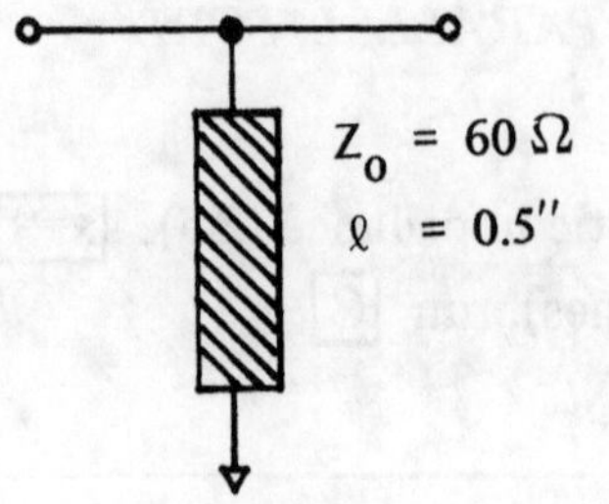


```
     TWO-PORT B
 .5777951709        S11
125.2956157
0.816181806         S12
35.29561566
0.816181806         S21
35.29561566
 .5777951709        S22
125.2956157
```

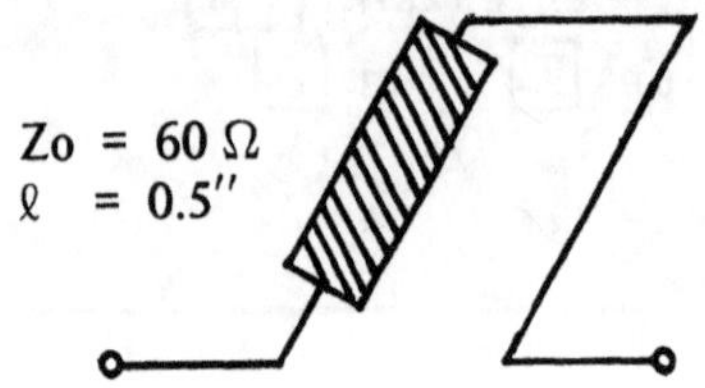


```
     TWO-PORT B
 .3329909797        S11
70.54958332
 .9429300119        S12
-19.45041668
 .9429300119        S21
-19.45041668
 .3329909797        S22
70.54958332
```

An overflow error will occur when the electrical length of the transmission line is very close to 90 degrees at the analysis frequency. The analysis frequency or transmission length should be changed slightly to prevent the error; 90.00000001 will not cause an overflow.

MP-40 SPECIAL TWO-PORTS

The special two-ports of this section include useful circuit elements which, when used in conjunction with the circuit elements of the previous section, provide the user with a wide catalogue of elements which may be used to analyze complex structures.

MP-41 Special Two-Ports

An analysis frequency is not required for this program.

The results of each calculation are stored as Two-Port B, Reg (22) thru Reg (29).

TRANSFORMER, IMPEDANCE RATIO

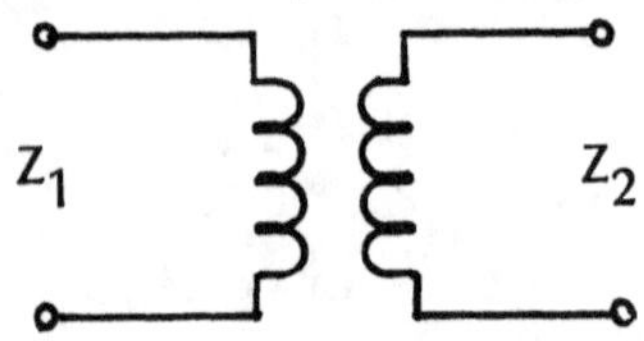

Enter Z_1, [x⇌t], enter Z_2, run [A].

TRANSFORMER, TURNS RATIO

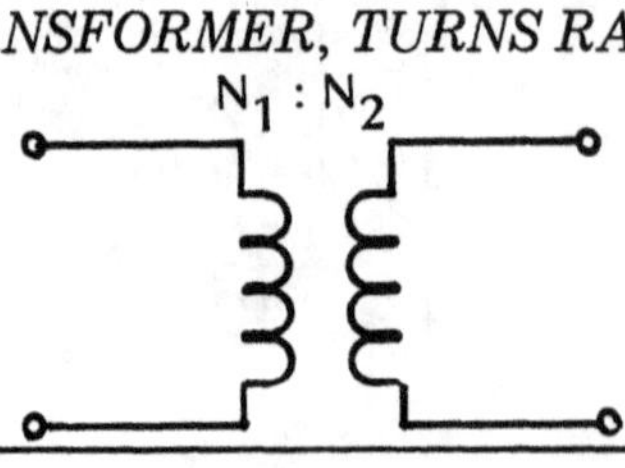

Enter N_1, [x⇌t], enter N_2, run [B].

SHORT

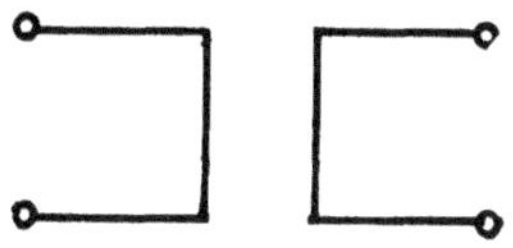

Run [C].

OPEN

Run [D].

IMAGE

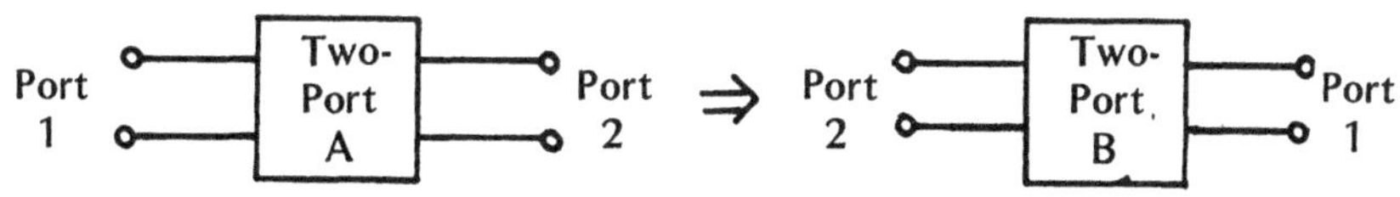

Run [E].

This program creates Two-Port B as a mirror image of Two-Port A.

EQUATE

Run [A′].

This program creates Two-Port B as a copy of Two-Port A.

CURRENT DEPENDENT VOLTAGE SOURCE, CDVS

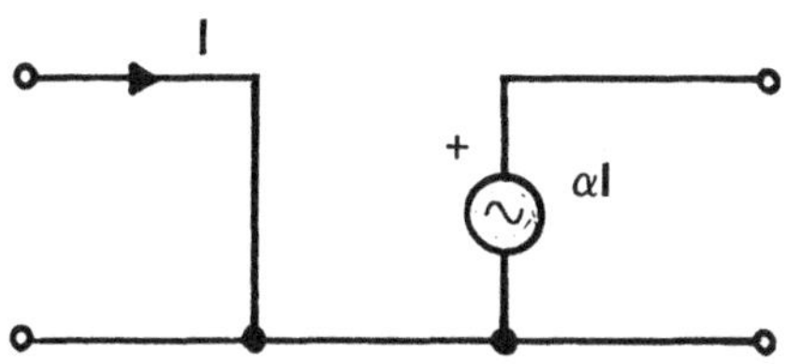

Enter $|\alpha|$, [x⇌t], enter angle of α
run [B′].

VOLTAGE DEPENDENT VOLTAGE SOURCE, VDVS

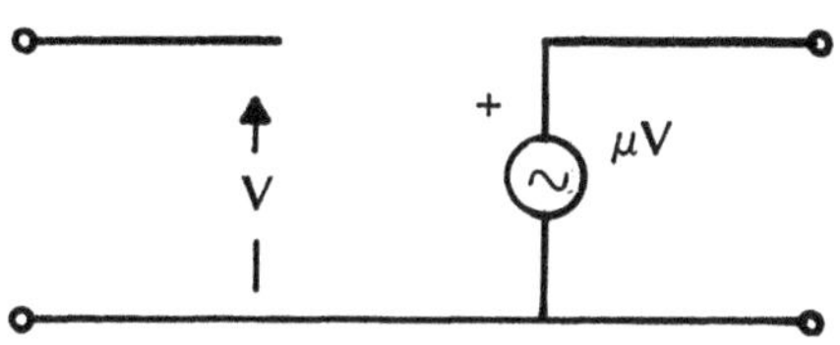

Enter $|\mu|$, [x⇌t], enter angle of μ
run [C′].

CURRENT DEPENDENT CURRENT SOURCE, CDCS

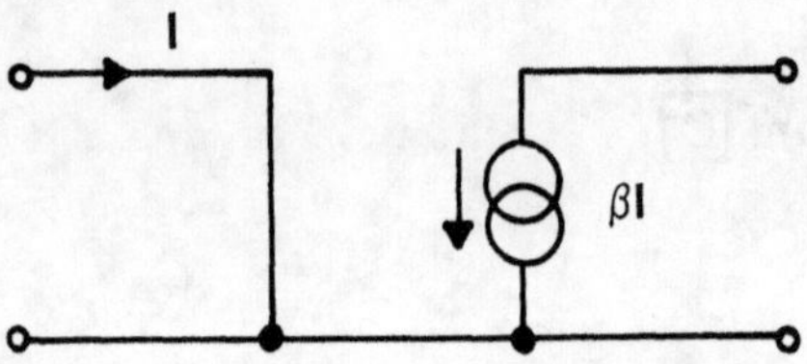

Enter $|\beta|$, [x⇌t], enter angle of β
run [D′] .

VOLTAGE DEPENDENT CURRENT SOURCE, VDCS

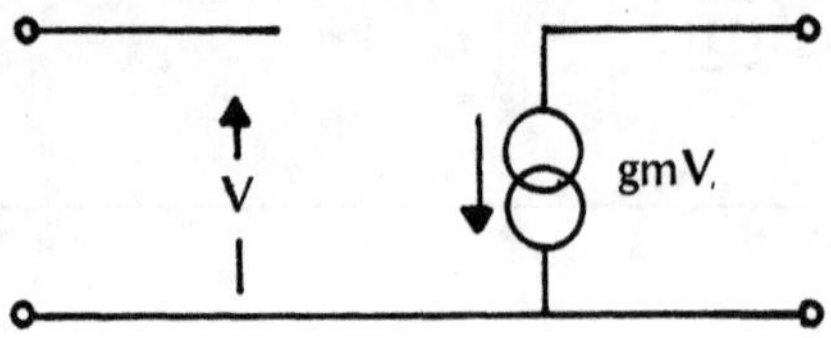

Enter |gm|, [x⇌t], enter angle of gm
run [E′] .

EXAMPLES:

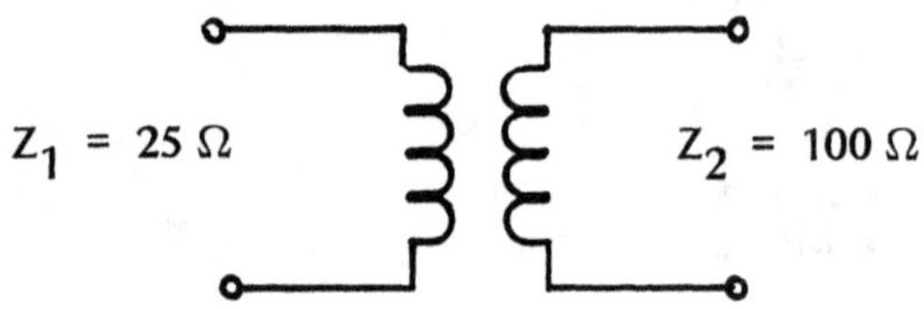


```
TWO-PORT B
      0.6        S11
    180.
      0.8        S12
      0.
      0.8        S21
      0.
      0.6        S22
-.0000000004
```

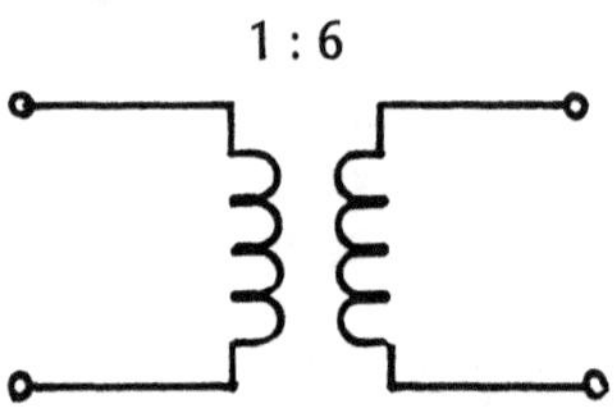


```
 TWO-PORT B
.9459459459      S11
   180.
.3243243243      S12
     0.
.3243243243      S21
     0.
.9459459459      S22
-.0000000004
```

SHORT

```
TWO-PORT B
   1.        S11
 180.
   0.        S12
   0.
   0.        S21
   0.
   1.        S22
 180.
```

OPEN

```
TWO-PORT B
   1.        S11
   0.
   0.        S12
   0.
   0.        S21
   0.
   1.        S22
   0.
```

Load Two-Port A into Reg (14) thru Reg (21).

```
TWO-PORT A
 0.979      S11
 -13.
 0.02       S12
 89.
 1.629      S21
 165.
 0.824      S22
 -2.
```

IMAGE

```
TWO-PORT B
 0.824      S11
 -2.
 1.629      S12
 165.
 0.02       S21
 89.
 0.979      S22
 -13.
```

EQUATE

```
TWO-PORT B
 0.979      S11
 -13.
 0.02       S12
 89.
 1.629      S21
 165.
 0.824      S22
 -2.
```

CDVS
$\alpha = 75\angle 30$

```
TWO-PORT B
 1.         S11
 180.
 0.         S12
 0.
 3.         S21
 30.
 1.         S22
 180.
```

VDVS
$\mu = 0.7\angle 45$

```
TWO-PORT B
 1.         S11
 0.
 0.         S12
 0.
 1.4        S21
 45.
 1.         S22
 180.
```

CDCS
$\beta = 0.5\angle 75$

```
TWO-PORT B
 1.         S11
 180.
 0.         S12
 0.
 1.         S21
 255.
 1.         S22
 0.
```

VDCS
$gm = 0.05\angle -20$

```
TWO-PORT B
 1.         S11
 0.
 0.         S12
 0.
 5.         S21
 160.
 1.         S22
 0.
```

MP-42 Coupled Lines

This program enables the user to analyze networks which contain coupled transmission lines. The user may terminate any two of the four ports associated with a pair of coupled lines to form a two-port network. The terminations may be arbitrary, which allows for analysis of inductance associated with grounding a shorted stub, or of capacitance associated with an open terminated line. As with MP-33, an overflow will be generated if the electrical length of the coupled line pair is 90 degrees at the analysis frequency. The analysis frequency or quarter-wavelength frequency should be changed slightly.

The analysis frequency should be stored in Reg (00).

The program assumes symmetrical coupled lines with equal even and odd mode phase velocity.

The program results are stored as Two-Port B, Reg (22) thru Reg (29).

The termination reflection coefficient must be referenced to the characteristic impedance of the coupler, $Zo = \sqrt{Zoe \cdot Zoo}$.

COUPLED PORT, CPL

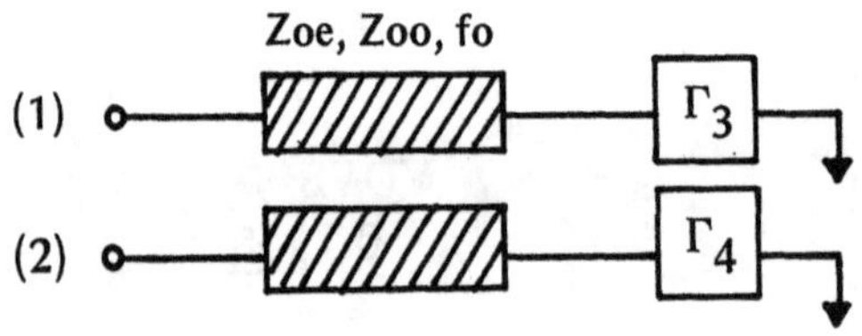

Load program MP-42.1

enter even mode impedance (Zoe), [x⇌t]

enter odd mode impedance (Zoo), run [A]

at stop, enter frequency at which the coupled line pair has an electrical length of 90° (fo), [R/S]

at "33", [CLR], enter $|\Gamma_3|$, [x⇌t], enter Θ_3, [R/S]

at "44", [CLR], enter $|\Gamma_4|$, [x⇌t], enter Θ_4, [R/S]

at "42.2", load Program MP-42.2, run [A]

at Program Stop, Two-Port B in Reg (22) thru Reg (29).

DIRECT PORT, DIR

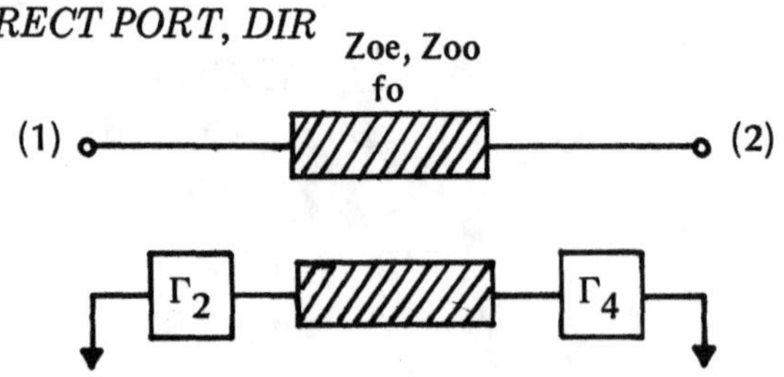

Load program MP-42.1

enter even mode impedance (Zoe), [x⇌t]

enter odd mode impedance (Zoo), run [B]

at stop, enter quarter-wavelength frequency (fo), [R/S]

at "22", [CLR], enter $|\Gamma_2|$, [x⇌t], enter Θ_2, [R/S]

at "44", [CLR], enter $|\Gamma_4|$, [x⇌t], enter Θ_4, [R/S]

at "42.2", load Program MP-42.2, run [A]

at Program Stop, Two-Port B in Reg (22) thru Reg (29).

ISOLATED PORT, ISO

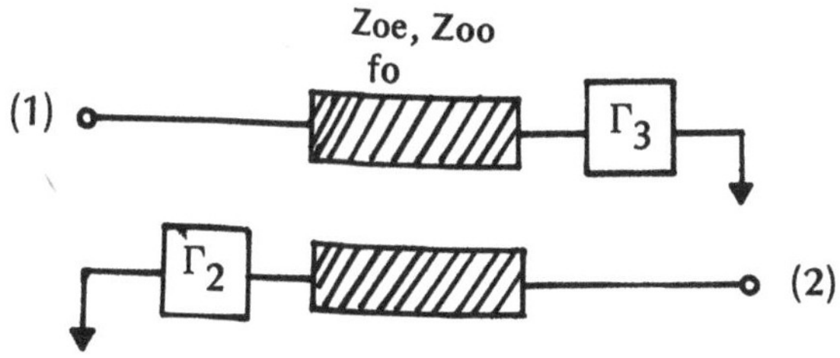

Load program MP-42.1

enter even mode impedance (Zoe), [x⇌t]

enter odd mode impedance (Zoo), run [C]

at stop, enter quarter-wavelength frequency (fo), [R/S]

at '42.2", load program MP-42.2, run [A]

at "22", [CLR], enter $|\Gamma_2|$, [x⇌t], enter Θ_2, [R/S]

at "33", [CLR], enter $|\Gamma_3|$, [x⇌t], enter Θ_3, [R/S]

at Program Stop, Two-Port B in Reg (22) thru Reg (29).

EXAMPLES: Analysis frequency is 2.0 GHz

COUPLED PORT

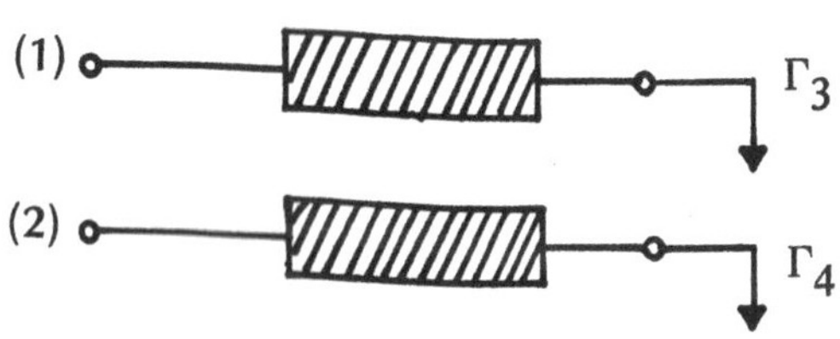

Zoe = 100Ω, Zoo = 30Ω
fo = 2.2 GHz
$\Gamma_3 = 1.0\angle 180$
$\Gamma_4 = 1.0\angle 180$

```
  TWO-PORT B
.9866750354     S11
17.58758429
.1627033326     S12
-72.4124157
.1627033326     S21
-72.4124157
.9866750354     S22
17.58758429
```

DIRECT PORT

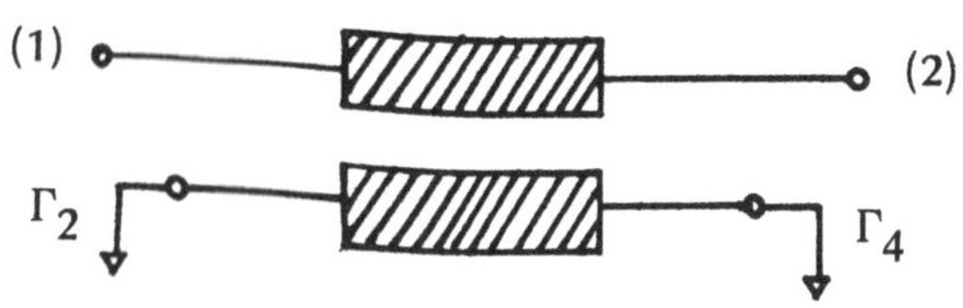

Zoe = 100Ω, Zoo = 30Ω
fo = 2.2 GHz
$\Gamma_2 = 1.0\angle 180$
$\Gamma_4 = 1.0\angle 180$

```
  TWO-PORT B
.0790643287     S11
188.1560304
0.996869516     S12
-81.84396957
0.996869516     S21
-81.84396957
.0790643287     S22
188.1560304
```

ISOLATED PORT

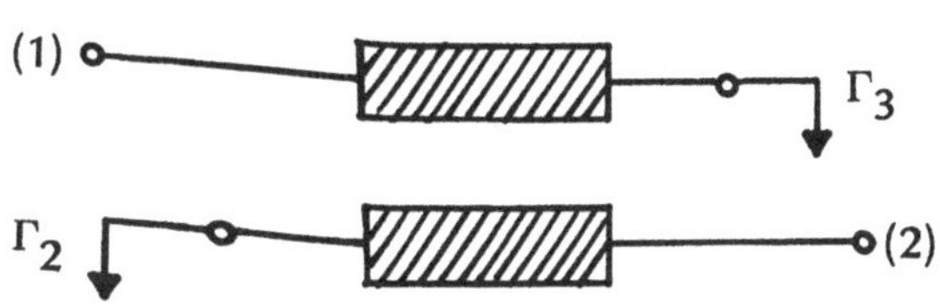

Zoe = 100Ω, Zoo = 30Ω
fo = 2.2 GHz
$\Gamma_2 = 1.0\angle 180$
$\Gamma_3 = 1.0\angle 180$

```
  TWO-PORT B
.4980383522     S11
13.24928168
.8671550033     S12
103.2492817
.8671550033     S21
103.2492817
.4980383522     S22
13.24928168
```

MP-50 MAPPING AND CONFIGURATION CONVERSION

The mapping programs provide the user with a powerful design tool. Any of the four S-parameters can be mapped as a function of series or shunt feedback, which allows the designer to investigate the effects of feedback elements on the circuit performance. Another feature of these programs is parameter conversions. The S-parameters which are measured in a particular configuration — for example, common emitter — can be converted to another configuration, common base or common collector. The map-source and map-load programs provide the user with a graphic presentation of S'_{11} or S'_{22} as a function of all possible load or source impedances.

MP-51 Two to Three-Port Conversion

The series and shunt feedback mappings, as well as the parameter conversions, require a transformation from two-port to three-port parameters. This program provides the transformation, and is essentially a preprocessor program for the parameter conversion of MP-52, and the three-port mapping of MP-54.

TWO-PORT TO THREE-PORT CONVERSIONS

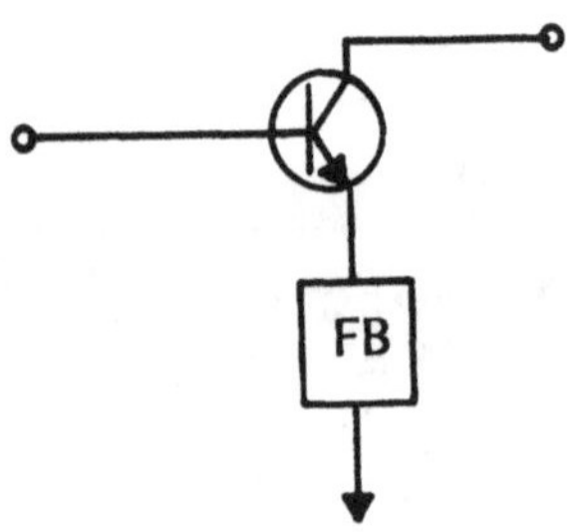

Series Feedback

Load Program MP-51.1

Two-Port A in Reg (14) thru Reg (21)

run [A]

at stop, three-port in Reg (14) thru Reg (31) .

DISPLAY THREE-PORT

Load MP-51.2, run [D]

at "11", [R/S] , at stop $|S_{11}|$ in display, [x⇌t] , Θ_{11} in display, [R/S]

at "12", [R/S] , at stop $|S_{12}|$ in display, [x⇌t] , Θ_{12} in display, [R/S]

.
.
.

continue to "99", Program Stop.

PRINT THREE-PORT

Load MP-51.2, run [E] .

SHUNT FEEDBACK

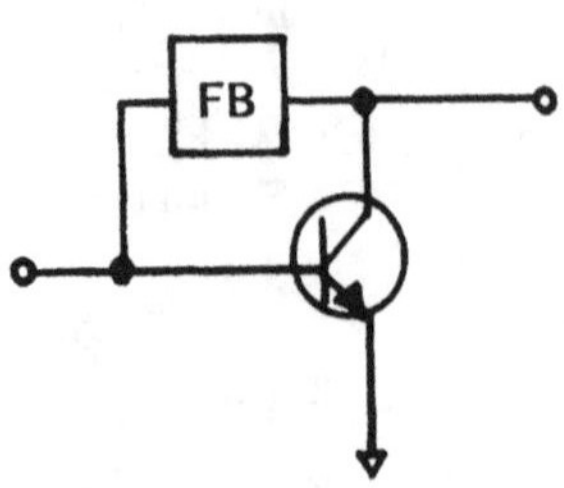

Load Program MP-51.1

Two-Port A in Reg (14) thru Reg (21)

run [B]

at "51.2", load Program MP-51.2, run [A]

at Program Stop, three-port in Reg (14) thru Reg (31) .

Display three-port: Same as above.

Print three-port: Same as above.

PARAMETER CONVERSIONS

Load Program MP-51.1

input Two-Port A in Reg (14) thru Reg (21)

run [C]

at Program Stop, three-port in Reg (14) thru Reg (31).

Display three-port: Same as above.

Print three-port: Same as above.

CHAIN WITH PROGRAM MP-52.

EXAMPLE:

Enter Two-Port A in Reg (14) thru Reg (21). Run the appropriate program function to obtain the output data. The three-port data for each example should be saved on a data card (bank 3 and bank 4) for use with the next examples.

```
TWO-PORT A
  0.979        S11
  -13.
  0.02         S12
  89.
  1.629        S21
  165.
  0.824        S22
  -2.
```

THREE-PORT PARAMETERS FOR SERIES FEEDBACK AND PARAMETER CONVERSIONS

```
 3-PORT S-PARAMETERS
 .9891514157        S11
-7.897244503
 .0294944412        S12
 87.02689694
 .1080815813        S13
 80.03678352
 .8595657011        S21
 166.6446516
 .9069545953        S22
-2.449167754
 .9438190139        S23
-9.747453397
 .8588371185        S31
-4.182930535
 .0928114295        S32
 5.752217771
 .0738734079        S33
 46.22627674
```

THREE-PORT PARAMETERS FOR SHUNT FEEDBACK

```
 3-PORT S-PARAMETERS
-.0291072713        S11
 130.2377038
-.4858743095        S12
 179.1825727
 .5331964923        S13
-1.643031836
-0.080010195        S21
 230.2417085
-.0336521753        S22
-50.90292859
 0.928282074        S23
 177.8150509
 .9712472344        S31
 -4.94538358
 0.494059426        S32
 183.8354638
 .4633642292        S33
-6.279113996
```

MP-52 Parameter Conversion

This program converts the three-port parameters of Program MP-51.1 label [C] , to three-port parameters with a specified common terminal. Program MP-53 is used as a post-processor to convert the three-port parameters to two-port parameters. These programs can be used to convert the two-port parameters of a three terminal network with one common terminal, to a two-port network with another common terminal. For example, one can convert common emitter to common base for a bipolar transitor, or from common source to common gate for a field effect transistor.

CONFIGURATION CONVERSION

Three-port parameters in Reg (14) thru Reg (31) from Program MP-51.1 [C]

run label corresponding to input data:

- [B′] for common emitter (or other convention)
- [C′] for common base (or other convention)
- [D′] for common collector (or other convention)

run label corresponding to desired transformation:

- [B] for common emitter (or other convention)
- [C] for common base (or other convention)
- [D] for common collector (or other convention)

at Program Stop, three-port parameters in Reg (14) thru Reg (31) .

Transformed three-port may be displayed or printed using Program MP-51.2, labels [D] and [E], respectively.

EXAMPLES: The three-port conversion example of MP-51 is based on the three-port parameters for an FET with common source (emitter), convert to common gate (base) and common drain (collector).

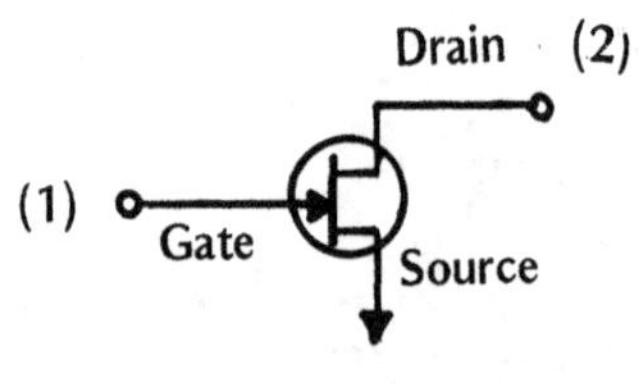

≡

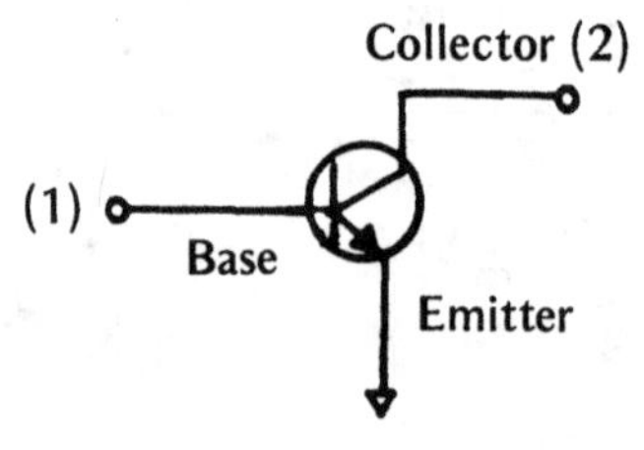

COMMON GATE THREE-PORT PARAMETERS

```
3-PORT S-PARAMETERS
.0738734079        S11
46.22627674
.0928114295        S12
5.752217771
.8588371185        S13
-4.182930535
.9438190139        S21
-9.747453397
.9069545953        S22
-2.449167754
.8595657011        S23
166.6446516
.1080815813        S31
80.03678352
.0294944412        S32
87.02689694
.9891514157        S33
-7.897244503
```

COMMON DRAIN THREE-PORT PARAMETERS

```
3-PORT S-PARAMETERS
.9891514157        S11
-7.897244503
.1080815813        S12
80.03678352
.0294944412        S13
87.02689694
.8588371185        S21
-4.182930535
.0738734079        S22
46.22627674
.0928114295        S23
5.752217771
.8595657011        S31
166.6446516
.9438190139        S32
-9.747453397
.9069545953        S33
-2.449167754
```

MP-53 Three to Two-Port Conversion

Program MP-53 is used in conjunction with MP-52 to convert the transformed three-port parameters to two-port parameters.

THREE-PORT TO TWO-PORT CONVERSION

The three-port parameters from MP-52 must be in Reg (14) thru Reg (31).

Run [A]

at stop, Two-Port A stored in Reg (14) thru Reg (21).

Display two-port : Run [B].

Print two-port : Run [E'], run [B].

EXAMPLE: Convert the three-port common gate and common drain parameters from MP-52 to two-port networks.

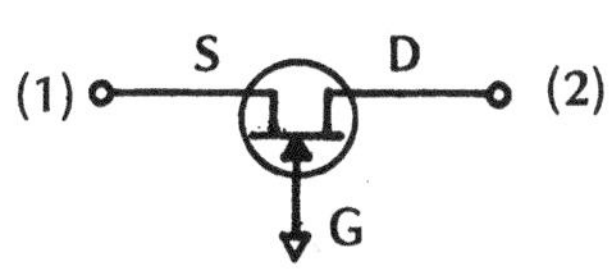

COMMON GATE PARAMETERS

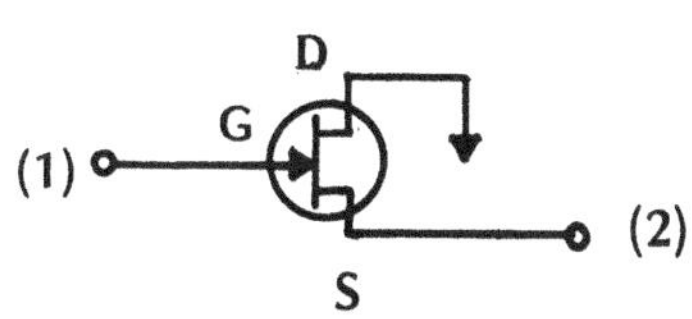

COMMON DRAIN PARAMETERS

COMMON GATE PARAMETERS		COMMON DRAIN PARAMETERS	
TWO-PORT A		TWO-PORT A	
0.043427102	S11	.9909211502	S11
9.687905182		-7.13451201	
.0916897351	S12	.0934869367	S12
-2.151618463		80.2854778	
.9527805267	S21	.9006507831	S21
-6.970853137		-4.287689152	
.9092496532	S22	.0558552634	S22
-1.656206334		84.64120529	

The above common gate and common drain parameters may be converted back to common source S-parameters using the same procedure followed above.

MP-54 Three-Port Mapping

Program MP-54 computes the contours of constant resistance or reactance for either series or shunt feedback. The contours, which are circles, can be plotted on the plane of the desired parameters; then, the results of a feedback element which corresponds to a point on the mapping can be determined directly. Program MP-54.1 computes three mapping constants from the input three-port parameters (from Program MP-51), and Program MP-54.2 computes the center and radius of the various contours.

SMITH CHART MAPPING:

No input required, run [B′]

at Program Stop, mapping constants in Reg (08) thru Reg (13).

S_{11} *MAPPING:*

Series or shunt three-port parameters from MP-51 in Reg (14) thru Reg (31)

run [B]

at Program Stop, mapping constants in Reg (08) thru Reg (13).

S_{12} *MAPPING:*

Series or shunt three-port parameters from MP-51 in Reg (14) thru Reg (31)

run [C]

at Program Stop, mapping constants in Reg (08) thru Reg (13).

S_{21} *MAPPING:*

Series or shunt three-port parameters from MP-51 in Reg (14) thru Reg (31)

run [D]

at Program Stop, mapping constants in Reg (08) thru Reg (13).

S_{22} MAPPING:

Series or shunt three-port parameters from MP-51 in Reg (14) thru Reg (31)

run [E]

at Program Stop, mapping constants in Reg (08) thru Reg (13).

After the mapping constants are determined, load MP-54.2.

CONTOURS OF CONSTANT RESISTANCE:

Enter ± R, ohms, run [A]

at stop, magnitude of center in display, [x⇌t]

angle of center in display, [R/S]

at stop, radius of contour in display, [x⇌t], input resistance in display.

CONTOURS OF CONSTANT REACTANCE:

Enter ± X, ohms, run [B]

at stop, magnitude of center in display, [x⇌t]

angle of center in display, [R/S]

at stop, radius of contour in display, [x⇌t], input reactance in display.

Print all output: Run [E′].

Turn print off: Run [E].

EXAMPLE: Use the three-port shunt feedback data of example MP-51 to compute mapping parameters. Output is registers (08) thru (13).

	SMITH CHART	S_{11} MAPPING
Reg (08)	0.	.3243654449
Reg (09)	0.	-.0238691972
Reg (10)	1.	.6295428485
Reg (11)	0.	-.1963578849
Reg (12)	1.	2.675275229
Reg (13)	0.	.3452987229

	S_{12} MAPPING	S_{21} MAPPING
Reg (08)	.3320829214	-0.480756739
Reg (09)	-.0298304716	.0694109769
Reg (10)	-.3317338733	-1.092736412
Reg (11)	.0498274255	.3522052476
Reg (12)	2.675275229	2.675275229
Reg (13)	.3452987229	.3452987229

Reg (08) S_{22} MAPPING

Reg (09)

Reg (10)

Reg (11)

Reg (12)

Reg (13)

```
 .2468037751
 .0634497025
 .5766942663
-.0922068870
 2.675275229
 .3452987229
```

Use the mapping constants from the S_{22} mappings to generate contours of constant 25 and 50 ohms resistance, along with contours for ± j25 reactance.

CONSTANT RESISTANCE

```
          25.        R
 .3411411843       CTR
 8.246427355
 .4920556163       RAD

          50.        R
 .4055340922       CTR
 5.427978024
 .4251142873       RAD
```

CONSTANT REACTANCE

```
          25.        X
  10.23388259      CTR
  76.33299717
  10.09954177      RAD

         -25.        X
  1.928660597      CTR
 -73.99806157
  1.848354868      RAD
```

MP-55 Map-Source and Map-Load

This program is similar to MP-54 above, except Two-Port A is used to generate the mapping constants. Map-load is a mapping of S'_{11} as a function of a mapping impedance connected as a load, while the map-source function maps S'_{22} as a function of source impedance. The mapping constants are the same as for Program MP-54. Program MP-55.2 is identical to Program MP-54.2, and is duplicated to allow Program MP-55 to stand alone.

MAP-LOAD

Store Two-Port A in Reg (14) thru Reg (21)

run [A]

at stop, mapping constants in Reg (08) thru Reg (13).

MAP-SOURCE

Store Two-Port A in Reg (14) thru Reg (21)

run [B]

at stop, mapping constants in Reg (08) thru Reg (13).

CONTOURS OF CONSTANT RESISTANCE AND REACTANCE:

See MP-54.2

Example: Load Two-Port A in Reg (19) thru Reg (21).

Calculate mapping constants for map-load and map-source. Then load MP-55.2 and compute contours of constant 25 and 100 Ω resistance, then constant ± j100 for the map-source.

```
TWO-PORT A
  0.979     S11
  -13.
  0.02      S12
  89.
  1.629     S21
  165.
  0.824     S22
  -2.
```

	MAP-LOAD		MAP-SOURCE
Reg (08)		Reg (08)	
Reg (09)	.9336773579	Reg (09)	.7833310923
Reg (10)	-.3013689328	Reg (10)	-.7951882324
Reg (11)	-.0574943816	Reg (11)	-.1042488842
Reg (12)	-.0836310786	Reg (12)	.7769833044
Reg (13)	10.03829995	Reg (13)	.8209280348
	1.798452774		8.70043003

CONSTANT REACTANCE

```
         25.    R
 .8968990032    CTR
-33.96483823
 .4872052064    RAD

        100.    R
 .7503118312    CTR
-19.00322178
 .2281387571    RAD
```

CONSTANT RESISTANCE

```
        100.    X
 .7742962567    CTR
-.4019924673
 .0960480168    RAD

       -100.    X
 .7387622325    CTR
-0.791647087
 .0601436591    RAD
```

MP-60 LADDER NETWORK ANALYSIS

The ladder network analysis program provides a simple and fast means of analyzing many types of circuits. This program includes the standard lumped and distributed elements which are most often used for matching networks and filters. The user stores a termination impedance in two data registers, enters input data for a circuit element, then executes the appropriate program for that element. The program computes the element impedance, and connects the impedance in either series or shunt with the termination impedance to form the input impedance at that point. A subroutine is included to convert impedance to reflection coefficient (referenced to 50 ohms), and another converts reflection coefficient to VSWR and insertion loss. Note that the insertion loss is calculated from the input reflection coefficient and assumes a lossless network, i.e., no resistive elements.

The major advantage of this program is the ability to actually "program" a circuit and allow the calculator to analyze the network over a specified frequency band, without operator intervention.

MP-61 Ladder Network Analysis

Reg (00) analysis frequency (GHz)
Reg (01) stop frequency (GHz) for frequency loop
Reg (02) step frequency (GHz) for frequency loop
Reg (03) magnitude of impedance at present node of network
Reg (04) angle of impedance at present node of network
Reg (05) thru Reg (08) used for scratch data memory.

The analysis frequency must be stored in Reg (00) before any calculations are made.

The program computes the impedance of each element and connects this impedance in either series or parallel with the nodal impedance in Reg (03) and Reg (04). The results are stored in Reg (03) and Reg (04) as the new nodal impedance. The results are also stored in the "x" and "t" registers for display. The program does not use the data in these registers, and can be viewed or modified by the user at any time.

SERIES RESISTOR

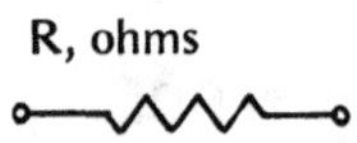

Enter R, ohms
run [A].

PARALLEL RESISTOR

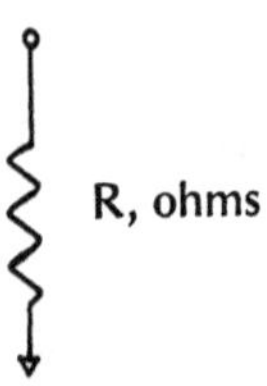

Enter R, ohms
run [E′], run [A].

SERIES INDUCTOR · CAPACITOR CONNECTED IN SERIES

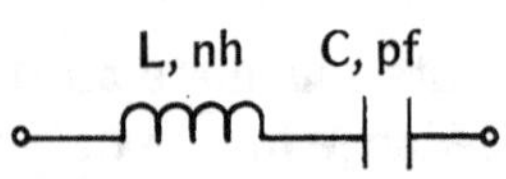

Enter L (nh), [x⇌t], enter C (pf),
run [B].

SERIES INDUCTOR · CAPACITOR CONNECTED IN PARALLEL

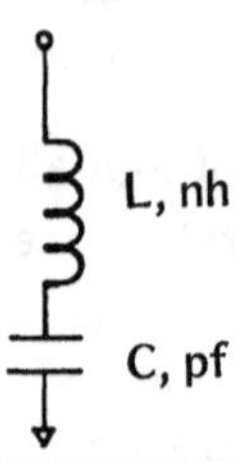

Enter L (nh), [x⇌t], enter C (pf)
run [E′], run [B].

PARALLEL INDUCTOR · CAPACITOR CONNECTED IN SERIES

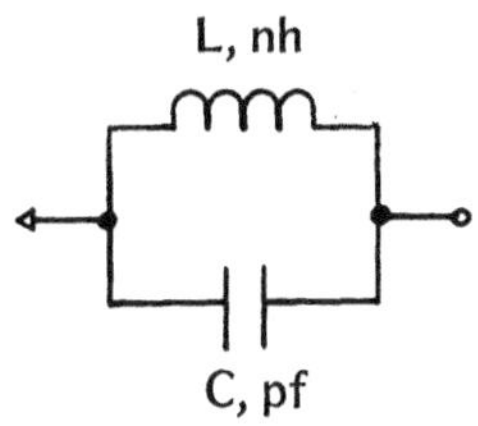

Enter L (nh), [x⇌t], enter C (pf)
run [C] .

PARALLEL INDUCTOR · CAPACITOR CONNECTED IN PARALLEL

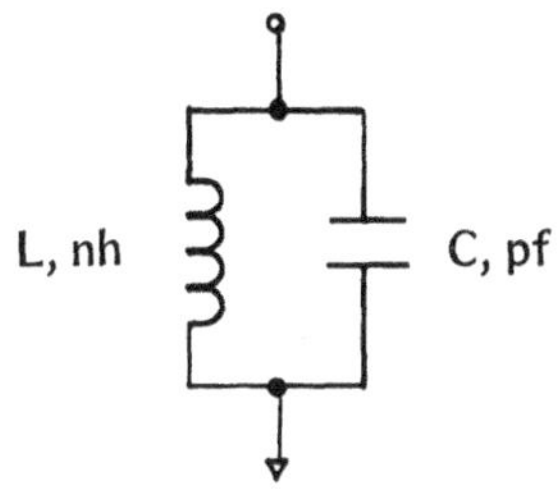

Enter L (nh), [x⇌t], enter C (pf)
run [E′] , run [C] .

SHORT-CIRCUIT TRANSMISSION LINE CONNECTED IN SERIES

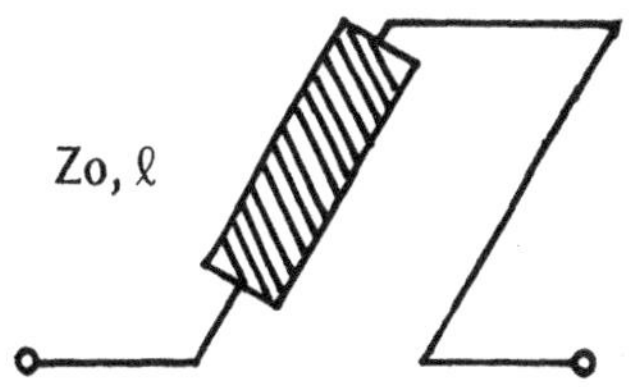

Enter characteristic impedance (Zo), [x⇌t],
enter length (inches) in air (ℓ)
run [D] .

SHORT-CIRCUIT TRANSMISSION LINE CONNECTED IN PARALLEL

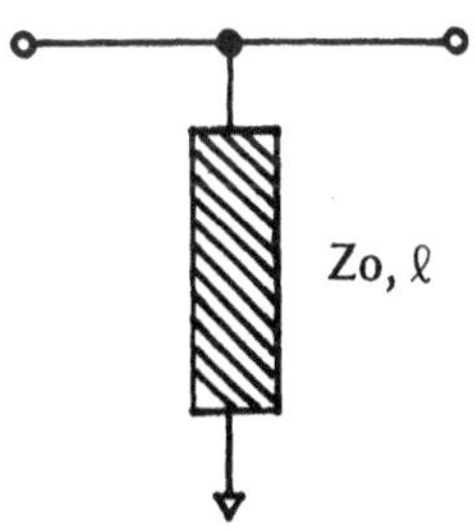

Enter characteristic impedance (Zo), [x⇌t],
enter length (inches) in air (ℓ)
run [E′] , run [D] .

OPEN-CIRCUIT TRANSMISSION LINE CONNECTED IN SERIES

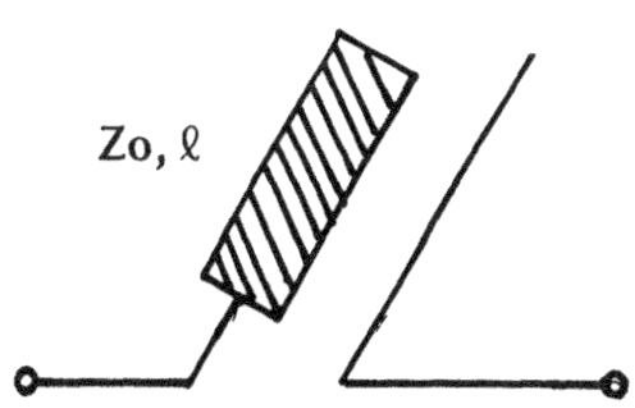

Enter characteristic impedance (Zo), [x⇌t]
enter length (inches) in air (ℓ)
run [E] .

OPEN-CIRCUIT TRANSMISSION LINE CONNECTED IN PARALLEL

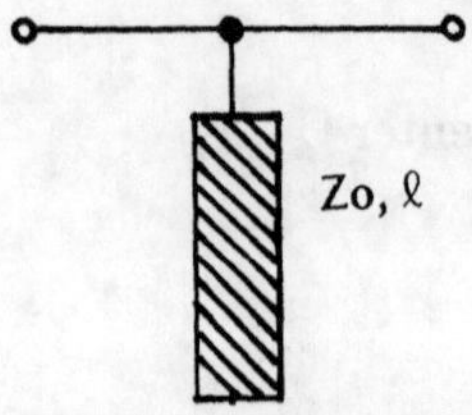

Enter characteristic impedance (Zo), [x⇌t]
enter length (inches) in air (ℓ)
run [E′] , run [E] .

TRANSMISSION LINE

Enter characteristic impedance (Zo), [x⇌t]
enter length (inches) in air (ℓ)
run [A′] .

CONVERT IMPEDANCE TO REFLECTION COEFFICIENT: Z ⇒ Γ

Magnitude of impedance in "x" register, angle of impedance in "t" register (this is the normal output from an element calculation)

run [B′]

at stop, magnitude in "x", angle is "t".

CONVERT REFLECTION COEFFICIENT TO VSWR AND INSERTION LOSS

Magnitude of reflection coefficient in "x", angle in "t" (normal output from [B′] above)

run [C′]

at stop, VSWR in "x", insertion loss (dB) in "t".

(The loss calculation assumes a lossless network.)

PRINT OUTPUT [D′]

The print output will print the frequency, then magnitude and angle of impedance, or reflection coefficient if [B′] , or VSWR and loss in dB if subroutine [C′] has been executed.

PROGRAMMING A NETWORK

Store start frequency (GHz) in Reg (00)
store stop frequency (GHz) in Reg (01)
store step frequency (GHz) in Reg (02) .

[GTO] [CLR]

Program network. The first portion of the network should be the termination impedance in Reg (03) and Reg (04). Any print statements should be included in the programming.

The last program steps should be [GTO] [=].

To execute the program: [GTO], [=], [R/S].

There are 425 program steps used for this program, and data registers (00) thru (08). The partition may be changed at any time to allocate more program memory.

LABEL "="

LABEL "CLR"

[TERMINATION IMPEDANCE]

[NETWORK TOPOLOGY] } USER PROGRAM STATEMENTS

[PRINT STATEMENTS]

GTO "="

LUMPED ELEMENT EXAMPLE

The 50 to 1000 ohm transformer shown below will be analyzed. This transformer is designed to operate from 1 to 2 GHz with a maximum of 0.1 dB ripple. The element values are in

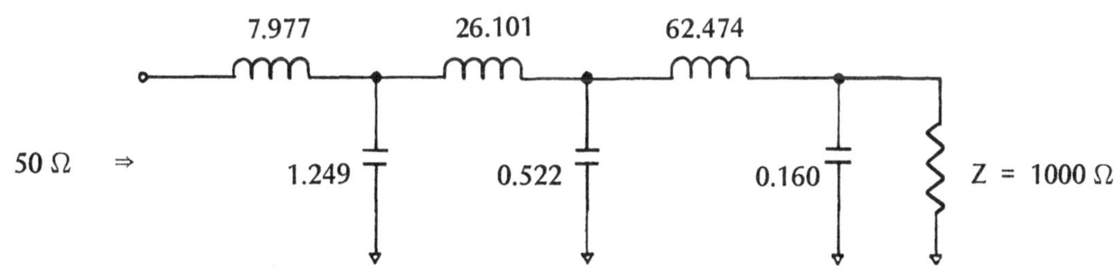

picofarads and nanohenrys. The circuit will be analyzed at 1.0 GHz, using the normal keyboard entry.

Store 1.0 GHz into Reg (00).

Store the termination impedance in Reg (03) and (04)

1000 ⇒ Reg (03), 0 ⇒ Reg (04).

Connect the 0.160 parallel capacitor. Note: use the SLC connected in parallel with L=0, and C = 0.160.

Connect the series 62.474 nh inductor using the PLC with L = 62.474 and C = 0.

.
.
.

Continue until the network is complete.

```
          1.    FREQ
       1000.    MAG     TERMINATION IMPEDANCE OF 1000 0 OHMS
          0.    ANG

          1.    FREQ
 705.2320352    MAG     SHUNT 0.160 pf CAPACITOR
-45.15170713    ANG

          1.    FREQ
 508.8283592    MAG     SERIES 62.474 nh INDUCTOR
-12.19184601    ANG

          1.    FREQ
 240.1310015    MAG     SHUNT 0.522 pf CAPACITOR
 -62.5299189    ANG

          1.    FREQ
  121.146866    MAG     SERIES 26.101 nh INDUCTOR
-23.88842185    ANG

          1.    FREQ
 74.08685727    MAG     SHUNT 1.249 pf CAPACITOR
-56.00239674    ANG

          1.    FREQ
 42.94020704    MAG     SERIES 7.977 nh INDUCTOR
-15.25959455    ANG

          1.    FREQ
 .1539878562    MAG     CONVERT Zin to Γin
-120.1383321    ANG

          1.    FREQ
  1.36403226    VSWR    CALCULATE VSWR AND INSERTION LOSS.
-.1042216437    DB
```

The network will now be programmed for analysis across the frequency band from 0.75 to 2.25 GHz in 0.25 GHz steps. The element values are stored in the following data registers:

ZT = 1000 Ω ⇌ Reg (09)

0.160 pf ⇌ Reg (10)

62.474 nh ⇌ Reg (11)

0.522 pf ⇌ Reg (12)

26.101 nh ⇌ Reg (13)

1.249 pf ⇌ Reg (14)

7.977 nh ⇌ Reg (15)

store the start frequency, 0.75, into Reg (00)

store the stop frequency, 2.25, into Reg (01)

store the step frequency, 0.25, into Reg (02)

GTO "CLR" and program the circuit as shown below. Note that in this example, the zero value for the undesired portion of the SLC or PLC is programmed as part of the network. A large value may also be used as a circuit value when only one element is desired. Should the normal program execution be interrupted for some reason, use [INV] [2nd] [ST FLG] [4] to clear flag four and allow a normal start frequency.

```
424  76 LBL
425  25 CLR
426  43 RCL   }
427  09  09   }
428  42 STO   }  TERMINATION IMPEDANCE
429  03  03   }
430  00  0    }
431  42 STO   }
432  04  04   }
433  00  0    }
434  32 X:T   }  PARALLEL CONNECTED SLC, L = 0, C = 0.160
435  43 RCL   }
436  10  10   }
437  10 E'    }
438  12  B    }
439  43 RCL   }
440  11  11   }
441  32 X:T   }  SERIES CONNECTED PLC, L = 62.474, C = 0
442  00  0    }
443  13  C    }
444  00  0    }
445  32 X:T   }
446  43 RCL   }  PARALLEL CONNECTED SLC, L = 0, C = 0.522
447  12  12   }
448  10 E'    }
449  12  B    }
450  43 RCL   }
451  13  13   }
452  32 X:T   }  SERIES CONNECTED PLC, L = 26.101, C = 0
453  00  0    }
454  13  C    }
455  00  0    }
456  32 X:T   }
457  43 RCL   }  PARALLEL CONNECTED SLC, L = 0, C = 1.249
458  14  14   }
459  10 E'    }
460  12  B    }
461  43 RCL   }
462  15  15   }
463  32 X:T   }  SERIES CONNECTED PLC, L = 7.977, C = 0
464  00  0    }
465  13  C    }
```

```
466  17 B'
467  18 C'     CONVERT Z = Γ
468  19 D'     CONVERT Γ ⇒ VSWR & LOSS
469  61 GTO    PRINT
470  95  =
```

If desired, a print statement may be used to print impedance, reflection coefficient, and insertion loss. If a printer is not available, insert a R/S at the desired output.

```
       0.75    FREQ
 3.964157379   VSWR
-1.914802038     DB

         1.    FREQ
  1.36403226   VSWR
-.1042216437     DB

       1.25    FREQ
 1.326520888   VSWR
-.0863982783     DB

        1.5    FREQ
 1.161172327   VSWR
-.0242212731     DB

       1.75    FREQ
 1.337466645   VSWR
-.0914789334     DB

         2.    FREQ
 1.363321409   VSWR
-.1038733231     DB

       2.25    FREQ
  23.7323841   VSWR
-8.091306051     DB
```

HP-67/97

INTRODUCTION

The network analysis programs used with the HP-67/97 calculators use the primary and secondary set of data registers to store Two-Ports A and B, respectively. Data register "A" is used to store the frequency, and the remaining data locations are shown below:

Reg (A) Frequency

TWO-PORT A (PRIMARY)		TWO-PORT B (SECONDARY)	
Reg (1)	$\lvert S_{11} \rvert$	Reg (1)	$\lvert S_{11} \rvert$
Reg (2)	$\lvert S_{12} \rvert$	Reg (2)	$\lvert S_{12} \rvert$
Reg (3)	$\lvert S_{21} \rvert$	Reg (3)	$\lvert S_{21} \rvert$
Reg (4)	$\lvert S_{22} \rvert$	Reg (4)	$\lvert S_{22} \rvert$
Reg (5)	Θ_{11}	Reg (5)	Θ_{11}
Reg (6)	Θ_{12}	Reg (6)	Θ_{12}
Reg (7)	Θ_{21}	Reg (7)	Θ_{21}
Reg (8)	Θ_{22}	Reg (8)	Θ_{22}

This arrangement enables the user to easily store or recall Two-Port A and B using magnetic data cards.

Complete program listings for the HP-67/97 network analysis programs can be found in Appendix B. A card labeling scheme is also shown for each program. Cards labeled in this manner should be sufficient to prompt a user who is moderately familiar with the programs for the necessary program inputs. The following pages contain a detailed description for the use for each program. Numerical examples of each program are included as well, which are intended to familiarize the user with the program execution, and provide a means to verify that each program is correct. The final chapter demonstrates the application of these programs to the analysis of a complex network.

MP-00 MISCELLANEOUS

This set of three programs is used to enter and recall the "S", "$\dot{Y}$", "Z", "H", and "G" parameters into the correct data storage registers. The last program contains some miscellaneous useful relations.

MP-01 Data Input/Output

This program is used to enter and recall normalized two-port data. The program card is labeled for "S" parameters; however, any of the parameter sets may be entered or recalled as long as normalization is not required. The instructions used to interchange Two-Port A and B, as well as the LOAD and SAVE functions, are included with this program — although they are calculator functions and are not programmed as part of MP-01.

ENTER TWO-PORT A:

Enter $|S_{11}|$, [ENTER ↑], enter Θ_{11}, run [f][A]
enter $|S_{12}|$, [ENTER ↑], enter Θ_{12}, run [f][B]
enter $|S_{21}|$, [ENTER ↑], enter Θ_{21}, run [f][C]
enter $|S_{22}|$, [ENTER ↑], enter Θ_{22}, run [f][D].

ENTER TWO-PORT B:

Enter $|S_{11}|$, [ENTER ↑], enter Θ_{11}, run [E][f][A]
enter $|S_{12}|$, [ENTER ↑], enter Θ_{12}, run [E][f][B]
enter $|S_{21}|$, [ENTER ↑], enter Θ_{21}, run [E][f][C]
enter $|S_{22}|$, [ENTER ↑], enter Θ_{22}, run [E][f][D].

RECALL TWO-PORT A:

Run [A], at stop $|S_{11}|$ in "x", Θ_{11} in "y"
run [B], at stop $|S_{12}|$ in "x", Θ_{12} in "y"
run [C], at stop $|S_{21}|$ in "x", Θ_{21} in "y"
run [D], at stop $|S_{22}|$ in "x", Θ_{22} in "y".

RECALL TWO-PORT B:

Run [E][A], at stop $|S_{11}|$ in "x", Θ_{11} in "y"
run [E][B], at stop $|S_{12}|$ in "x", Θ_{12} in "y"
run [E][C], at stop $|S_{21}|$ in "x", Θ_{21} in "y"
run [E][D], at stop $|S_{22}|$ in "x", Θ_{22} in "y".

EXCHANGE A & B:

Execute [f][P⇌S]

Two-Port A is exchanged with Two-Port B.

SAVE A:

Execute [f][W/DATA]

load one side of data card, press any key to interrupt load.

SAVE B:

Execute [f][P⇌S] (exchange A & B)

save data as above

execute [f][P⇌S] (restore A & B).

LOAD A:

Load data card, press any key to stop data load into secondary.

A second method may be used:

store 8 into the index register

enter 8, [h] [STI]

execute [g] [MERGE].

The second method does not load the frequency register "A". If this method is used, ensure that the correct frequency is in Reg (A).

LOAD B:

Execute [f] [P⇌S]	(exchange A & B)
load data as above	
execute [f] [P⇌S]	(restore A & B).

Program verification is accomplished by entering a known set of values for Two-Port A and B, then recalling the register set corresponding to A and B. The following set is entered:

$A_{11} = 1\angle 2$ $\quad A_{12} = 3\angle 4$
$A_{21} = 5\angle 6$ $\quad A_{22} = 7\angle 8$

$B_{11} = 9\angle 10$ $\quad B_{12} = 11\angle 12$
$B_{21} = 13\angle 14$ $\quad B_{22} = 15\angle 16$

The convention $A_{11} = 1\angle 2$ is used throughout this book to indicate the complex number associated with S_{11} of Two-Port A. The "1" is the polar magnitude, $|S_{11}|$, and the "2" is the angle, Θ_{11}.

```
R01=  1.00000
R02=  3.00000
R03=  5.00000
R04=  7.00000
R05=  2.00000
R06=  4.00000
R07=  6.00000
R08=  8.00000

R11=  9.00000
R12= 11.00000
R13= 13.00000
R14= 15.00000
R15= 10.00000
R16= 12.00000
R17= 14.00000
R18= 16.00000
```

The register numbers associated with the primary set are 0 through 9, and those associated with the secondary set are 10 through 19. The secondary registers may be accessed through the index register using the [STO] [(i)] and [RCL] [(i)] functions, as well as the [f] [P⇌S] command. Most of the listings shown in this chapter were made using the HP-41C printer, where "R15=" corresponds to data register 15, or register 5 if the [f] [P⇌S] function is used.

MP-02 Parameter Normalization

This program is used to enter and recall non-normalized values of "Z", "Y", "H", and "G" into and from Two-Port A. Two-Port A is always in normalized form. The normalization impedance is a real 50 ohms, or 20 mmho in the case of "Y" parameters.

ENTER Z: Run [A]

at "11", enter $|Z_{11}|$, [ENTER ↑], enter Θ_{11}, [R/S]
at "12", enter $|Z_{12}|$, [ENTER ↑], enter Θ_{12}, [R/S]
at "21", enter $|Z_{21}|$, [ENTER ↑], enter Θ_{21}, [R/S]
at "22", enter $|Z_{22}|$, [ENTER ↑], enter Θ_{22}, [R/S]
at stop, normalized "Z" parameters are in Two-Port A.

RECALL Z: Run [f] [A]

at "11", $|Z_{11}|$ in "y", Θ_{11} in "z", [R/S]
at "12", $|Z_{12}|$ in "y", Θ_{12} in "z", [R/S]
at "21", $|Z_{21}|$ in "y", Θ_{21} in "z", [R/S]
at "22", $|Z_{22}|$ in "y", Θ_{22} in "z".

ENTER Y (mmho): Run [B]

at "11", enter $|Y_{11}|$, [ENTER ↑], enter Θ_{11}, [R/S]
at "12", enter $|Y_{12}|$, [ENTER ↑], enter Θ_{12}, [R/S]
at "21", enter $|Y_{21}|$, [ENTER ↑], enter Θ_{21}, [R/S]
at "22", enter $|Y_{22}|$, [ENTER ↑], enter Θ_{22}, [R/S]
at stop, normalized "Y" parameters are in Two-Port A.

RECALL Y (mmho): Run [f] [B]

at "11", $|Y_{11}|$ is in "y", Θ_{11} in "z", [R/S]
at "12", $|Y_{12}|$ is in "y", Θ_{12} in "z", [R/S]
at "21", $|Y_{21}|$ is in "y", Θ_{21} in "z", [R/S]
at "22", $|Y_{22}|$ is in "y", Θ_{22} in "z".

ENTER H: Run [C]

at "11", enter $|h_{11}|$, [ENTER ↑], Θ_{11} [R/S]
at "12", enter $|h_{12}|$, [ENTER ↑], Θ_{12} [R/S]
at "21", enter $|h_{21}|$, [ENTER ↑], Θ_{21} [R/S]
at "22", enter $|h_{22}|$, [ENTER ↑], Θ_{22} [R/S]
at stop, normalized "H" parameters are in Two-Port A.

RECALL H: Run [f][C]	at "11", $\|h_{11}\|$ in "y", Θ_{11} in "z", [R/S] at "12", $\|h_{12}\|$ in "y", Θ_{12} in "z", [R/S] at "21", $\|h_{21}\|$ in "y", Θ_{21} in "z", [R/S] at "22", $\|h_{22}\|$ in "y", Θ_{22} in "z".
ENTER G: Run [D]	at "11", enter $\|g_{11}\|$, [ENTER ↑], Θ_{11}, [R/S] at "12", enter $\|g_{12}\|$, [ENTER ↑], Θ_{12}, [R/S] at "21", enter $\|g_{21}\|$, [ENTER ↑], Θ_{21}, [R/S] at "22", enter $\|g_{22}\|$, [ENTER ↑], Θ_{22}, [R/S] at stop, normalized "G" parameters are in Two-Port A.
RECALL G: Run [f][D]	at "11", $\|g_{11}\|$ in "y", Θ_{11} in "z", [R/S] at "12", $\|g_{12}\|$ in "y", Θ_{12} in "z", [R/S] at "21", $\|g_{21}\|$ in "y", Θ_{21} in "z", [R/S] at "22", $\|g_{22}\|$ in "y", Θ_{22} in "z".

Program verification is accomplished by entering a particular set of the parameters which results in the simple normalized values of the previous example. Enter the following parameters, recall each Two-Port A register to verify the correct normalization and storage, then recall the parameter set to obtain the input values. The storage registers should be cleared before each parameter set check to ensure that the data is not from the previous example. The parameter sets are shown below:

```
   TWO-PORT A          TWO-PORT A          TWO-PORT A          TWO-PORT A
 Z-PARAMETERS        Y-PARAMETERS        H-PARAMETERS        G-PARAMETERS
Z11= 50.00000       Y11= 20.00000       H11= 50.00000       G11= 0.02000
      2.00000             2.00000             2.00000             2.00000
Z12= 150.00000      Y12= 60.00000       H12= 3.00000        G12= 3.00000
      4.00000             4.00000             4.00000             4.00000
Z21= 250.00000      Y21= 100.00000      H21= 5.00000        G21= 5.00000
      6.00000             6.00000             6.00000             6.00000
Z22= 350.00000      Y22= 140.00000      H22= 0.14000        G22= 350.00000
      8.00000             8.00000             8.00000             8.00000
```

MP-03 Miscellaneous Relations

This program provides the user with the common relations between reflection coefficient and impedance, admittance, VSWR, return loss, and reflection loss. Equivalence between a series and parallel load is also included.

REFLECTION COEFFICIENT ⇒ IMPEDANCE, Γ ⇒ Z

Enter |Γ|, [ENTER ↑], enter Θ, run [A]
at Program Stop, |Z| in "x", Θ in "y".

IMPEDANCE ⇒ REFLECTION COEFFICIENT, Z ⇒ Γ

Enter |Z|, [ENTER ↑], enter Θ, run [f] [A]
at Program Stop, |Γ| in "x", Θ in "y".

REFLECTION COEFFICIENT ⇒ ADMITTANCE, Γ ⇒ Ymmho

Enter |Γ|, [ENTER ↑], enter Θ, run [B]
at Program Stop, |Y|mmho in "x", Θ in "y".

ADMITTANCE ⇒ REFLECTION COEFFICIENT, Ymmho ⇒ Γ

Enter |Y|mmho, [ENTER ↑], enter Θ, run [f] [B]
at Program Stop, |Γ| in "x", Θ in "y".

RETURN LOSS ⇌ REFLECTION COEFFICIENT, RTN LOSS ⇌ Γ

Enter return loss (dB), run [C]
at Program Stop, |Γ| in "x"

or, enter |Γ|, run [C]
at Program Stop, return loss (dB) in "x".

REFLECTION LOSS ⇌ REFLECTION COEFFICIENT, RFL LOSS ⇌ Γ

Enter reflection loss (dB), run [f] [C]
at Program Stop, |Γ| in "x"

or enter |Γ|, run [f] [C]
at Program Stop, reflection loss (dB) in "x".

VSWR ⇌ REFLECTION COEFFICIENT

Enter VSWR, run [D]

at Program Stop, $|\Gamma|$ in "x"

or enter $|\Gamma|$, run [D]

at Program Stop, VSWR in "x".

MILLIWATT ⇒ dBm

Enter MW, run [f] [D]

at Program Stop, dBm in "x".

SERIES ⇒ SHUNT EQUIVALENT LOAD, Rs ± jXs ⇒ Rp ± jXp

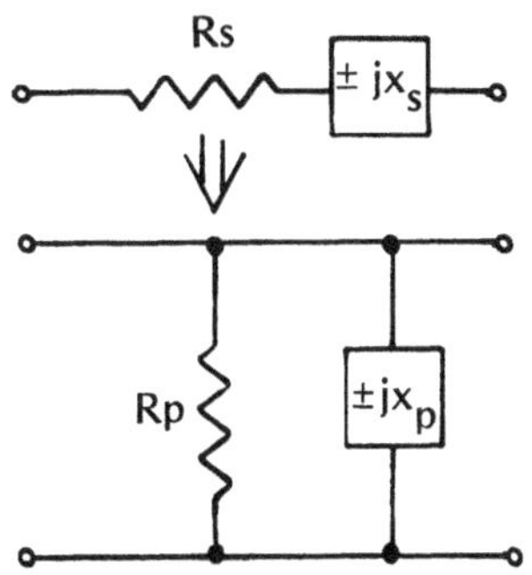

Enter Rs, [ENTER ↑], enter Xs, run [E]

at Program Stop, Rp in "x", Xp in "y".

SHUNT ⇒ SERIES EQUIVALENT LOAD, Rp ± jXp ⇒ Rs ± jXs

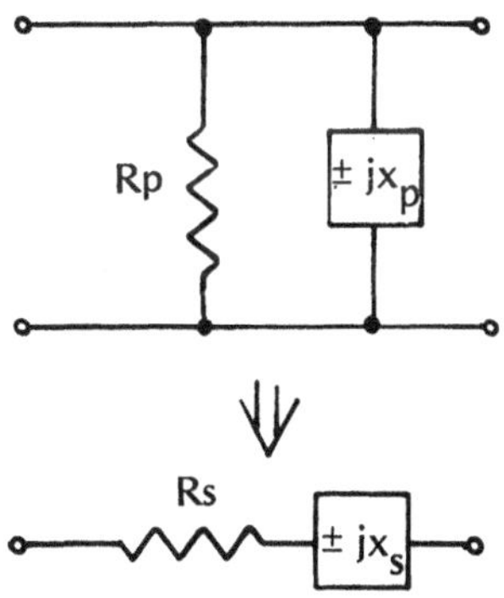

Enter Rp, [ENTER ↑], enter Xp, run [f] [E]

at Program Stop, Rs in "x", and Xs in "y".

The following examples can be used for program verification.

```
Γ=   0.500000000       |Γ|=  0.500000000           Γ=   0.500000000       Rs + jXs=  69.07435702
     45.00000000°      RTN LOSS= -6.020599914dB         45.00000000°                j 65.12392843
Z=   94.93362340                                   Y=   10.53367568
     43.31385670°      |Γ|=  0.500000000               -43.31385670°
 =   69.07435702       RFL LOSS= -1.249387366dB     =   7.664374840       Rp + jXp=  130.4737856
   j 65.12392843                                      j-7.226041907                 j 138.3883477
                       100 MW = +20dBM
```

MP-10 TWO-PORT ANALYSIS

The next group of four programs is used to characterize Two-Port A. These programs provide the user with useful design information such as the stability factor and maximum available gain — along with simultaneous matched load and source if stable — if not, maximum stable gain can be calculated. S'_{11} and S'_{22} and transducer gain can be calculated as well. Stability circles, mismatch circles, and contours of constant gain can also be determined using the programs in this section. Two-Port A remains unchanged by all of these programs.

MP-11 Two-Port Characterization

This program is used to analyze Two-Port A for K, the stability factor; GA max, the maximum available gain; and Γms and Γml, the simultaneous matched source and load. If the stability factor is less than 1, the maximum available gain is not defined, and a simultaneous matched source and load do not exist (positive real). Maximum stable gain, MSG, must be calculated under these conditions.

Two-Port A must be in primary registers (1) thru (8). Run [A] before any other label. At Program Stop, K is in "x".	
STABILITY FACTOR: Run [A] as above. To recall K, run [E] .	
MAXIMUM AVAILABLE GAIN, GA max Run [B] at Program Stop, GA max (dB) in "x".	$K \geqslant 1$
SIMULTANEOUS MATCHED LOAD, Γml Run [C] at Program Stop, |Γml| in "x", Θ in "y" .	$K \geqslant 1$
SIMULTANEOUS MATCHED SOURCE, Γms Run [D] at Program Stop, |Γms| in "x", Θ in "y" .	$K \geqslant 1$
MAXIMUM STABLE GAIN, MSG Run [f][A] at Program Stop, MSG (dB) in "x".	$K < 1$

As an example of the two-port characterization, the S-parameters for a one-half micron gate-length GaAs FET measured at 14 GHz will be used. These parameters are shown below. This data will be used in several examples, and should be saved on a data card to eliminate additional data entry. Label [A] is executed first, and the stability factor is greater than 1, thus GA max, Γms, and Γml all exist and may be calculated as shown.

```
    TWO-PORT A
  S-PARAMETERS
S11=  0.604000000
     -114.0000000
S12=  0.076000000
      121.0000000
S21=  1.297000000
      89.00000000
S22=  0.738000000
     -83.00000000

    FREQ=  14.00000000 GHZ
       K=  1.212408635
  Ga max=  9.538470017 dB
     Γml=  0.876520075
           78.11401102°
     Γms=  0.815474933
           104.6931174°
```

An example of an unstable device can be obtained by increasing the feedback term, S_{12}. Set S_{12} (register ②) to 0.15 and again run label [A]. This time the stability factor is less than 1, and the MSG must be calculated as shown.

```
    TWO-PORT A
  S-PARAMETERS
S11=  0.604000000
     -114.0000000
S12=  0.150000000
      121.0000000
S21=  1.297000000
      89.00000000
S22=  0.738000000
     -83.00000000

    FREQ=  14.00000000 GHZ
       K=  0.536616083
     MSG=  9.368487170 dB
```

MP-12 Stability

This program computes the center and radius of the stability circle on the input and output planes. This information is necessary when designing with potentially unstable devices. This program should be used in conjunction with MP-11, even though the stability factor is greater than 1. A stability factor greater than 1 is a necessary, but not sufficient condition for absolute stability.

Two-Port A must be in primary registers ① thru ⑧.

Run [A] before any other label. At Program Stop, K is in "x".

INPUT PLANE STABILITY CIRCLE

Run [B]

at Program Stop, center in "x", angle of center in "y", radius of circle in "z", stability code in "t".

OUTPUT PLANE STABILITY CIRCLE

Run [C]

at Program Stop, center in "x", angle of center in "y", radius of circle in "z", stability code in "t".

STABILITY CODE

"0" stable region outside circle

"1" stable region inside circle

"10" unconditional stability

The center and radius of the stability circle are with respect to the coordinate system of the input or output plane. Thus, with a standard Smith Chart, the center of the stability circle would be determined in the same manner as any reflection coefficient, and the radius determined from the magnitude scale of the Smith Chart. In those cases where the stability circle is too large for a standard Smith Chart, program MP-54 may be used to construct a Smith Chart with a more convenient scale factor.

The S-parameters from the previous example are used for program verification of the stability program.

```
TWO-PORT A
S-PARAMETERS
S11=  0.604000000
     -114.0000000
S12=  0.076000000
      121.0000000
S21=  1.297000000
      89.00000000
S22=  0.738000000
     -83.00000000
```

```
INPUT PLANE
  FREQ=  14.00000000
     K=  1.212408635
CENTER=  1.520675968
         104.6931174°
RADIUS=  0.455641153
  CODE=  10.00000000
         (UNCONDITIONAL)

OUTPUT PLANE
  FREQ=  14.00000000
     K=  1.212408635
CENTER=  1.290442363
         78.11401103°
RADIUS=  0.248815574
  CODE=  10.00000000
         (UNCONDITIONAL)
```

```
TWO-PORT A
S-PARAMETERS
S11=  0.604000000
     -114.0000000
S12=  0.150000000
      121.0000000
S21=  1.297000000
      89.00000000
S22=  0.738000000
     -83.00000000
```

```
INPUT PLANE
  FREQ=  14.00000000
     K=  0.536616083
CENTER=  1.571453031
         98.27532476°
RADIUS=  0.789061649
  CODE=  0.000000000
         (STABLE OUTSIDE)

OUTPUT PLANE
  FREQ=  14.00000000
     K=  0.536616083
CENTER=  1.303026410
         74.10166667°
RADIUS=  0.456275977
  CODE=  0.000000000
         (STABLE OUTSIDE)
```

MP-13 Constant Gain, S'_{11} and S'_{22}

This program computes the contours of constant gain on either the input or output plane for the condition of a conjugate match at the opposite port of the network. The calculations of S'_{11} and S'_{22} are included to provide a convenient means of computing the conjugate match impedance. Stability factor, and maximum available gain or maximum stable gain are included as well.

Two-Port A must be in primary registers (1) thru (8).

INPUT PLANE CONSTANT GAIN: Run [A]

at "21" enter gain in dB, R/S

at stop, center of circle in "x", angle of center in "y", radius of circle in "z".

To compute another circle, enter gain in dB, [R/S]

output as above.

OUTPUT PLANE CONSTANT GAIN: Run [B]

at "21" enter gain in dB, [R/S]

at stop, center of circle in "x", angle of center in "y", radius of circle in "z".

To compute another circle, enter gain in dB, [R/S] .

CALCULATE S'_{11}:

Enter $|\Gamma L|$, [ENTER ↑], enter Θ , run [C]

at stop, $|S'_{11}|$ in "x", Θ in "y".

CALCULATE S'_{22}:

Enter $|\Gamma S|$, [ENTER ↑], enter Θ, run [D]

at stop, $|S'_{22}|$ in "x", Θ in "y".

CALCULATE K:

Run [E]

at stop, K in "x".

CALCULATE GA max: $K \geq 1$

Run [f] [A]

at stop, GA max (dB) in "x".

CALCULATE MSG: $K < 1$

Run [f] [B]

at stop, MSG (dB) in "x".

As an example of this program, the S-parameters of the previous examples will be used. These are the unstable S-parameters for the example of MP-11,

```
TWO-PORT A
S-PARAMETERS
S11=  0.604000000                FREQ=  14.00000000
     -114.0000000                   K=  0.536616083
S12=  0.150000000                 MSG=  9.368487170
      121.0000000
S21=  1.297000000
      89.00000000
S22=  0.738000000
     -83.00000000
```

and agree exactly with these calculations. The actual device parameters at 14 GHz will be used for the remaining examples of MP-13. The stability factor is greater than 1, so that the maximum available gain can be calculated as shown. Constant gain contours can be computed for the input and output planes as shown. The desired gain must be less than or equal to the maximum available gain. Note that if the gain is equal to the maximum available gain, the constant gain contour reduces to a single point equal to the matched source or load reflection coefficient of MP-11 for the input and output planes, respectively. Based on these values, S'_{11} and S'_{22} can be computed, and will result in the matched condition for the opposite plane. The example shown here for the calculation of S'_{11} and S'_{22} is based on an input or output reflection coefficient which falls on the 8 dB constant gain contour. Thus, a source reflection coefficient which falls on the 8 dB gain input plane contour, and a load reflection coefficient which presents the conjugate of S'_{22} to the device, will result in 8 dB gain. These calculations can be made using the transducer gain program of MP-14.

```
TWO-PORT A                       INPUT PLANE
S-PARAMETERS                       FREQ=  14.00000000
S11=  0.604000000                  GAIN=  8.000000000
     -114.0000000                CENTER=  0.681186407
S12=  0.076000000                         104.6931174°
      121.0000000                RADIUS=  0.270554460
S21=  1.297000000
      89.00000000
S22=  0.738000000                OUTPUT PLANE
     -83.00000000                  FREQ=  14.00000000
                                   GAIN=  8.000000000
                                 CENTER=  0.771342292
   FREQ=  14.00000000                     78.11401102°
      K=  1.212408635            RADIUS=  0.197146267
Ga max=  9.538470017


   FREQ=  14.00000000              FREQ=  14.00000000
     Γs=  0.410631947                Γl=  0.574196025
          104.6931174°                    78.11401102°
  S22'=  0.782617146               S11'=  0.633194292
         -80.77155743°                   -108.8845410°
```

MP-14 Transducer Gain and Mismatch

MP-14 can be used to compute the transducer gain of a two-port network with arbitrary source and load terminations. Mismatch loss for a given source and load reflection coefficient can be calculated, as well as contours of constant mismatch for a given reflection coefficient.

Two-Port A must be in primary registers (1) thru (8).
TRANSDUCER GAIN, GT: Run [A] at "11", enter $\|\Gamma s\|$, [ENTER ↑], Θ, [R/S] at "22", enter $\|\Gamma l\|$, [ENTER ↑], Θ, [R/S] at stop, transducer gain (dB) in "x".
MISMATCH CIRCLES: Enter $\|\Gamma\|$, [ENTER ↑], Θ run [B] at "21", enter desired mismatch (+dB), [R/S] at stop, center of circle in "x", angle of center in "y", radius in "z". To compute another circle for the same Γ, enter desired mismatch (+dB), [R/S].
MISMATCH LOSS: Run [C] at "11", enter $\|\Gamma_1\|$, [ENTER ↑], Θ, [R/S] at "22", enter $\|\Gamma_2\|$, [ENTER ↑], Θ, [R/S] at stop, mismatch (dB) in "x".

The constant gain, along with the S'_{11} and S'_{22} calculations of the previous examples, will be used to verify the MP-14 program. The transducer gain can be calculated based on a source reflection coefficient which is on the input plane 8 dB constant gain contour and the corresponding S'_{22} conjugate. A second example based on the output plane and S'_{11} conjugate is also shown. Thus, the constant gain calculations of the previous example are correct. The transducer gain program provides a check for the previous program and should be used during the design process before the matching networks are synthesized to ensure that the desired load and source impedances are correct.

```
FREQ=  14.00000000          FREQ=  14.00000000
  Γs=  0.410631947            Γs=  0.683194292
       104.6931174°                108.8845410°
  Γl=  0.782617146            Γl=  0.574196025
       80.77155743°                78.11401102°
  Gt=  7.999999993            Gt=  7.999999999
```

Thus, a constant gain contour can be constructed on one plane, a reflection coefficient selected which falls on this contour and is more desirable than other points (lower VSWR, better stability, lower noise figure, simple matching network, etc.), then the resulting reflection coefficient on the opposite plane can be computed using S'_{11} or S'_{22}. This procedure will indeed result in the desired gain; however, the designer is restricted to a single point on one of the planes resulting from the conjugate of S'_{11} or S'_{22}. Another method — one which provides more design flexibility — is to increase the gain for the constant gain contour, then construct a mismatch circle around the corresponding S'_{11} or S'_{22}. For example, assume that 6dB is desired from the device at 14 GHz. An 8 dB gain contour is constructed on the input plane, a source reflection coefficient selected, and S'_{22} calculated. This is the input plane example of MP-13, and the first example above where Γl is set equal to S'^{*}_{22}. A 2 dB mismatch circle is constructed for S'_{22}. The load reflection coefficient may now be selected from any point which falls on the 2 dB mismatch contour. The point which is closest to the origin — magnitude of the center minus the radius — will be used, along with the source reflection coefficient from the previous example, to check the transducer gain which does, in fact, result in the desired 6 dB gain.

```
  FREQ=  14.00000000
     Γ=  0.782617146
        -80.77155743°
   GMM=  2.00000000[illegible]
CENTER=  0.63801063[illegible]
         80.77155743°
RADIUS=  0.304158675

  [illegible]EQ=  14.0000000[illegible]
   [illegible]s=  0.4106319[illegible]
         104.693[illegible]
     [illegible]=  0.33385[illegible]57
         80.77155743°
   Gt=  5.999999988
```

The mismatch loss portion of the program can be checked by using the S'_{22} calculation from MP-13 and the 2 dB load reflection coefficient from the previous example.

```
  Γ1=  0.78261714[illegible]
      -80.77155743°
  Γ2=  0.333851957
       80.77155743°
 GMM= -2.000000009
```

MP-20 INTERCONNECTIONS AND PARAMETER CONVERSIONS

This group of six programs is used to interconnect two-ports in cascade, series, parallel, series · parallel, or parallel · series. The hybrid connection is also included for amplifier designs. One-port networks connected in series or parallel and the various parameter set conversions are also included in this section. The examples of the various interconnections use the same Two-Ports A and B. These two-ports may be saved on a data card and reloaded for each example.

MP-21 Cascade Connection

Run [A]

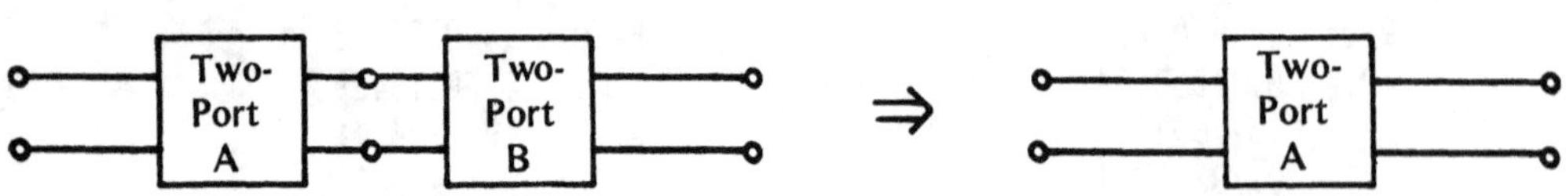

Two-Port A in primary (1) thru (8)

Two-Port B in secondary (1) thru (8)

resulting Two-Port A in primary (1) thru (8).

```
  TWO-PORT A
S-PARAMETERS
S11=  0.604000000
     -114.0000000
S12=  0.076000000
      121.0000000
S21=  1.297000000
      89.00000000
S22=  0.738000000
     -83.00000000
```

```
  TWO-PORT B
S-PARAMETERS
S11=  0.400000000
      30.00000000
S12=  0.200000000
     -50.00000000
S21=  1.700000000
      25.00000000
S22=  0.900000000
     -70.00000000
```

```
  TWO-PORT A
S-PARAMETERS
S11=  0.646965298
     -115.5290470
S12=  0.017767980
      55.00297470
S21=  2.577409131
      98.00297465
S22=  1.098354246
     -82.47638562
```

RESULTS OF CASCADE OF TWO-PORT A AND TWO-PORT B ABOVE

MP-22 Series and Parallel Connection

SERIES CONNECTION, SER: Run [A]

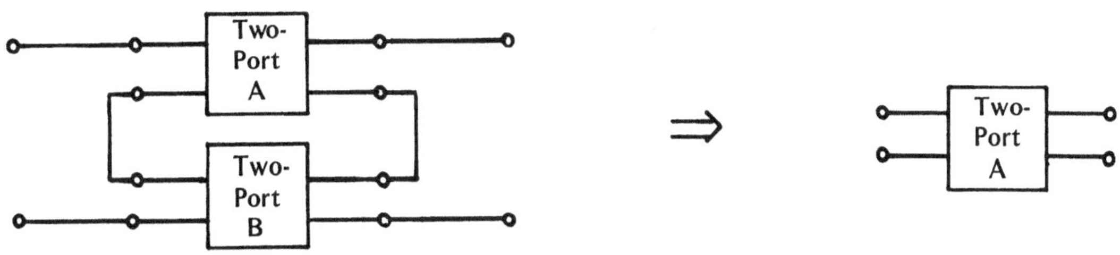

Two-Port A in primary (1) thru (8)

Two-Port B in secondary (1) thru (8)

resulting Two-Port A in primary (1) thru (8).

```
TWO-PORT A
S-PARAMETERS
S11=  0.604000000
     -114.0000000
S12=  0.076000000
      121.0000000
S21=  1.297000000
      89.00000000
S22=  0.738000000
     -83.00000000
```

```
TWO-PORT B
S-PARAMETERS
S11=  0.400000000
      30.00000000
S12=  0.200000000
     -50.00000000
S21=  1.700000000
      25.00000000
S22=  0.900000000
     -70.00000000
```

```
TWO-PORT A
S-PARAMETERS
S11=  0.356149662
     -368.6787656
S12=  0.110304189
     -388.2000589
S21=  1.267553654
     -301.7136300
S22=  0.963492491
     -407.2223713
```

RESULTS OF SERIES CONNECTION OF TWO-PORTS A AND B ABOVE

PARALLEL CONNECTION, PAR: Run [B]

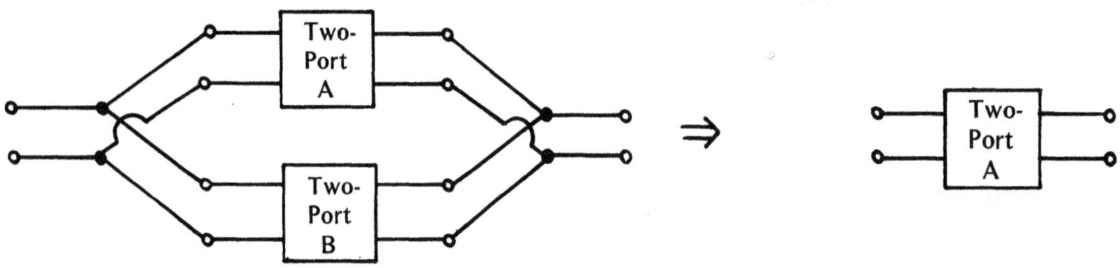

Two-Port A in primary (1) thru (8)

Two-Port B in secondary (1) thru (8)

resulting Two-Port A in primary (1) thru (8).

```
TWO-PORT A
S-PARAMETERS
S11=  0.604000000
     -114.0000000
S12=  0.076000000
      121.0000000
S21=  1.297000000
      89.00000000
S22=  0.738000000
     -83.00000000
```

```
TWO-PORT B
S-PARAMETERS
S11=  0.400000000
      30.00000000
S12=  0.200000000
     -50.00000000
S21=  1.700000000
      25.00000000
S22=  0.900000000
     -70.00000000
```

```
  TWO-PORT A
S-PARAMETERS
S11=  0.371502200
     -135.9171204
S12=  0.046622723
     -118.5723825
S21=  0.844157744
     -317.4861182
S22=  0.739149252
     -117.4158935
```

RESULTS OF PARALLEL CONNECTION OF TWO-PORTS A AND B ABOVE

MP-23 Hybrid Connection

HYBRID CONNECTION, HYB:

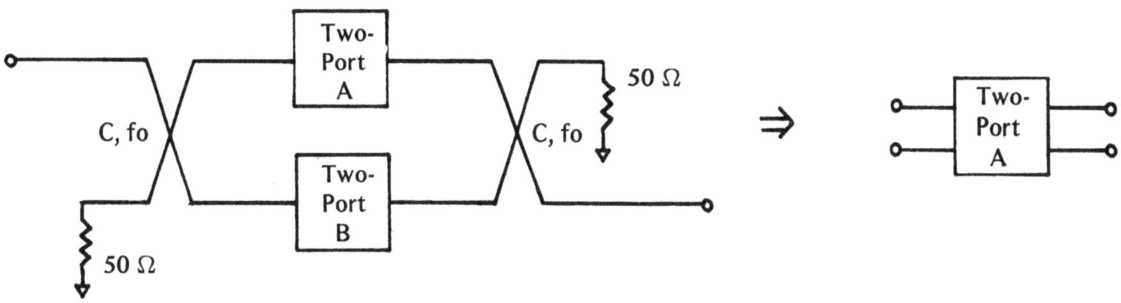

Frequency (GHz) in Reg Ⓐ

enter coupling at fo (+dB), [ENTER ↑], center frequency, fo, [A]

Two-Port A in primary ① thru ⑧

Two-Port B in secondary ① thru ⑧

resulting Two-Port A in primary ① thru ⑧.

NOTE: Offset analysis frequency slightly from center frequency to avoid overflow due to TAN 90°.

```
  TWO-PORT A
S-PARAMETERS
S11=  0.604000000
     -114.0000000
S12=  0.076000000
      121.0000000
S21=  1.297000000
      89.00000000
S22=  0.738000000
     -83.00000000
```

```
  TWO-PORT B
S-PARAMETERS
S11=  0.400000000
      30.00000000
S12=  0.200000000
     -50.00000000
S21=  1.700000000
      25.00000000
S22=  0.900000000
     -70.00000000
```

```
  TWO-PORT A
 S-PARAMETERS
S11=  0.488330376
     -146.7524036
S12=  0.062493013
     -154.6681445
S21=  1.270054437
     -57.90709036
S22=  0.180232049
     -65.33813020
```

RESULTS OF ANALYSIS AT 14 GHz FOR TWO-PORTS A AND B ABOVE CONNECTED USING TWO HYBRID COUPLERS WITH 2.5 dB COUPLING AT CENTER FREQUENCY, fo, OF 12 GHz

MP-24 Series · Parallel and Parallel · Series Connection

SERIES · PARALLEL CONNECTION, SER · PAR: Run [A]

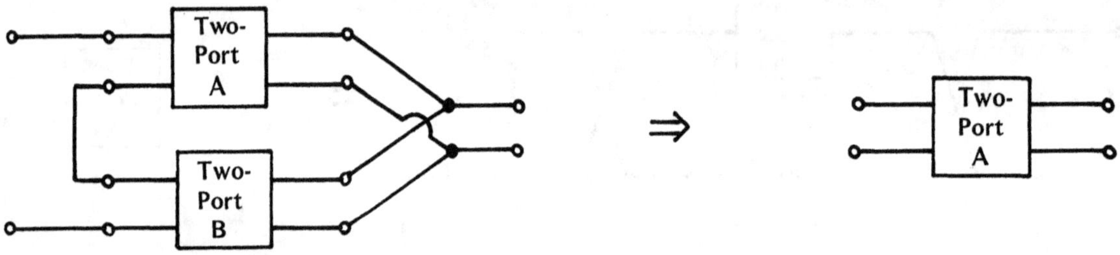

Two-Port A in primary (1) thru (8)

Two-Port B in secondary (1) thru (8)

resulting Two-Port A in primary (1) thru (8).

```
  TWO-PORT A
 S-PARAMETERS
S11=  0.604000000
     -114.0000000
S12=  0.076000000
      121.0000000
S21=  1.297000000
      89.00000000
S22=  0.738000000
     -83.00000000
```

```
  TWO-PORT B
 S-PARAMETERS
S11=  0.400000000
      30.00000000
S12=  0.200000000
     -50.00000000
S21=  1.700000000
      25.00000000
S22=  0.900000000
     -70.00000000
```

```
  TWO-PORT A
 S-PARAMETERS
S11=  0.340239764
     -353.4533316
S12=  0.084290004
     -404.6369351
S21=  1.402604081
     -332.0755075
S22=  0.925604168
     -119.8455740
```

RESULTS OF SERIES · PARALLEL CONNECTION OF TWO-PORTS A AND B ABOVE

PARALLEL · SERIES CONNECTION, PAR · SER: Run [B]

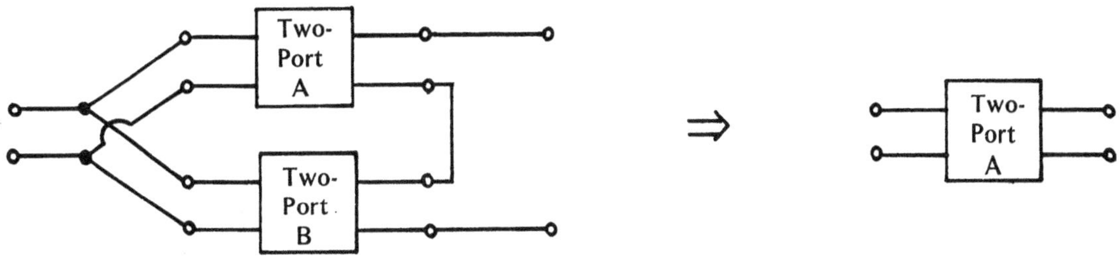

Two-Port A in primary (1) thru (8)

Two-Port B in secondary (1) thru (8)

resulting Two-Port A in primary (1) thru (8).

```
TWO-PORT A
S-PARAMETERS
S11=  0.604000000
     -114.0000000
S12=  0.076000000
      121.0000000
S21=  1.297000000
      89.00000000
S22=  0.738000000
     -83.00000000
```

```
TWO-PORT B
S-PARAMETERS
S11=  0.400000000
      30.00000000
S12=  0.200000000
     -50.00000000
S21=  1.700000000
      25.00000000
S22=  0.900000000
     -70.00000000
```

```
TWO-PORT A
S-PARAMETERS
S11=  0.542627951
     -494.7156706
S12=  0.041814463
     -473.2475215
S21=  0.692881188
     -302.0563839
S22=  0.797074684
     -43.02759580
```

RESULTS OF PARALLEL · SERIES CONNECTION OF TWO-PORTS A AND B ABOVE

MP-25 Parameter Conversions

This program converts a normalized parameter set in primary registers (1) through (8), Two-Port A, to another normalized parameter set which is also Two-Port A.

S ⇒ Y CONVERSION: Run [A]

normalized S in Two-Port A at start

normalized Y in Two-Port A st atop.

(Continued)

Y ⇒ S CONVERSION: Run [f][A]

normalized Y in Two-Port A at start

normalized S in Two-Port A at stop.

S ⇒ Z CONVERSION: Run [B]

normalized S in Two-Port A at start

normalized Z in Two-Port A at stop.

Z ⇒ S CONVERSION: Run [f][B]

normalized Z in Two-Port A at start

normalized S in Two-Port A at stop.

S⇒H CONVERSION: Run [C]

normalized S in Two-Port A at start

normalized H in Two-Port A at stop.

H⇒S CONVERSION: Run [f][C]

normalized H in Two-Port A at start

normalized S in Two-Port A at stop.

S ⇒ G CONVERSION: Run [D]

normalized S in Two-Port A at start

normalized G in Two-Port A at stop.

G⇒S CONVERSION: Run [f][D]

normalized G in Two-Port A at start

normalized S in Two-Port A at stop.

Z ⇌ Y CONVERSION: Run [E]

normalized Z(Y) in Two-Port A at start

normalized Y(Z) in Two-Port A at stop.

G⇌H CONVERSION: Run [f][E]

normalized G(H) in Two-Port A at start

normalized H(G) in Two-Port A at stop.

The two-port S-parameters of the previous examples are shown below with the equivalent Y, Z, H, and G parameters. Conversions between the various parameters should be used to verify the correct operation of this program.

```
TWO-PORT A
S-PARAMETERS
S11=  0.604000000
      -114.0000000
S12=  0.076000000
      121.0000000
S21=  1.297000000
      89.00000000
S22=  0.738000000
      -83.00000000
```

```
TWO-PORT A
Z-PARAMETERS
Z11=  0.600042141
      -60.83608760
Z12=  0.090725470
      -299.8991413
Z21=  1.548301763
      -331.8991413
Z22=  1.023528023
      -73.02902420
```

```
TWO-PORT A
Y-PARAMETERS
Y11=  1.412584368
      -306.6270897
Y12=  0.125211403
      6.502793200
Y21=  2.136831447
      -25.49720680
Y22=  0.828125981
      -294.4341530
```

```
TWO-PORT A
H-PARAMETERS
H11=  0.707922319
      -53.37291030
H12=  0.088639947
      -226.8701171
H21=  1.512710672
      -78.87011710
H22=  0.977012820
      -286.9709758
```

```
TWO-PORT A
G-PARAMETERS
G11=  1.666549616
      -299.1639127
G12=  0.151198496
      -419.0630540
G21=  2.580321709
      -271.0630540
G22=  1.207545740
      -65.56584715
```

MP-26 One-Port Used in Series or Parallel

This program is used to convert a one-port to a two-port network with the one-port connected in either series or parallel. This program is useful for analyzing networks with parallel or series branches, such as amplifier bias networks, power dividers, or switches. When using a cascade of two-ports as the network branch, the user should recall that the input to the two-port is calculated by connecting a 50 ohm reference impedance at the output port. Therefore, unless the branch has a 50 ohm impedance as the last element, a short or open should be cascaded as the last element to obtain the desired results.

This program uses a reflection coefficient stored as S_{11} of Two-Port A (primary (1) and (5)) to compute Two-Port B as the series or parallel connection of the one-port. Measured one-port data may be stored directly.

USE SERIES: Run [A]

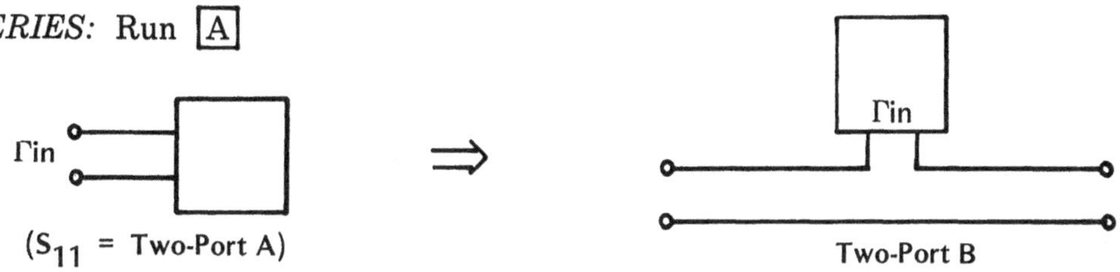

(Continued)

Γin in primary (1) and (5) at start
Two-Port B in secondary (1) thru (8) at stop.

USE PARALLEL: Run [B]

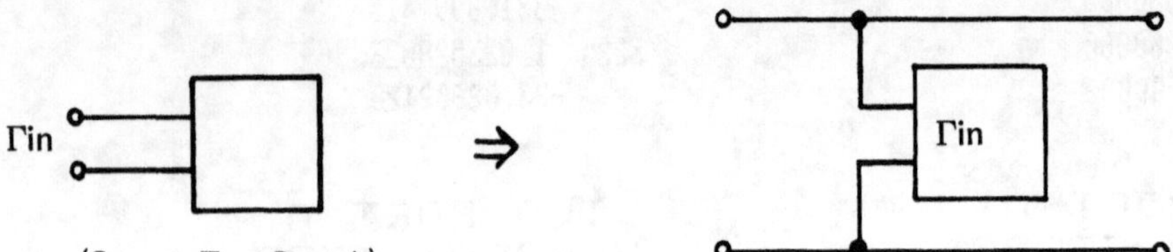

(S_{11} = Two-Port A)

Γin in primary (1) and (5) at start
Two-Port B in secondary (1) thru (8) at stop.

ENTER Z/Θ:

Enter |Z|, ENTER ↑ , enter Θ, run [C]

Zin converted to Γin and stored in primary (1) and (5)
[A] or [B] above can be used to form Two-Port B.

TERMINATE WITH SHORT:

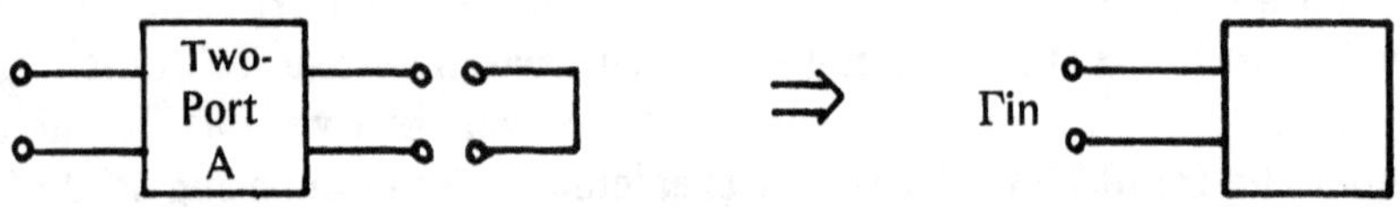

Two-Port A in primary (1) thru (8) at start
Γin (S'_{11} with Γl = 1/180) in primary (1) and (5) at stop
[A] or [B] above can be used to form Two-Port B.

TERMINATE WITH OPEN:

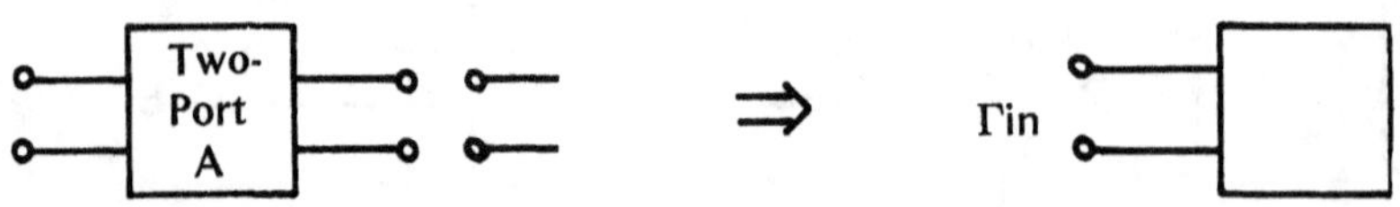

(Continued)

Two-Port A in primary (1) thru (8) at start
Γin (S'_{11} with Γl = 1∠0) in primary (1) and (5) at stop
[A] or [B] above can be used to form Two-Port B.

EXAMPLES: *CONNECT Γin = 0.6∠30 IN SERIES*

Store |Γ| = 0.6 in primary (1)

store Θ = 30 in primary (5)

run [A] to connect Γin in series and form Two-Port B.

```
 TWO-PORT B
S-PARAMETERS
S11=  0.619959572
      18.06400394
S12=  0.453371195
     -25.08838580
S21=  0.453371195
     -25.08838580
S22=  0.619959572
      18.06400394
```

CONNECT Γin = 0.6∠30 IN PARALLEL

Store |Γ| = 0.6 in primary (1)

store Θ = 30 in primary (5)

run [B] to connect Γin in parallel and form Two-Port B.

```
 TWO-PORT B
S-PARAMETERS
S11= -0.160335401
     -36.85667472
S12=  0.876998519
      6.295715019
S21=  0.876998519
      6.295715019
S22= -0.160335401
     -36.85667472
```

CONVERT Z = 60∠−45 TO Γin:

Enter Zin = 60∠−45, run [C]

Γin in primary (1) and (5).

```
Zin=  60.00000000
     -45.00000000°
Γin=  0.423771988
     -75.46483480°
```

Terminate Two-Port A with an open and short to form Γin. The two-port parameters from the previous examples will be used.

```
 TWO-PORT A
S-PARAMETERS
S11=  0.604000000
     -114.0000000
S12=  0.076000000
      121.0000000
S21=  1.297000000
      89.00000000
S22=  0.738000000
     -83.00000000
```

Run [D] to calculate Γin from Two-Port A above with a short-circuit termination.

Γin in primary (1) and (5).

```
   SHORT
Γin=  0.528995571
     -113.7025484°
```

Run [E] to calculate Γin from Two-Port B above with an open-circuit termination.

Γin in primary (1) and (5).

```
   OPEN
Γin=  0.631353977
     -121.4111066°
```

MP-30 LUMPED AND DISTRIBUTED ELEMENTS

These three programs provide the user with a wide range of standard lumped and distributed elements. The two-port S-parameters for each element are calculated at the analysis frequency specified by register (A) and are stored at Two-Port B, which can then be interconnected with Two-Port A and another element calculated.

MP-31 Series Resistor, Inductor, and Capacitor

Frequency (GHz) in Reg (A).

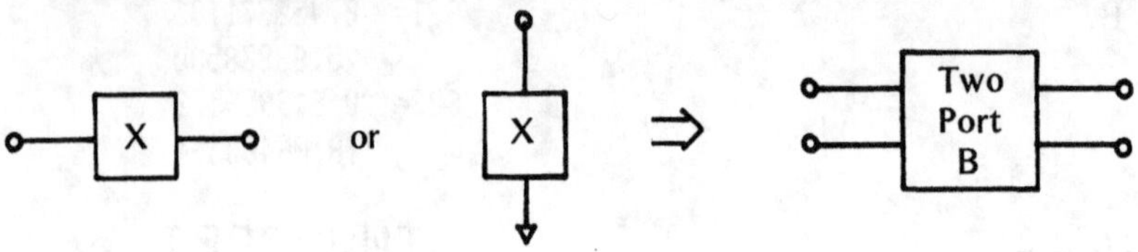

Secondary (1) thru (8)

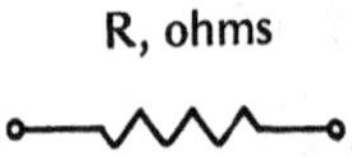

Enter R, run [A].

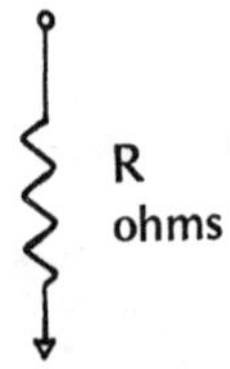

Run [f][E] enter R, run [A].

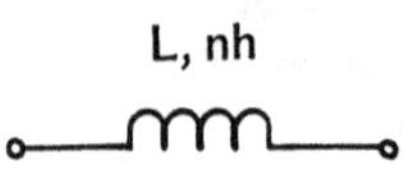

Enter L, run [B].

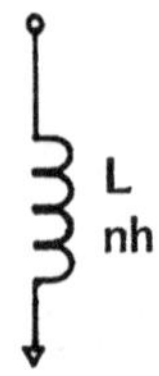

Run [f][E] enter L, run [B].

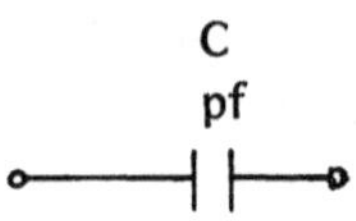

Enter C, run [C].

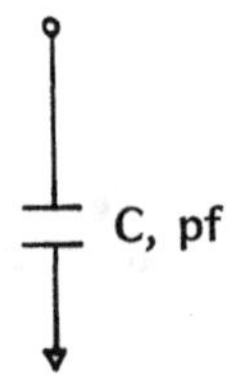

Run [f][E] enter C, run [C].

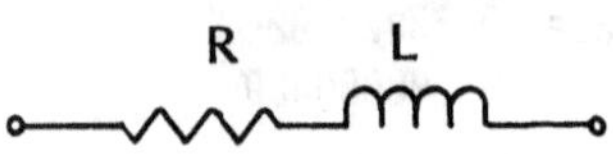

Enter R, [ENTER ↑], enter L run [D].

Run [f][E] enter R, [ENTER ↑], enter L run [D].

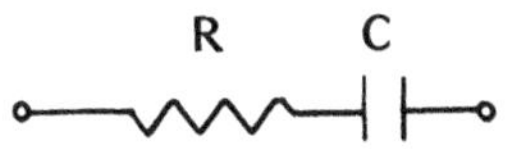 Enter R, [ENTER ↑], enter C run [E] .	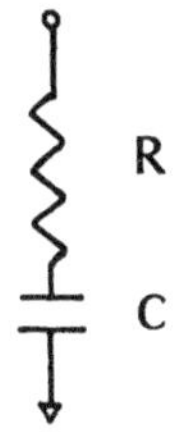 Run [f][E] enter R, [ENTER ↑], enter C run [E] .
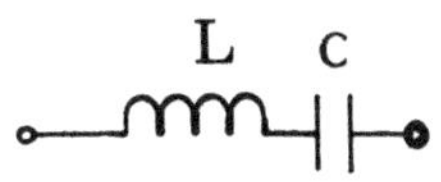 Enter L, [ENTER ↑], enter C run [f] [A] .	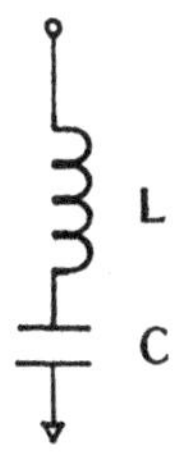 Run [f][E] enter L, [ENTER ↑], enter C run [f] [A] .
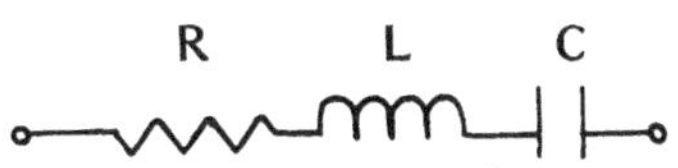 Enter R, [ENTER ↑], enter L, [ENTER ↑], enter C run [f] [B] .	 Run [f] [E] enter R, [ENTER ↑], enter L [ENTER ↑], enter C run [f] [B] .

The following examples are various combinations of a 20 ohm resistor, a 1 nanohenry inductor, and a 0.5 picofarad capacitor. The analysis frequency is 10 GHz. Store 10 in Register (A), then use the table above to generate the S-parameters of the desired element for program verification.

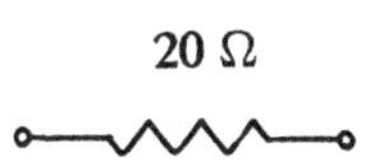


```
TWO-PORT B
S-PARAMETERS
S11=  0.166666667
       1.120045562-48
S12=  0.833333333
      -2.240091124-49
S21=  0.833333333
      -2.240091124-49
S22=  0.166666667
       1.120045562-48
```

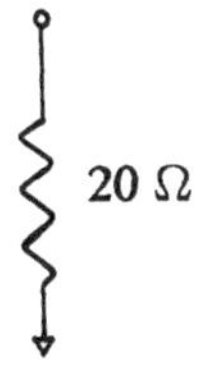


```
TWO-PORT B
S-PARAMETERS
S11=  0.555555556
       180.0000000
S12=  0.444444444
       7.466970412-49
S21=  0.444444444
       7.466970412-49
S22=  0.555555556
       180.0000000
```

TWO-PORT B
S-PARAMETERS
S11= 0.532018045
57.85809235
S12= 0.846733016
-32.14190765
S21= 0.846733016
-32.14190765
S22= 0.532018045
57.85809235

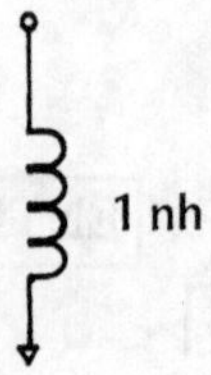

TWO-PORT B
S-PARAMETERS
S11= 0.369697848
111.6969840
S12= 0.929152034
21.69698397
S21= 0.929152034
21.69698397
S22= 0.369697848
111.6969840

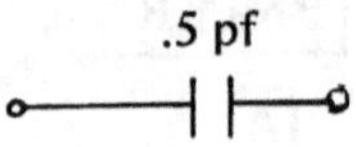

TWO-PORT B
S-PARAMETERS
S11= 0.303314471
-72.34321285
S12= 0.952890514
17.65678715
S21= 0.952890514
17.65678715
S22= 0.303314471
-72.34321285

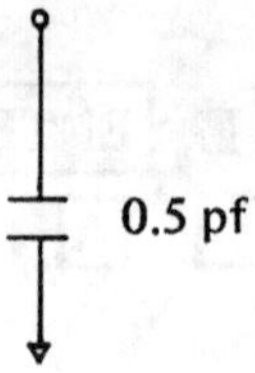

TWO-PORT B
S-PARAMETERS
S11= 0.617667825
-128.1460260
S12= 0.786439100
-38.14602599
S21= 0.786439100
-38.14602599
S22= 0.617667825
-128.1460260

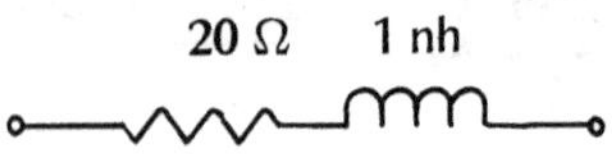

TWO-PORT B
S-PARAMETERS
S11= 0.486793053
44.70671351
S12= 0.738256887
-27.63649934
S21= 0.738256887
-27.63649934
S22= 0.486793053
44.70671351

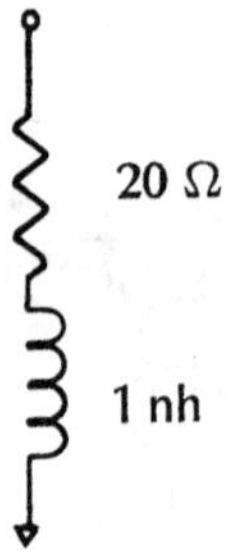

TWO-PORT B
S-PARAMETERS
S11= 0.323481542
125.6101340
S12= 0.853191186
17.95334681
S21= 0.853191186
17.95334681
S22= 0.323481542
125.6101340

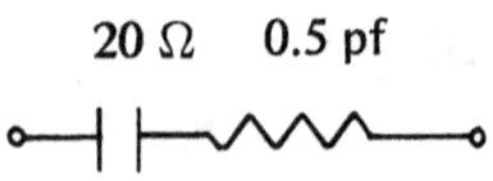

TWO-PORT B
S-PARAMETERS
S11= 0.302800829
-43.00204108
S12= 0.805477524
14.85605129
S21= 0.805477524
14.85605129
S22= 0.302800829
-43.00204108

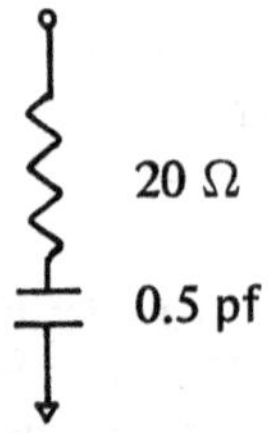

TWO-PORT B
S-PARAMETERS
S11= 0.453556069
-144.7261186
S12= 0.682016069
-22.58421096
S21= 0.682016069
-22.58421096
S22= 0.453556069
-144.7261186

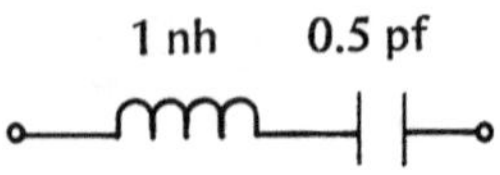

TWO-PORT B
S-PARAMETERS
S11= 0.296106304
72.77611194
S12= 0.955154991
-17.22388806
S21= 0.955154991
-17.22388806
S22= 0.296106304
72.77611194

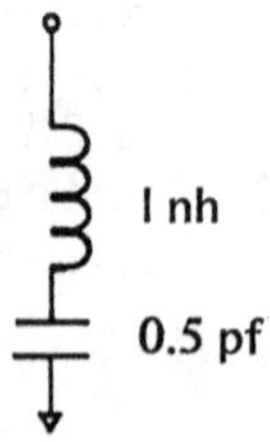

TWO-PORT B
S-PARAMETERS
S11= 0.627741844
128.8837157
S12= 0.778421593
38.88371571
S21= 0.778421593
38.88371571
S22= 0.627741844
128.8837157

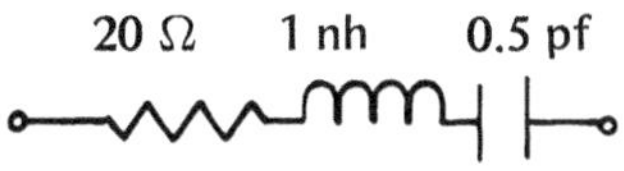


```
  TWO-PORT B
S-PARAMETERS
S11= 0.297664547
     42.68706555
S12= 0.806843859
    -14.48512049
S21= 0.806843859
    -14.48512049
S22= 0.297664547
     42.68706555
```

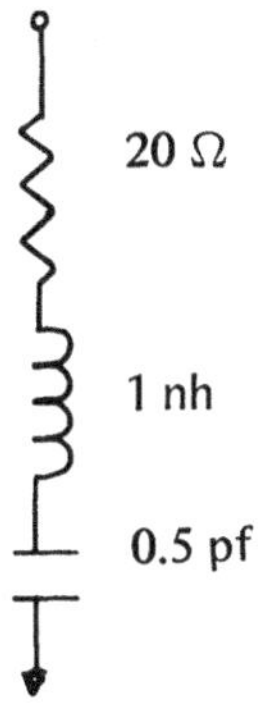


```
  TWO-PORT B
S-PARAMETERS
S11= 0.457500117
     145.4367289
S12= 0.675132188
     22.60891497
S21= 0.675132188
     22.60891497
S22= 0.457500117
     145.4367289
```

MP-32 Parallel Resistor, Inductor, and Capacitor

Frequency (GHz) in Reg (A)

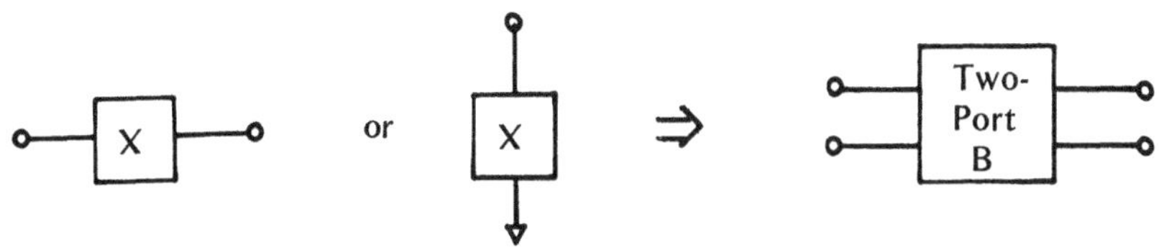

Secondary (1) thru (8)

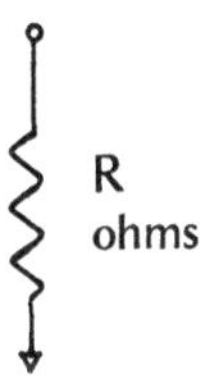 Enter R, run [A] .	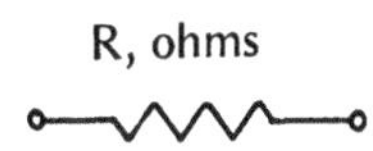 Run [f][E] enter R, run [A] .
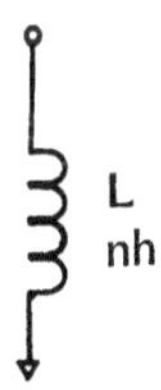 Enter L, run [B] .	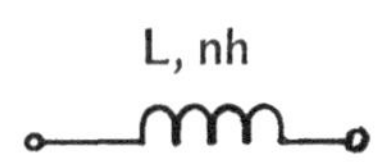 Run [f][E] enter L, run [B] .
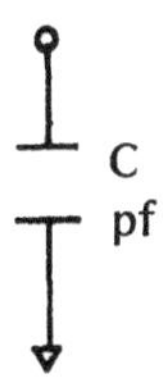 Enter C, run [C] .	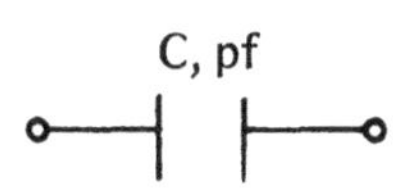 Run [f][E] enter C, run [C] .

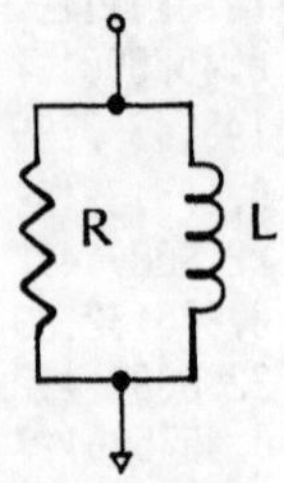	Enter R, [ENTER ↑], enter L run [D] ..	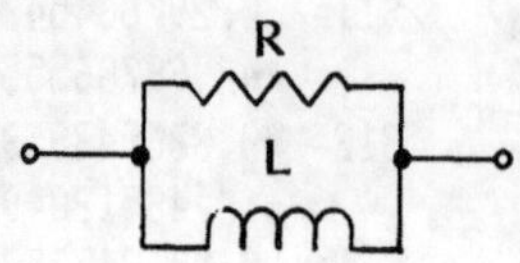	Run [f] [E] enter R, [ENTER ↑], enter L run [D] .
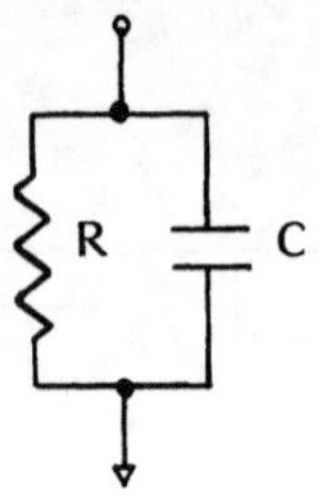	Enter R, [ENTER ↑], enter C run [E] .	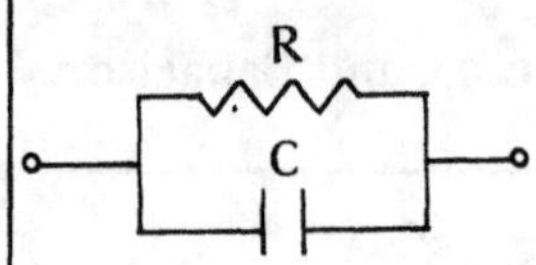	Run [f] [E] enter R, [ENTER ↑], enter C run [E] .
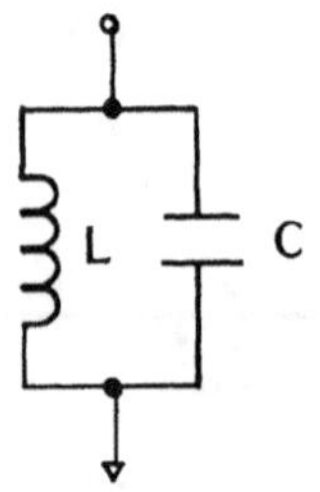	Enter L, [ENTER ↑], enter C run [f] [A] .	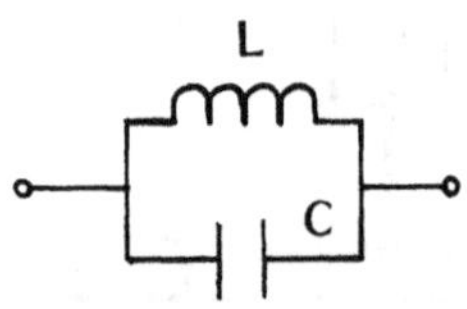	Run [f] [E] enter L, [ENTER ↑], enter C run [f] [A] .
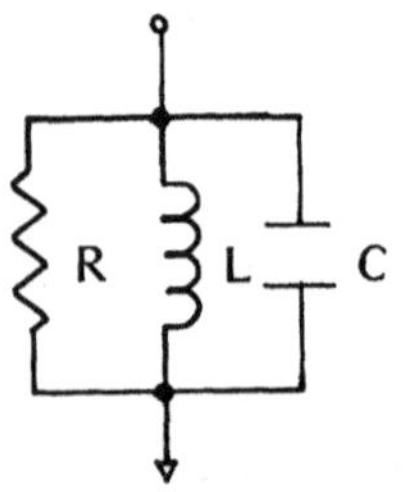	Enter R, [ENTER ↑], enter L [ENTER ↑], enter C run [f] [B] .	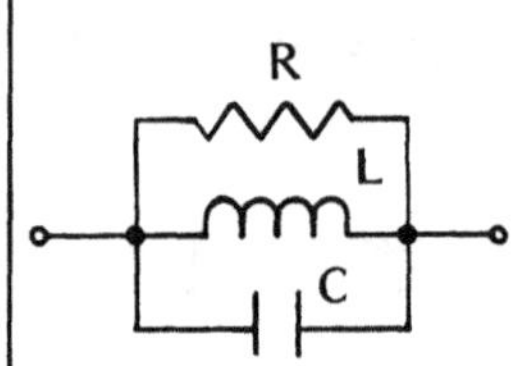	Run [f] [E] enter R, [ENTER ↑], enter L [ENTER ↑], enter C run [f] [B] .

Store 10 in register (A) (analysis frequency is 10 GHz). Use the above table to calculate Two-Port B for the elements.

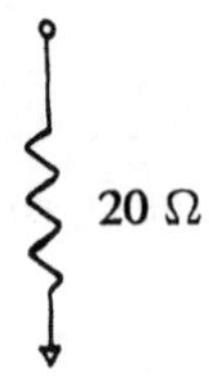


```
TWO-PORT B
S-PARAMETERS
S11=  0.555555556
     -180.0000000
S12=  0.444444444
     -2.986788163-49
S21=  0.444444444
     -2.986788163-49
S22=  0.555555556
     -180.0000000
```

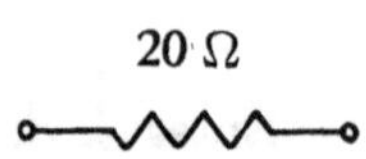


```
TWO-PORT B
S-PARAMETERS
S11=  0.166666667
     -4.480182245-49
S12=  0.833333333
      8.960364488-50
S21=  0.833333333
      8.960364488-50
S22=  0.166666667
     -4.480182245-49
```

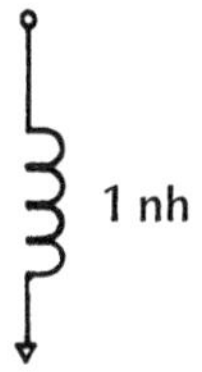


```
TWO-PORT B
S-PARAMETERS
S11=  0.369697848
       111.6969840
S12=  0.929152034
       21.69698398
S21=  0.929152034
       21.69698398
S22=  0.369697848
       111.6969840
```

```
TWO-PORT B
S-PARAMETERS
S11=  0.532018044
       57.85809237
S12=  0.846733016
      -32.14190763
S21=  0.846733016
      -32.14190763
S22=  0.532018044
       57.85809237
```

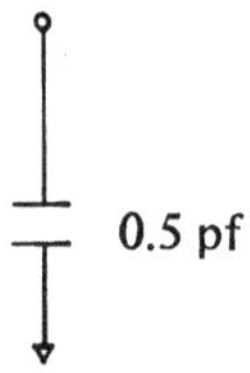


```
TWO-PORT B
S-PARAMETERS
S11=  0.617667825
      -128.1460260
S12=  0.786439100
      -38.14602599
S21=  0.786439100
      -38.14602599
S22=  0.617667825
      -128.1460260
```

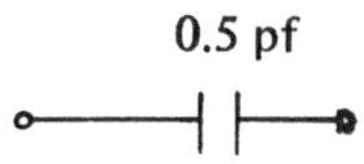


```
TWO-PORT B
S-PARAMETERS
S11=  0.303314471
      -72.34321285
S12=  0.952890514
       17.65678715
S21=  0.952890514
       17.65678715
S22=  0.303314471
      -72.34321285
```

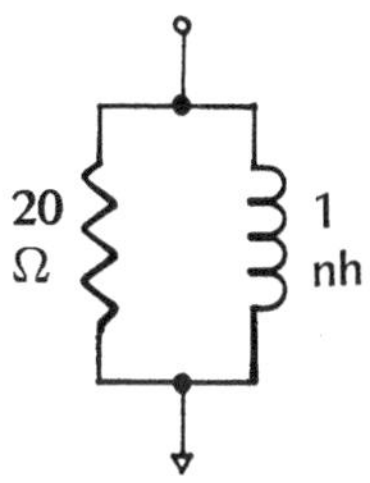


```
TWO-PORT B
S-PARAMETERS
S11=  0.574113666
       172.3716526
S12=  0.437653973
       10.02843976
S21=  0.437653973
       10.02843976
S22=  0.574113666
       172.3716526
```

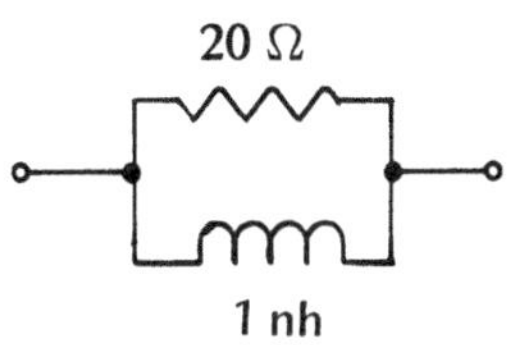


```
TWO-PORT B
S-PARAMETERS
S11=  0.161095505
       14.85605128
S12=  0.845299131
      -2.800735871
S21=  0.845299131
      -2.800735871
S22=  0.161095505
       14.85605128
```

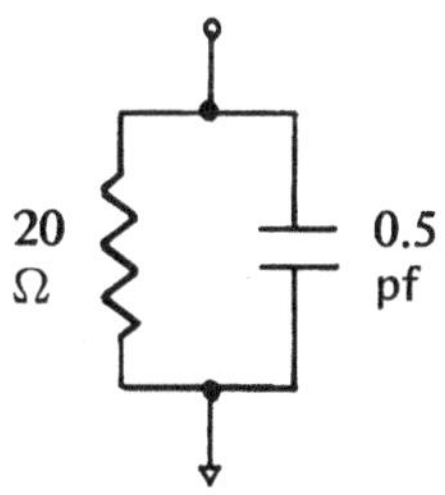


```
TWO-PORT B
S-PARAMETERS
S11=  0.619461317
      -167.1004429
S12=  0.419614679
      -19.24235053
S21=  0.419614679
      -19.24235053
S22=  0.619461317
      -167.1004429
```

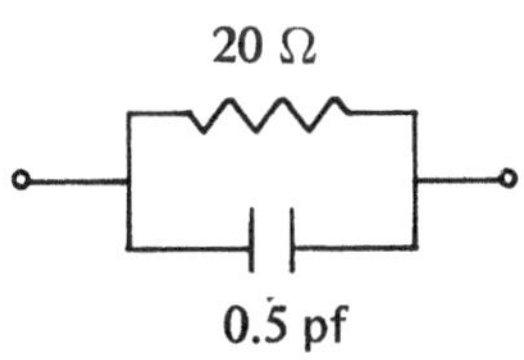


```
TWO-PORT B
S-PARAMETERS
S11=  0.147651377
      -27.63649933
S12=  0.871888627
       4.505408301
S21=  0.871888627
       4.505408301
S22=  0.147651377
      -27.63649933
```

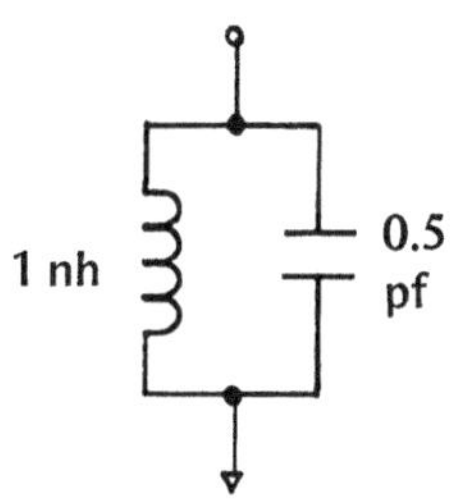


```
TWO-PORT B
S-PARAMETERS
S11=  0.361329828
      -111.1818878
S12=  0.932438071
      -21.18188778
S21=  0.932438071
      -21.18188778
S22=  0.361329828
      -111.1818878
```

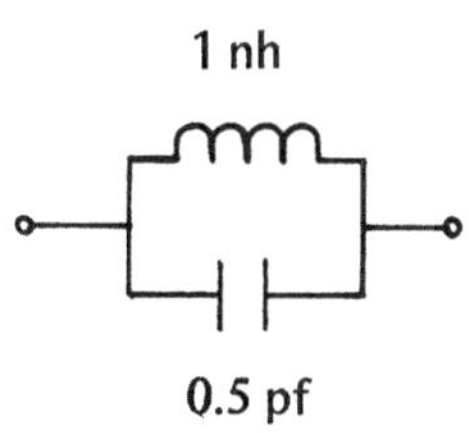


```
TWO-PORT B
S-PARAMETERS
S11=  0.542116196
      -57.17218602
S12=  0.840303535
       32.82781398
S21=  0.840303535
       32.82781398
S22=  0.542116196
      -57.17218602
```

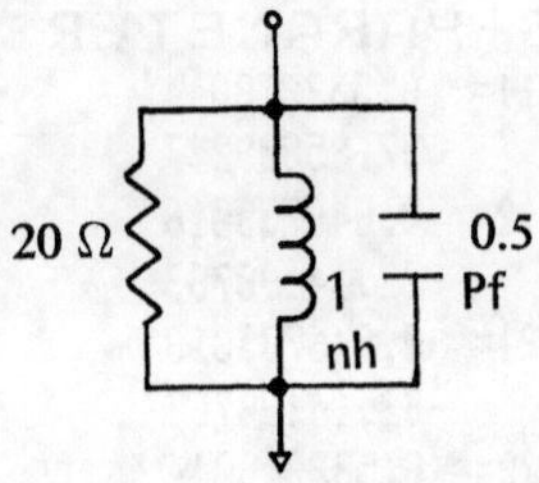


```
TWO-PORT B
S-PARAMETERS
S11=  0.573200119
     -172.5481265
S12=  0.437995963
     -9.772014583
S21=  0.437995963
     -9.772014583
S22=  0.573200119
     -172.5481265
```

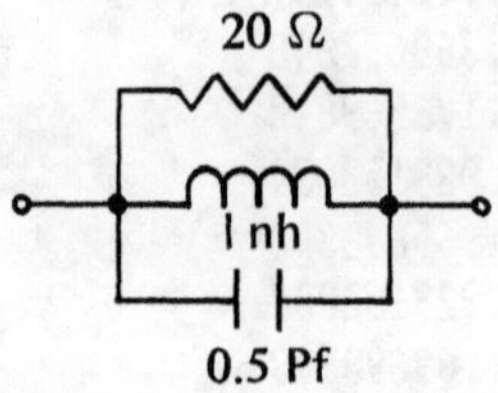


```
TWO-PORT A
S-PARAMETERS
S11=  0.161368772
     -14.48512047
S12=  0.844725586
      2.738767576
S21=  0.844725586
      2.738767576
S22=  0.161368772
     -14.48512047
```

MP-33 Transmission Line Elements

Analysis frequency (GHz) in Reg (A)
characteristic impedance in ohms, length in inches with air dielectric
transmission line element stored as Two-Port B, secondary (1) thru (8).

TRANSMISSION LINE: TRL

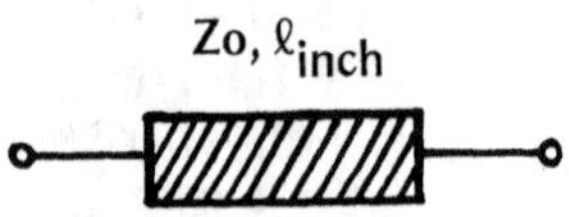

Enter characteristic impedance (Zo), [ENTER ↑]
enter length (inches, $\epsilon r = 1$), ℓ
run [A] .

OPEN CIRCUIT TRANSMISSION LINE CONNECTED IN PARALLEL: OTRL

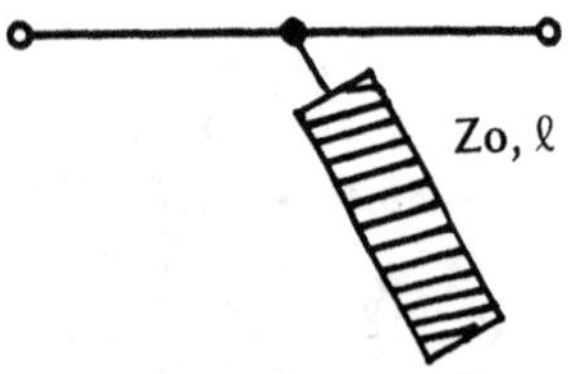

Enter Zo, [ENTER ↑] , enter ℓ
run [B] .

OPEN-CIRCUIT TRANSMISSION LINE CONNECTED IN SERIES: OTRL

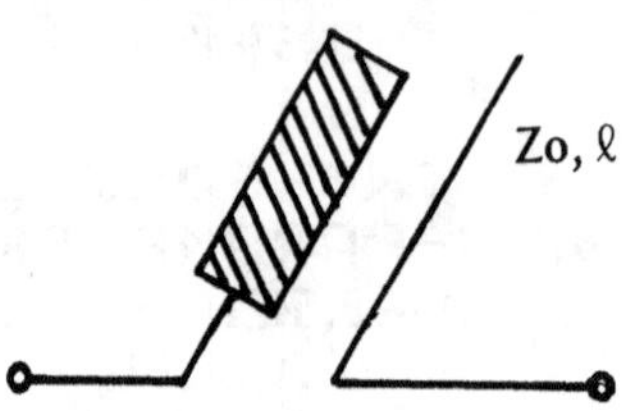

Run [f][E]
enter Zo, [ENTER ↑] , enter ℓ
run [B] .

SHORT-CIRCUIT TRANSMISSION LINE CONNECTED IN PARALLEL: STRL

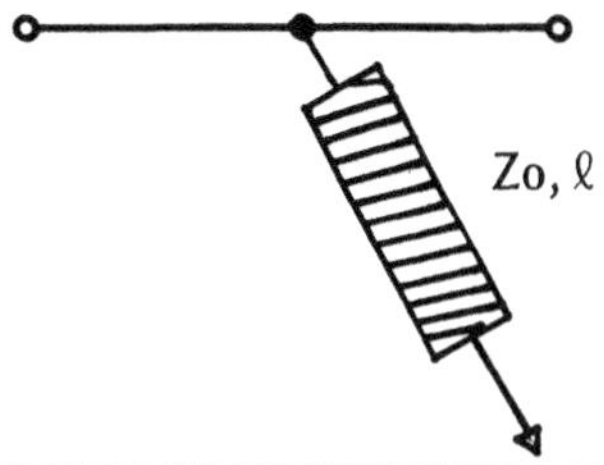

Enter Zo, [ENTER ↑], enter ℓ
run [C] .

SHORT-CIRCUIT TRANSMISSION LINE CONNECTED IN SERIES: STRL

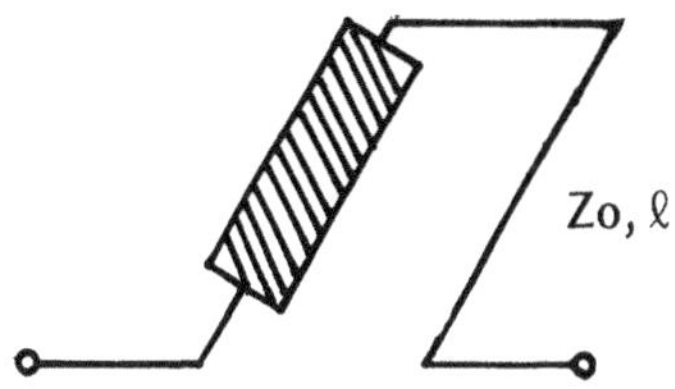

Run [f][E]
enter Zo, [ENTER ↑], enter ℓ
run [C] .

The following examples are for a 25 ohm transmission line, 0.1 inch long. The analysis frequency is 10 GHz.

Zo = 25 Ω
ℓ = 0.1″

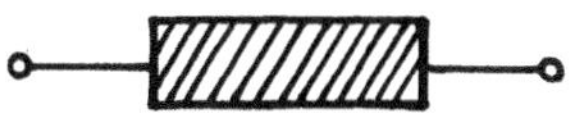

f = 10 GHz

```
    TWO-PORT B
  S-PARAMETERS
S11=  0.355567433
     -126.3426113
S12=  0.934650630
     -36.34261125
S21=  0.934650630
     -36.34261125
S22=  0.355567433
     -126.3426113
```

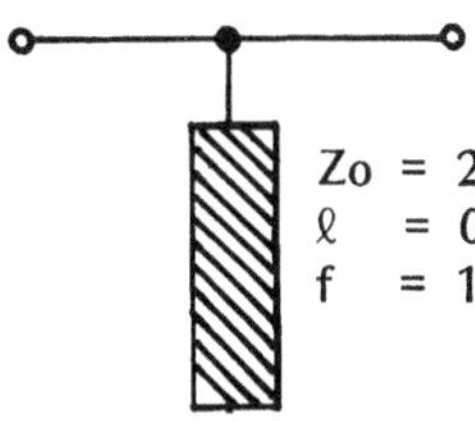

Zo = 25 Ω
ℓ = 0.1″
f = 10 GHz

```
    TWO-PORT B
  S-PARAMETERS
S11=  0.507237566
     -120.4800000
S12=  0.861806272
     -30.47999998
S21=  0.861806272
     -30.47999998
S22=  0.507237566
     -120.4800000
```

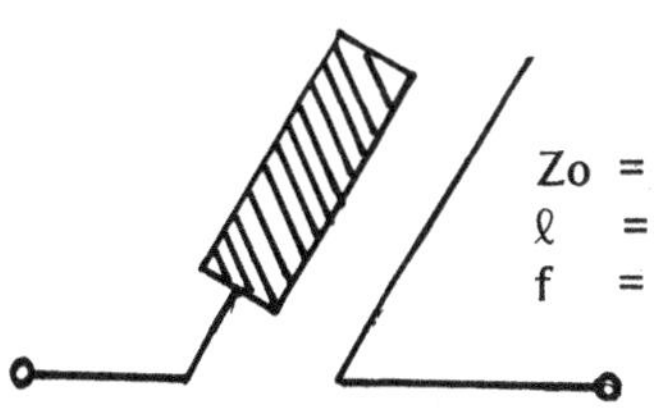

Zo = 25 Ω
ℓ = 0.1″
f = 10 GHz

```
    TWO-PORT B
  S-PARAMETERS
S11=  0.390949439
     -66.98641081
S12=  0.920412155
      23.01358919
S21=  0.920412155
      23.01358919
S22=  0.390949439
     -66.98641081
```

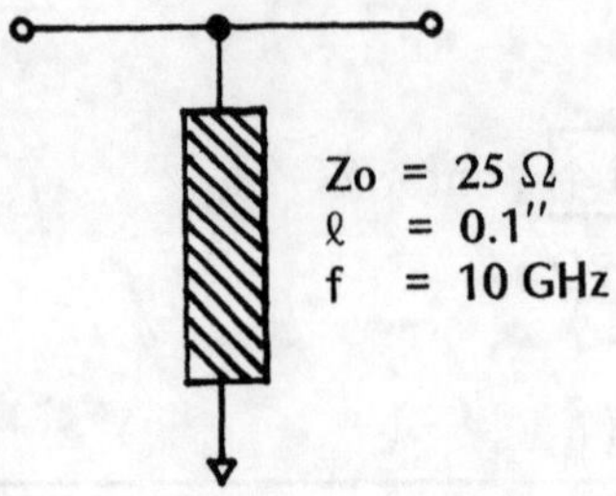


```
TWO-PORT B
S-PARAMETERS
S11=  0.861806272
      149.5200000
S12=  0.507237566
      59.52000002
S21=  0.507237566
      59.52000002
S22=  0.861806272
      149.5200000
```

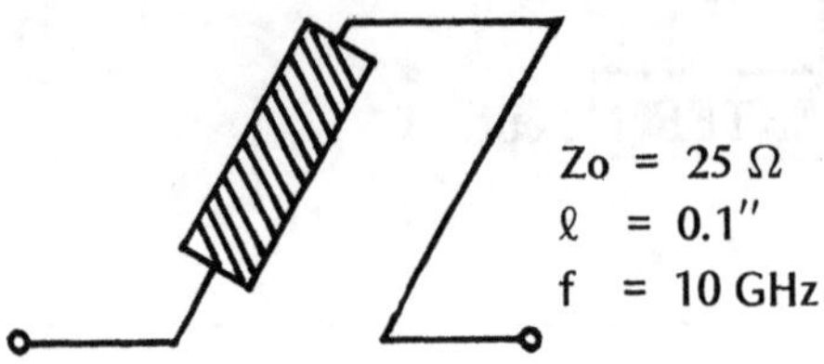


```
TWO-PORT B
S-PARAMETERS
S11=  0.145576216
      81.62935199
S12=  0.989347040
     -8.370648008
S21=  0.989347040
     -8.370648008
S22=  0.145576216
      81.62935199
```

An overflow error will occur when the electrical length of the transmission line approaches 90 degrees at the analysis frequency. The analysis frequency or electrical length should be changed slightly to prevent this error.

MP-40 SPECIAL TWO-PORTS

The special two-ports of this section include useful circuit elements which, when used in conjunction with the circuit elements of the previous section, provide the user with a wide range of catalog elements which may be used to analyze complex structures.

MP-41 Special Two-Ports

An analysis frequency is not required for this program.

The S-parameters of each element are stored as Two-Port B, sec (1) thru (8)

IDEAL TRANSFORMER IMPEDANCE RATIO:

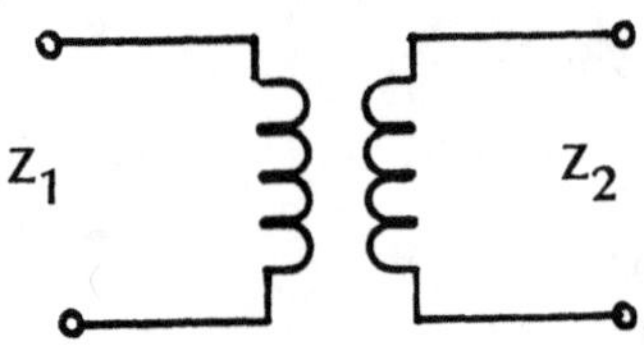

Enter Z_1 (ohms), [ENTER ↑], enter Z_2 run [A].

IDEAL TRANSFORMER TURNS RATIO:

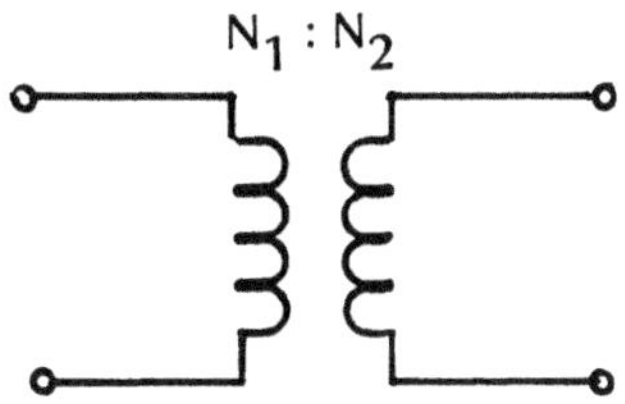

Enter N_1, ENTER ↑ , enter N_2
run B .

SHORT

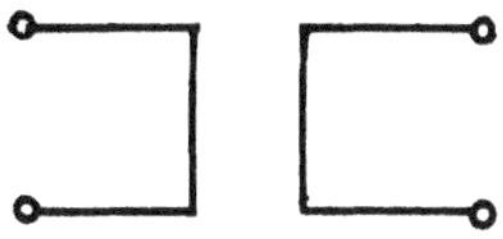

Run C .

OPEN

Run D .

IMAGE

Run E .

This program creates Two-Port B as the mirror-image of Two-Port A.

EQUATE:

run f A .

This program creates Two-Port B as a copy of Two-Port A.

CURRENT DEPENDENT VOLTAGE SOURCE: CDVS

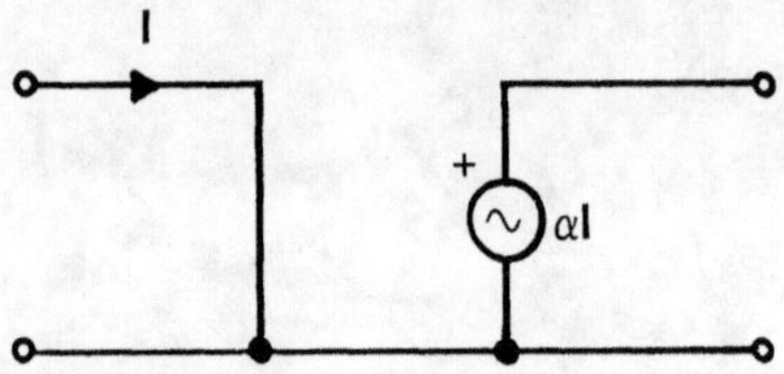

enter |α|, [ENTER ↑], enter angle of α

run [f] [B].

VOLTAGE DEPENDENT VOLTAGE SOURCE: VDVS

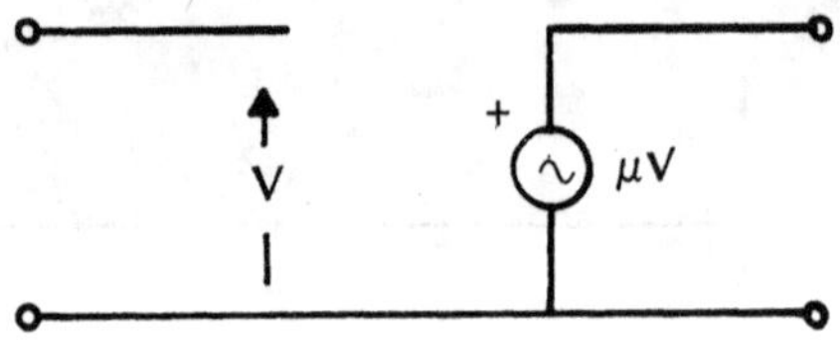

enter |μ|, [ENTER ↑], enter angle of μ

run [f] [C].

CURRENT DEPENDENT CURRENT SOURCE: CDCS

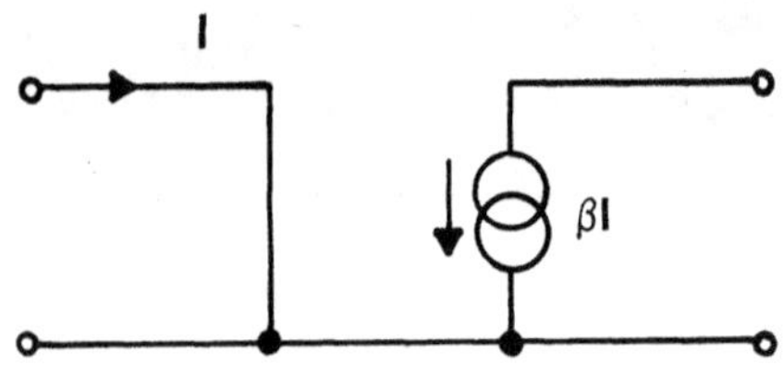

enter |β|, [ENTER ↑], enter angle of β

run [f] [D].

VOLTAGE DEPENDENT CURRENT SOURCE: VDCS

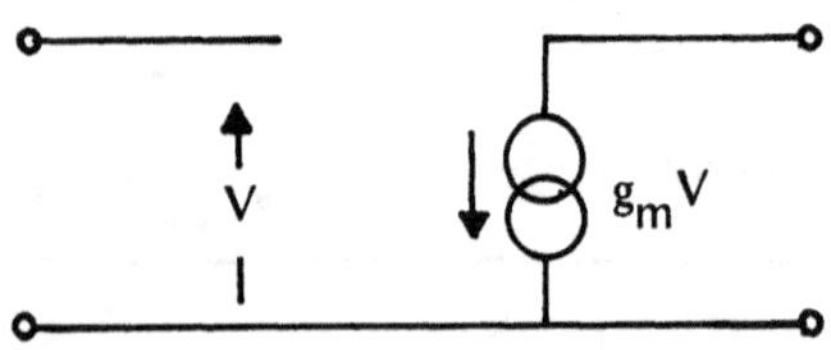

enter |gm|, [ENTER ↑], enter angle of gm

run [f] [E].

EXAMPLES:

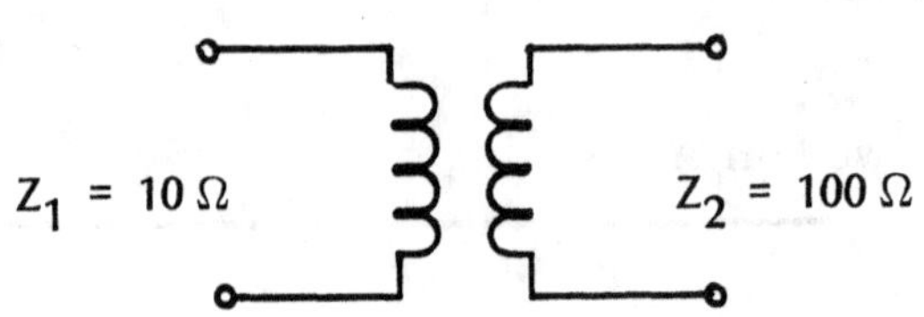


```
TWO-PORT B
S-PARAMETERS
S11= -0.818181818
      0.000000000
S12=  0.574959575
      0.000000000
S21=  0.574959575
      0.000000000
S22=  0.818181818
      0.000000000
```

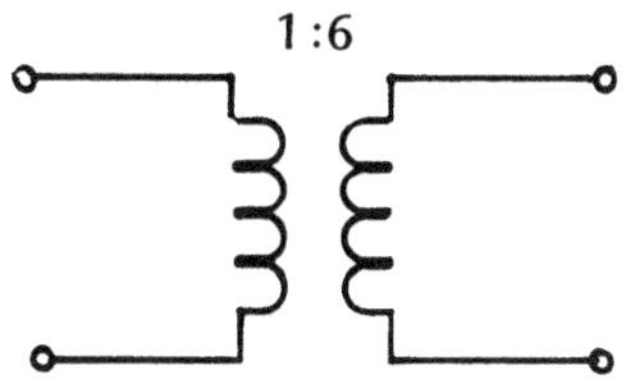


```
  TWO-PORT B
S-PARAMETERS
S11= -0.945945946
      0.000000000
S12=  0.324324324
      0.000000000
S21=  0.324324324
      0.000000000
S22=  0.945945946
      0.000000000
```

SHORT

```
  TWO-PORT B
S-PARAMETERS
S11=  1.000000000
      180.0000000
S12=  0.000000000
      0.000000000
S21=  0.000000000
      0.000000000
S22=  1.000000000
      180.0000000
```

OPEN

```
  TWO-PORT B
S-PARAMETERS
S11=  1.000000000
      0.000000000
S12=  0.000000000
      0.000000000
S21=  0.000000000
      0.000000000
S22=  1.000000000
      0.000000000
```

Two-Port A from the previous examples will be used for the Equate and Image examples.

```
  TWO-PORT A
S-PARAMETERS
S11=  0.604000000
     -114.0000000
S12=  0.076000000
      121.0000000
S21=  1.297000000
      89.00000000
S22=  0.738000000
     -83.00000000
```

IMAGE

```
  TWO-PORT B
S-PARAMETERS
S11=  0.738000000
     -83.00000000
S12=  1.297000000
      89.00000000
S21=  0.076000000
      121.0000000
S22=  0.604000000
     -114.0000000
```

EQUATE

```
  TWO-PORT B
S-PARAMETERS
S11=  0.604000000
     -114.0000000
S12=  0.076000000
      121.0000000
S21=  1.297000000
      89.00000000
S22=  0.738000000
     -83.00000000
```

CDVS

$\alpha = 75 \angle 30$

```
  TWO-PORT B
S-PARAMETERS
S11=  1.000000000
      180.0000000
S12=  0.000000000
      0.000000000
S21=  3.000000000
      30.00000000
S22=  1.000000000
      180.0000000
```

VDVS

$\mu = 0.7 \angle 45$

```
  TWO-PORT B
S-PARAMETERS
S11=  1.000000000
      0.000000000
S12=  0.000000000
      0.000000000
S21=  1.400000000
      45.00000000
S22=  1.000000000
      180.0000000
```

CDCS

$\beta = 0.5 \angle 75$

```
  TWO-PORT B
S-PARAMETERS
S11=  1.000000000
      180.0000000
S12=  0.000000000
      0.000000000
S21=  1.000000000
      255.0000000
S22=  1.000000000
      0.000000000
```

VDVS

$gm = 0.05 \angle -20$

```
  TWO-PORT B
S-PARAMETERS
S11=  1.000000000
      0.000000000
S12=  0.000000000
      0.000000000
S21=  5.000000000
      160.0000000
S22=  1.000000000
      0.000000000
```

MP-42 Coupled Lines

This program allows the user to analyze networks which contain coupled transmission lines. The user may terminate any two of the four ports associated with a pair of coupled lines to form a two-port network. The terminations may be arbitrary, which allows for analysis of the inductance associated with grounding a shorted stub, or the capacitance of an open ended line. As with MP-33, an overflow may occur if the electrical length of the coupled line pair is 90 degrees at the analysis frequency. The analysis frequency or quarter-wavelength frequency should be changed slightly.

The analysis frequency should be stored in Reg (A).

The program assumes symmetrical coupled lines with equal even and odd mode phase velocity.

Program results are stored as Two-Port B, secondary registers (0) thru (8).

The termination reflection coefficient must be referenced to the characteristic impedance of the coupler, $Zo = \sqrt{Zoe \cdot Zoo}$.

COUPLED PORT: CPL

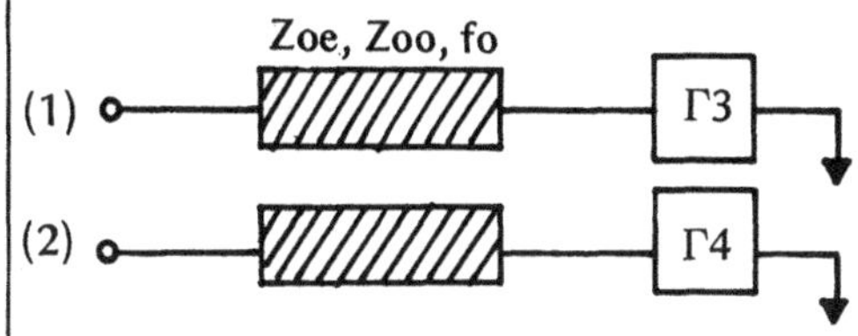

Enter even mode impedance (Zoe), [ENTER ↑]
enter odd mode impedance (Zoo), [ENTER ↑]
enter quarter-wavelength frequency of the coupled line pair (fo)
run [A]

at "3", enter $|\Gamma_3|$, [ENTER ↑], enter Θ_3, [R/S] *
at "4", enter $|\Gamma_4|$, [ENTER ↑], enter Θ_4, [R/S] *
at Program Stop, load Program MP-42.2 and [R/S] (or insert Program MP-42.2 into card reader after Γ_4 entry; MP-42.2 will be automatically loaded and execution will continue)

at MP-42.2 Program Stop, Two-Port B will contain the two-port S-parameters for the coupled line pair.

*Refer to SHO and OPE for automatic short and open termination.

DIRECT PORT: DIR

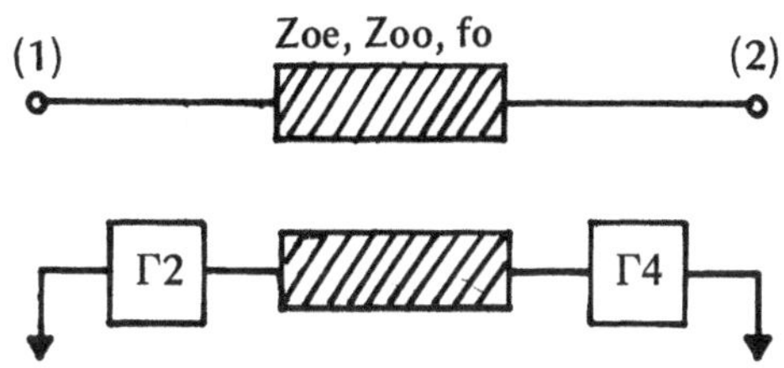

Enter even mode impedance (Zoe), [ENTER ↑]
enter odd mode impedance (Zoo), [ENTER ↑]
enter quarter-wavelength frequency of the coupled line pair (fo)
run [B]

at "2", enter $|\Gamma_2|$, [ENTER ↑], enter Θ_2, [R/S] *
at "4", enter $|\Gamma_4|$, [ENTER ↑], enter Θ_4, [R/S] *
at Program Stop, load program MP-42.2 and [R/S] (or insert Program MP-42.2 into card reader after Γ_4 entry; MP-42.2 will be automatically loaded and execution will continue)

at MP-42.2 Program Stop, Two-Port B will contain the two-port S-parameters for the coupled line pair.

*Refer to SHO and OPE for automatic short and open termination.

ISOLATED PORT: ISO

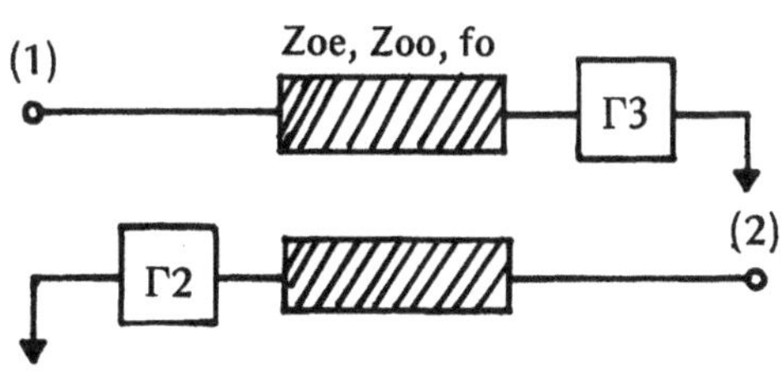

Enter even mode impedance (Zoe), [ENTER ↑]
enter odd mode impedance (Zoo), [ENTER ↑]
enter quarter-wavelength frequency of the coupled line pair (fo)
Run [C]

at "2", enter $|\Gamma_2|$, ENTER ↑, enter Θ_2, R/S *

at "3", enter $|\Gamma_3|$, ENTER ↑, enter Θ_3, R/S *

at Program Stop, load program MP-42.2 and R/S (or insert Program MP-42.2 into card reader after Γ_3 entry; MP-42.2 will be automatically loaded and execution will continue)

at MP-42.2 Program Stop, Two-Port B will contain the two-port S-parameters for the coupled line pair.

*Refer to SHO and OPE for automatic short and open termination.

SHO

Run D for automatic shorted termination.

OPE

Run E for automatic open termination.

EXAMPLES: Store the analysis frequency of 2.4 GHz into Reg Ⓐ

COUPLED PORT

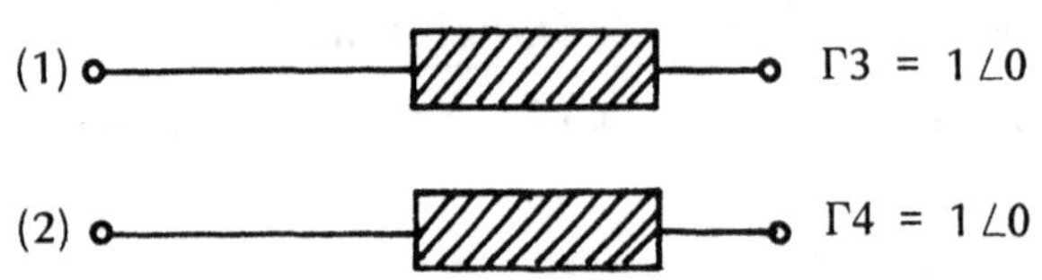

Zoe = 100 Ω, Zoo = 30 Ω, fo = 2.2 GHz

$\Gamma_3 = \Gamma_4 = 1\angle 0$, Open.

```
TWO-PORT B
S-PARAMETERS
S11=  0.981250938
      159.0265352
S12=  0.192734524
      69.02653523
S21=  0.192734524
      69.02653523
S22=  0.981250938
      159.0265352
```

DIRECT PORT

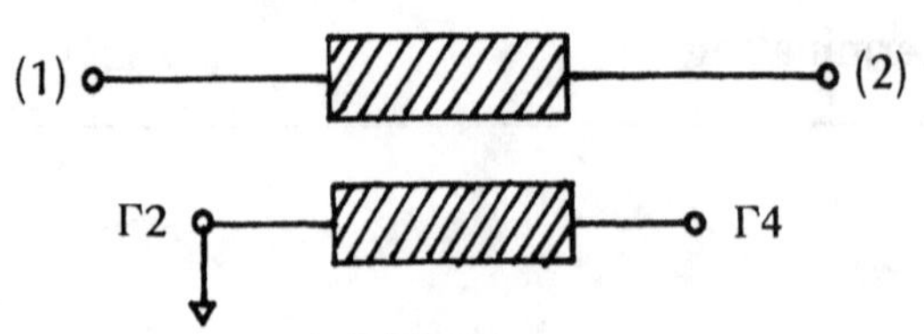

Zoe = 100 Ω, Zoo = 30 Ω, fo = 2.2 GHz

$\Gamma_2 = 1\angle 180$, Short

$\Gamma_4 = 1\angle 0$, Open

```
TWO-PORT B
S-PARAMETERS
S11=  0.815613759
      132.5998639
S12=  0.578596748
     -43.03746193
S21=  0.578596748
     -43.03746193
S22=  0.815613759
     -38.67478776
```

ISOLATED PORT

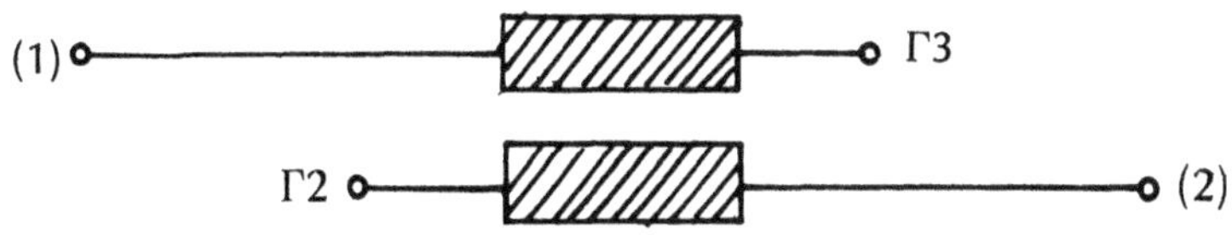

Zoe = 100 Ω, Zoo = 30 Ω, fo = 2.2 GHz

$\Gamma_2 = \Gamma_3 = 1\angle 0$, Open

```
TWO-PORT B
S-PARAMETERS
S11=  0.353677801
      165.6871322
S12=  0.935367315
     -104.3128677
S21=  0.935367315
     -104.3128677
S22=  0.353677801
      165.6871322
```

MP-50 MAPPING AND CONFIGURATION CONVERSION

The mapping programs provide the user with a powerful design tool. Any of the four S-parameters of a two-port network can be mapped as a function of series or shunt feedback impedance, which allows the designer to investigate the effects of feedback elements on circuit performance. Another feature of these programs is the parameter conversions. The S-parameters which are measured in a particular configuration - for example, common emitter - can be converted to another configuration, common base or common collector. The map-source and map-load programs provide the user with a graphic presentation of S'_{11} or S'_{22} as a function of all possible load and source impedances.

MP-51 Two to Three-Port Conversion

The series and shunt feedback mappings, as well as the parameter conversions, require a transformation from two-port to three-port parameters. This program provides the transformation, and is essentially a preprocessor program for the parameter conversion of MP-52 and the three-port mapping of MP-54.

TWO-PORT TO THREE-PORT CONVERSION: SERIES FEEDBACK

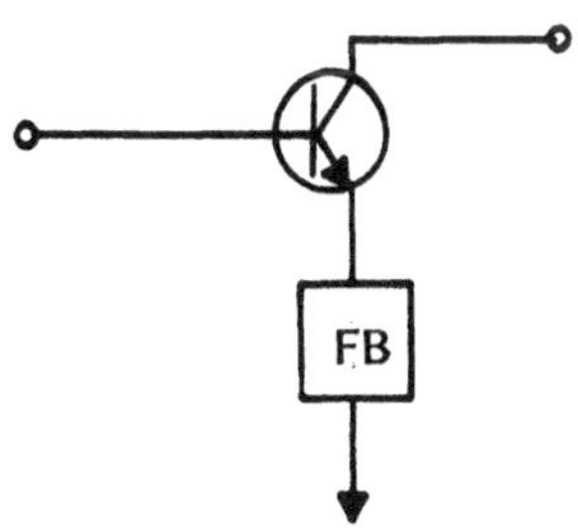

Load Program MP-51.1

Two-Port A in primary (1) thru (8)

run [A]

at Program Stop, "51.2", load Program MP-51.2

run [A]

at Program Stop, THREE-PORT in primary (0) thru (9) and secondary (0) thru (7).

See "RECALL THREE-PORT" to display three-port data.

TWO-PORT TO THREE-PORT CONVERSION: SHUNT FEEDBACK

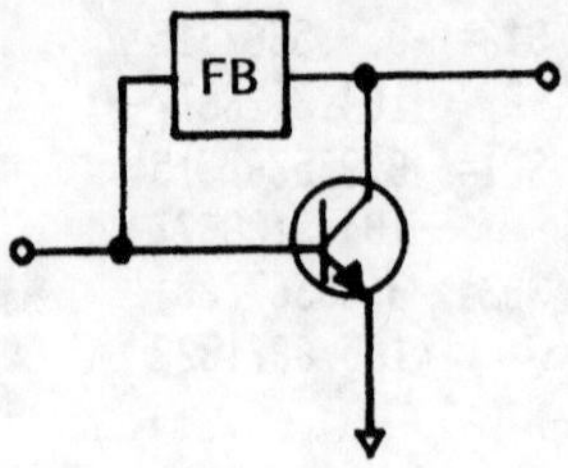

Load Program MP-51.1

Two-Port A in primary (1) thru (8)

run [B]

at Program Stop, "51.2", load Program MP-51.2

run [B]

at Program Stop, THREE-PORT in primary (0) thru (9) and secondary (0) thru (7).

See "RECALL THREE-PORT" to display three-port data.

PARAMETER CONVERSIONS:

Load Program MP-51.1

Two-Port A in primary (1) thru (8)

run [C]

at Program Stop, "51.2", load Program MP-51.2

run [C]

at Program Stop, THREE-PORT in primary (0) thru (9) and secondary (0) thru (7).

See "RECALL THREE-PORT" to display three-port data.

RECALL THREE-PORT:

Run after any of the above, Program MP-51.2

run [D]

at "11", $|S_{11}|$ in "y", Θ_{11} in "z", [R/S]

at "13", $|S_{13}|$ in "x", Θ_{13} in "y", [R/S]

⋮

at "22", $|S_{22}|$ in "y", Θ_{22} in "z".

THE THREE-PORT DATA IS STORED AS SHOWN BELOW:

PRIMARY		SECONDARY	
(0)	$\|S_{11}\|$	(0)	$\|S_{32}\|$
(1)	Θ_{11}	(1)	Θ_{32}
(2)	$\|S_{13}\|$	(2)	$\|S_{21}\|$
(3)	Θ_{13}	(3)	Θ_{21}
(4)	$\|S_{12}\|$	(4)	$\|S_{23}\|$
(5)	Θ_{12}	(5)	Θ_{23}
(6)	$\|S_{31}\|$	(6)	$\|S_{22}\|$
(7)	Θ_{31}	(7)	Θ_{22}
(8)	$\|S_{33}\|$		
(9)	Θ_{33}		

EXAMPLE:

The Two-Port A data from the previous examples will be used. Run the appropriate program function to obtain the three-port parameters shown below. The three-port parameters from each example should be saved on a data card for use with following examples.

```
TWO-PORT A
S-PARAMETERS
S11=  0.604000000
     -114.0000000
S12=  0.076000000
      121.0000000
S21=  1.297000000
      89.00000000
S22=  0.738000000
     -83.00000000
```

SERIES FEEDBACK
AND
PARAMETER CONVERSION

```
THREE-PORT
S-PARAMETERS
S11=  0.666984677
     -430.8038291
S12=  0.420829125
     -293.7311723
S13=  0.658469791
      21.81118281
S21=  0.993936005
     -280.9017271
S22=  0.809205669
     -421.8522486
S23=  0.504029229
     -31.38537959
S31=  0.686364058
     -30.28159300
S32=  0.556115032
      36.17626465
S33=  0.045276709
      156.7943864
```

SHUNT FEEDBACK

```
THREE-PORT
S-PARAMETERS
S11=  0.405914726
     -158.9011771
S12=  0.227120793
     -32.00337670
S13=  0.429468002
     -3.437822850
S21=  0.339015093
     -255.8002529
S22=  0.252108741
     -149.2927337
S23=  0.979726187
      152.1692145
S31=  0.849514977
     -33.97839943
S32=  0.590697132
      179.1878385
S33=  0.566141282
     -58.58439462
```

MP-52 Parameter Conversion

This program converts the three-port parameters of Program MP-51.1 label [C], to three-port parameters with a specified common terminal. Program MP-53 is used as a post-processor to convert the three-port parameters to two-port parameters. These programs are used to convert the two-port parameters of a three terminal network with one common terminal, to a two-port network with another common terminal — for example, common emitter to common base for a bipolar transistor, or from common source to common gate for a field effect transistor.

CONFIGURATION CONVERSION:

Three-port parameters in primary (0) thru (9) and secondary (0) thru (7) from Program MP-51.1 [C]

run label corresponding to input data:

[f][B] for common emitter (or other convention)

[f][C] for common base (or other convention)

[f][D] for common collector (or other convention)

run label corresponding to desired transformation:

[B] for common emitter (or other convention)

[C] for common base (or other convention)

[D] for common collector (or other convention)

if program stops with display "52.2", load Program MP-52.2 and [R/S]. At Program Stop, three-port parameters are stored in the same location as MP-51.

Program MP-51.2 [D] may be used to display the transformed three-port parameters.

EXAMPLES: The three-port conversion example of MP-51 is based on the two-port parameters for an FET with common source (emitter). Convert these parameters to common gate (base) and common drain (collector).

The three-port parameters for each configuration can be converted into the other configurations as a check for this program.

COMMON GATE THREE-PORT PARAMETERS		COMMON DRAIN THREE-PORT PARAMETERS	
THREE-PORT S-PARAMETERS		THREE-PORT S-PARAMETERS	
S11=	0.045276709	S11=	0.666984677
	156.7943864		-430.8038291
S12=	0.556115032	S12=	0.658469791
	36.17626465		21.81118281
S13=	0.686364058	S13=	0.420829125
	-30.28159300		-293.7311723
S21=	0.504029229	S21=	0.686364058
	-31.38537959		-30.28159300
S22=	0.809205669	S22=	0.045276709
	-421.8522486		156.7943864
S23=	0.993936005	S23=	0.556115032
	-280.9017271		36.17626465
S31=	0.658469791	S31=	0.993936005
	21.81118281		-280.9017271
S32=	0.420829125	S32=	0.504029229
	-293.7311723		-31.38537959
S33=	0.666984677	S33=	0.809205669
	-430.8038291		-421.8522486

MP-53 Three to Two-Port Conversion

Program MP-53 is used in conjunction with MP-52 to convert the transformed three-port parameters into two-port parameters.

THREE-PORT TO TWO-PORT CONVERSION:

The three-port parameters from MP-52 must be in the appropriate data registers

run [A]

at Program Stop, Two-Port A in primary (1) thru (8).

RECALL TWO-PORT:

Run [A] as above

run [B] , at "11", $|S_{11}|$ in "y", Θ_{11} in "z", [R/S]

⋮

at "22", $|S_{22}|$ in "y", Θ_{22} in "z".

Example: Convert the three-port common source parameters MP-51 and the common gate and drain parameters of MP-52, into two-port parameters.

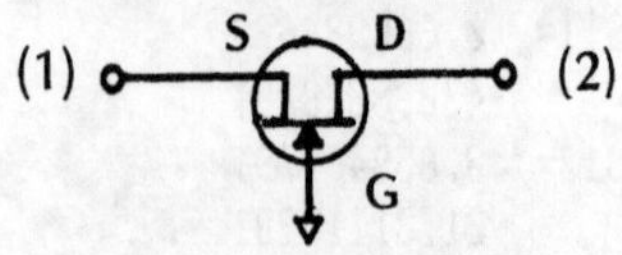

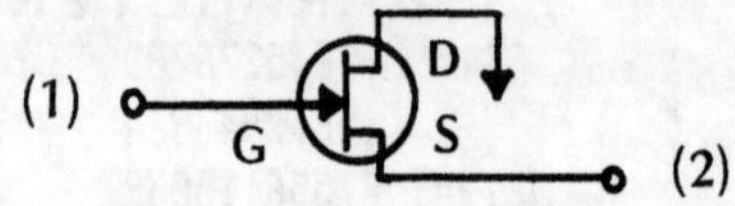

COMMON GATE PARAMETERS

```
 TWO-PORT A
S-PARAMETERS
S11=  0.364195984
     -165.9262780
S12=  0.381094868
      21.58890965
S21=  0.965447348
     -41.29237523
S22=  1.016786698
     -47.72059987
```

COMMON DRAIN PARAMETERS

```
 TWO-PORT A
S-PARAMETERS
S11=  0.823051470
     -53.80100465
S12=  0.561568790
      12.75716157
S21=  1.039991947
     -32.71387050
S22=  0.209352803
     -158.1410197
```

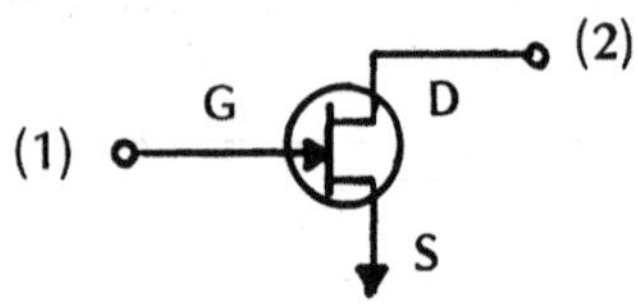

COMMON SOURCE PARAMETERS

```
 TWO-PORT A
S-PARAMETERS
S11=  0.604000002
     -114.0000002
S12=  0.076000000
      121.0000012
S21=  1.297000002
      89.00000004
S22=  0.738000001
     -83.00000010
```

These two-port parameters may again be transformed using MP-51, MP-52, and MP-53.

MP-54 Three-Port Mapping

Program MP-54 is used to compute the contours of constant resistance or reactance for either series or shunt feedback. The contours, which are circles, can be plotted on the plane of the desired parameter, then the results of a feedback element which corresponds to a point on the mapping can be determined directly. Program MP-54.1 computes three mapping constants from the input three-port parameters (from Program MP-51), and Program MP-54.2 computes the center and radius of the various contours.

SMITH CHART MAPPING:

No input required, run [f] [B]

at Program Stop, mapping constants in Reg (B) thru (E) and secondary (8) and (9).

S_{11} MAPPING:

Series or shunt three-port parameters from Program MP-51 in primary (0) thru (9) and secondary (0) thru (7)

run [B]

at Program Stop, mapping constants in Reg (B) thru (E) and secondary (8) and (9).

S_{12} MAPPING:

Series or shunt three-port parameters from Program MP-51 in primary (0) thru (9) and secondary (0) thru (7)

run [C]

at Program Stop, mapping constants in Reg (B) thru (E) and secondary (8) and (9).

S_{21} MAPPING:

Series or shunt three-port parameters from Program MP-51 in primary (0) thru (9) and secondary (0) thru (7)

run [D]

at Program Stop, mapping constants in Reg (B) thru (E) and secondary (8) and (9).

S_{22} MAPPING:

Series or shunt three-port parameters from Program MP-51 in primary (0) thru (9) and secondary (0) thru (7)

run [E]

at Program Stop, mapping constants in Reg (B) thru (E) and secondary (8) and (9).

NOTE: Program MP-54 does not modify the input three-port parameters. All parameters may be mapped in any order without additional input.

After the mapping constants have been calculated, load Program MP-54.2.

CONTOURS OF CONSTANT RESISTANCE:

Enter ± R, ohms, run [A]

at stop, ± R in "x", magnitude of center in "y", angle of center in "z", radius of contour in "t".

Enter another ±R and run as above for another contour.

CONTOURS OF CONSTANT REACTANCE:

Enter ± x, ohms, run [B]

at stop, ± x in "x", magnitude of center in "y", angle of center in "z", radius of contour in "t".

Enter another ±x and run as above for another contour.

EXAMPLE: Use the shunt feedback three-port parameters from the example of MP-51 to compute the mapping constants for the four parameters. The Smith Chart constants can be verified as well; however, no input is required for this data. The HP-41C is used to print the output data. "R18=" corresponds to data register 18, or secondary register 8 for the HP-67. These register numbers correspond to the [RCL] [(i)] function; secondary (0) through (9) is 10 through 19, and (A) through (I) is 20 through 25.

SMITH CHART

```
R18=  1.000000000
R19=  0.000000000

R21=  1.000000000
R22=  0.900000000
R23=  0.000000000
R24=  0.000000000
```

S_{11} MAPPING

```
R18= -0.150436727
R19= -0.515431025

R21=  0.930388885
R22= -1.323113847
R23= -0.095232206
R24= -0.036350432
```

S_{12} MAPPING

```
R18= -0.108505763
R19=  0.357235053

R21=  0.930388885
R22= -1.323113847
R23=  0.069362868
R24= -0.292090339
```

S_{21} MAPPING

```
R18=  0.798158981
R19=  0.929135948

R21=  0.930388885
R22= -1.323113847
R23= -0.775523212
R24=  0.367666514
```

S_{22} MAPPING

```
R18= -0.111132482
R19= -0.844425665

R21=  0.930388885
R22= -1.323113847
R23=  0.201072059
R24=  0.111926606
```

The mapping constants from the S_{22} mapping are used with Program MP-54.2 to determine the contours of constant 25 and 50 ohm resistance, and ±j100 constant reactance.

```
      R=  50.00000000
 CENTER=  0.355739484
         -66.20966967°
 RADIUS=  0.410497028

      R=  25.00000000
 CENTER=  0.244737401
         -48.48211666°
 RADIUS=  0.553988436

      X=  100.0000000
 CENTER=  1.532222559
         -35.29292706°
 RADIUS=  1.170682687

      X= -100.0000000
 CENTER=  0.716516929
         -101.7962127°
 RADIUS=  0.238456742
```

MP-55 Map-Source and Map-Load

This program is similar to MP-54 above, except Two-Port A is used to generate the mapping constants. Map-load is a mapping of S'_{11} as a function of a mapping impedance connected as the load, while the map-source function maps S'_{22} as a function of source impedance. The mapping constants are the same as for Program MP-54. Program MP-55.2 is identical to Program MP-54.2, and is duplicated to allow Program MP-55 to stand alone.

MAP-LOAD: Two-Port A in primary (1) thru (8). run [A] at stop, mapping constants in Reg (B) thru (E) and secondary (8) and (9).
MAP-SOURCE: Two-Port A in primary (1) thru (8) run [B] at stop, mapping constants in Reg (B) thru (E) and secondary (8) and (9).
CONTOURS OF CONSTANT RESISTANCE AND REACTANCE: Load Program MP-55.2, execute as MP-54.2.

EXAMPLE: The two-port data from the previous examples will be used for this example. Calculate the mapping constants for map-load and map-source. Then load MP-55.2 and compute contours of constant 25 and 50 ohm resistance, then contours of ±j100 ohms.

```
TWO-PORT A
S-PARAMETERS
S11=  0.604000000
     -114.0000000
S12=  0.076000000
      121.0000000
S21=  1.297000000
      89.00000000
S22=  0.738000000
     -83.00000000
```

MAPPING CONSTANTS MAP-LOAD	MAPPING CONSTANTS MAP-SOURCE
R18= -0.145798912	R18= -0.147771849
R19= 0.160009285	R19= 0.047395562
R21= 0.333651618	R21= 0.342204389
R22= -1.073443619	R22= -0.594542799
R23= -0.183247048	R23= 0.165770893
R24= -0.698838202	R24= -0.787593737

CONSTANT R AND X MAP-LOAD	CONSTANT R AND X MAP-SOURCE
R= 25.00000001	R= 25.00000000
CENTER= 0.660852535	CENTER= 0.763455186
-114.1804510°	-84.13287993°
RADIUS= 0.086638543	RADIUS= 0.063055372
R= 50.00000000	R= 50.00000000
CENTER= 0.648595438	CENTER= 0.754339357
-116.8131655°	-85.76739888°
RADIUS= 0.054156843	RADIUS= 0.039565890
X= 100.0000000	X= 100.0000000
CENTER= 0.706528206	CENTER= 0.776204925
-123.1802521°	-89.52321805°
RADIUS= 0.077951394	RADIUS= 0.037785221
X= -100.0000000	X= -100.0000001
CENTER= 0.608828115	CENTER= 0.721115837
-120.7922581°	-88.07285076°
RADIUS= 0.023500142	RADIUS= 0.020468158

MP-60 LADDER NETWORK ANALYSIS

The ladder network analysis program provides a simple and fast means of analyzing many types of circuits. This program includes the standard lumped and distributed elements which are most often used for matching networks and filters. The user stores a termination impedance in two-data registers, enters input data for a circuit element, then executes the appropriate program for that element. The program computes the element impedance, and connects the impedance in either series or shunt with the termination impedance to form the input impedance at that point. A subroutine is included to convert impedance to reflection coefficient (referenced to 50 ohms). The VSWR and insertion loss (for a lossless network) may easily be calculated from the input reflection coefficient.

The major advantage of this program is the ability to "program" a circuit topology and analyze the network over a specified frequency range, without operator intervention.

MP-61 Ladder Network Analysis

Data is stored in the following locations:

Reg (A)	The analysis frequency (GHz), or start frequency for frequency loop
Reg (B)	stop frequency (GHz) for frequency loop
Reg (C)	step frequency (GHz) for frequency loop
Reg (1)	magnitude of termination impedance or impedance at present node
Reg (2)	angle of termination impedance or impedance at present node

The program computes the impedance of the specified element and connects this impedance in series or parallel, as instructed, with the impedance in Reg (1) and Reg (2). The results are stored in Reg (1) and Reg (2) as the new nodal impedance, and duplicated in the "x" and "y" registers for display. The program does not use the data in the "x" and "y" registers so that they may be viewed at any time. All impedance calculations are based on the frequency (GHz) in Reg (A).

SERIES RESISTOR:

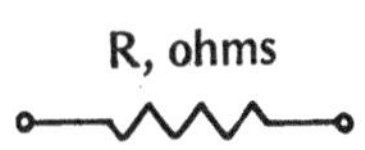

Enter R, ohms

run [A] .

PARALLEL RESISTOR:

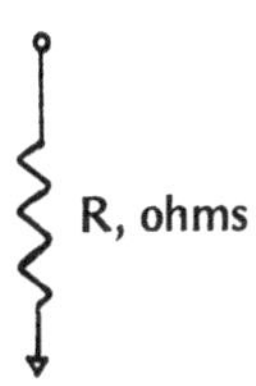

Enter R, ohms

run [f] [E] [A] .

SERIES INDUCTOR · CAPACITOR; CONNECTED IN SERIES:

Enter L (nh), [ENTER ↑], enter C (pf)
run [B].

SERIES INDUCTOR · CAPACITOR; CONNECTED IN PARALLEL:

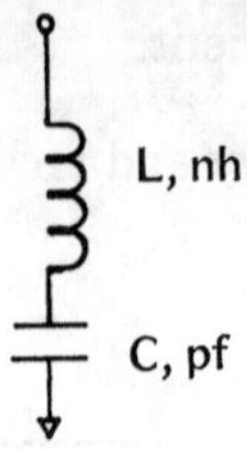

Enter L (nh), [ENTER ↑], enter C (pf)
run [f] [E] [B].

PARALLEL INDUCTOR · CAPACITOR; CONNECTED IN SERIES:

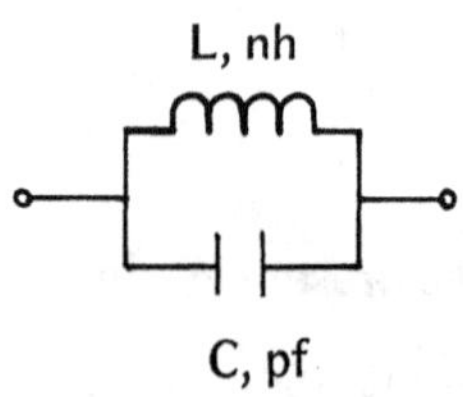

Enter L (nh), [ENTER ↑], enter C (pf)
run [C].

PARALLEL INDUCTOR · CAPACITOR; CONNECTED IN PARALLEL:

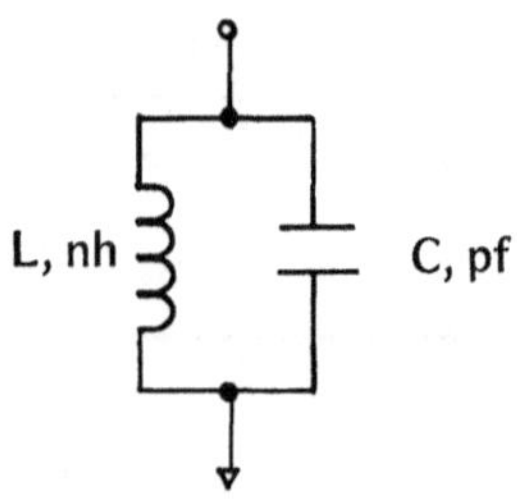

Enter L (nh), [ENTER ↑], enter C (pf)
run [f] [E] [C].

SHORT-CIRCUIT TRANSMISSION LINE CONNECTED IN SERIES:

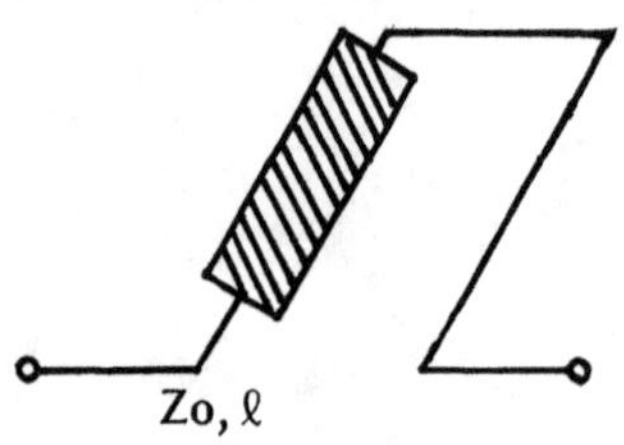

Enter characteristic impedance (Zo), [ENTER ↑], enter length (inches in air)
run [D].

SHORT-CIRCUIT TRANSMISSION LINE CONNECTED IN PARALLEL:

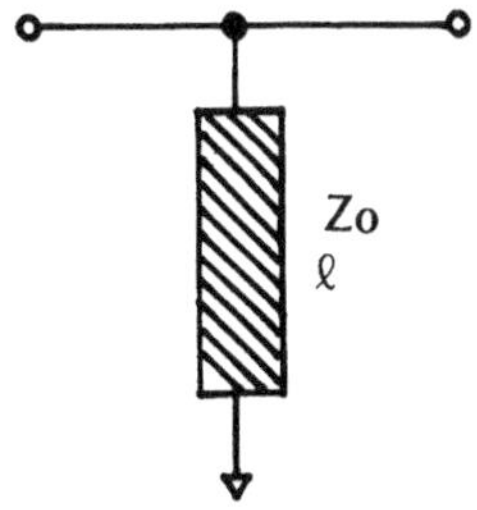

Enter characteristic impedance (Zo), [ENTER ↑], enter length (inches in air)

run [f] [E] [D] .

OPEN-CIRCUIT TRANSMISSION LINE CONNECTED IN SERIES:

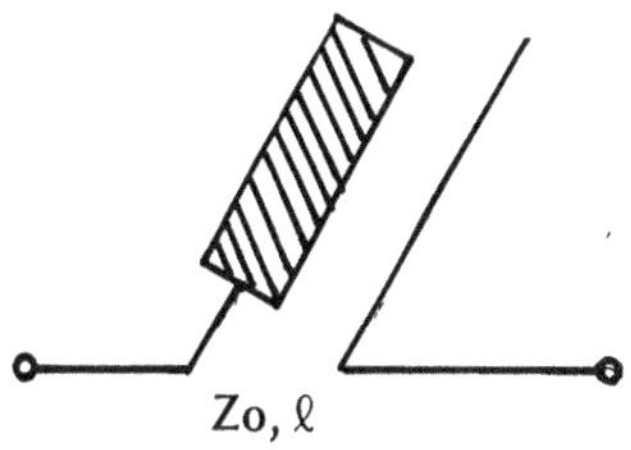

Enter characteristic impedance (Zo), [ENTER ↑], enter length (inches in air)

run [E] .

OPEN-CIRCUIT TRANSMISSION LINE CONNECTED IN PARALLEL:

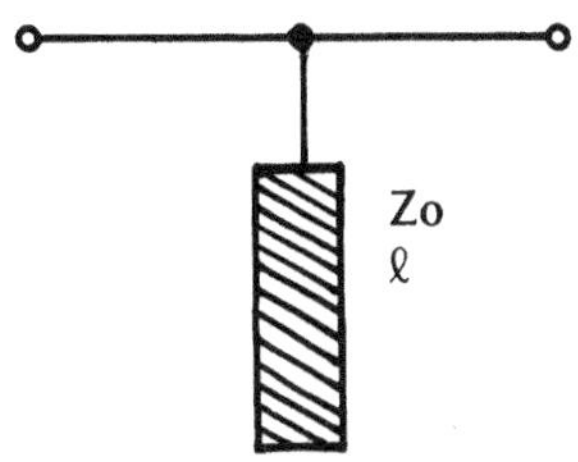

Enter characteristic impedance (Zo), [ENTER ↑], enter length (inches in air)

run [f] [E] [E] .

TRANSMISSION LINE

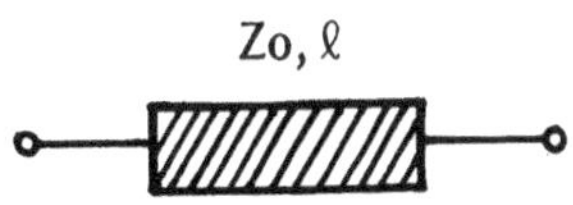

Enter characteristic impedance (Zo), [ENTER ↑], enter length (inches in air)

run [f] [A] .

CONVERT IMPEDANCE TO REFLECTION COEFFICIENT:

Enter angle of impedance (Θ), [ENTER ↑], enter magnitude of impedance, |Z|

run [f] [B] .

NOTE: The input format for this subroutine is the same as a normal calculation, the magnitude is in "x", and the angle in "Y". To convert to reflection coefficient following any element calculation, execute [f][B].

USE Z∠Θ IN SERIES:

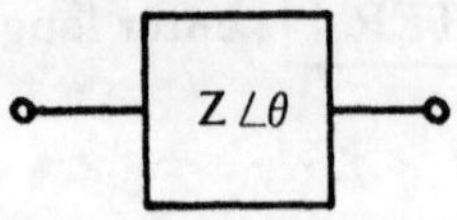

Enter |Z|, [ENTER ↑], enter Θ

run [f] [C] .

USE Z∠Θ IN PARALLEL:

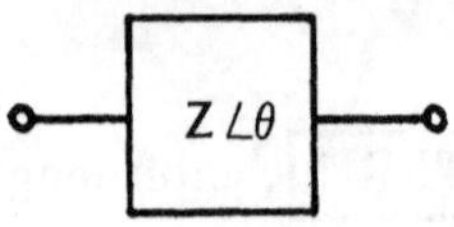

Enter |Z|, [ENTER ↑], enter Θ

run [f] [E] [f] [C] .

PROGRAMMING A NETWORK:

Store start (and analysis) frequency (GHz) in Reg (A)

store stop frequency (GHz) in Reg (B)

store step frequency (GHz) in Reg (C)

execute [GTO][1]

switch to [W/PRGM]

program network using the same keystrokes that would be used for normal calculation. Termination impedance should be stored in Reg (1) and (2) as the first steps. Data for each element can be recalled from specified data storage registers.

Switch to [RUN]

execute [h][RTN] [R/S]

the program will stop with frequency in "x", magnitude of reflection coefficient in "y", and angle in "z". Program steps 002 through 004 convert the calculated impedance to reflection coefficient, recall the frequency and stops. These steps may be eliminated if impedance is desired.

Program steps 171 through 224 are available for the network topology. If more memory is required, unused elements may be deleted from the program (see Appendix A). All data registers except those specified above are available.

Additional elements may be programmed by calculating the impedance with magnitude in "y" and angle in "x", then transfering to LBL 6. Flag 0 is the parallel connection flag. When set, the calculated impedance is paralleled with the impedance in Reg (1) and (2).

VSWR may be calculated from the reflection coefficient as:

$$\text{VSWR} = \frac{1 + |\Gamma|}{1 - |\Gamma|}$$

Insertion loss (for a lossless network only) is:

$$\text{IL} = 10 \text{ Log } (1 - |\Gamma|^2) \text{ dB.}$$

DISTRIBUTED ELEMENT EXAMPLE:

A 150 ohm to 50 ohm distributed element transformer network is shown below. This network is calculated for the 2 to 8 GHz frequency band, and has a theoretical ripple of 0.17 dB. Note that the line lengths are commensurate — that is, they are all the same.

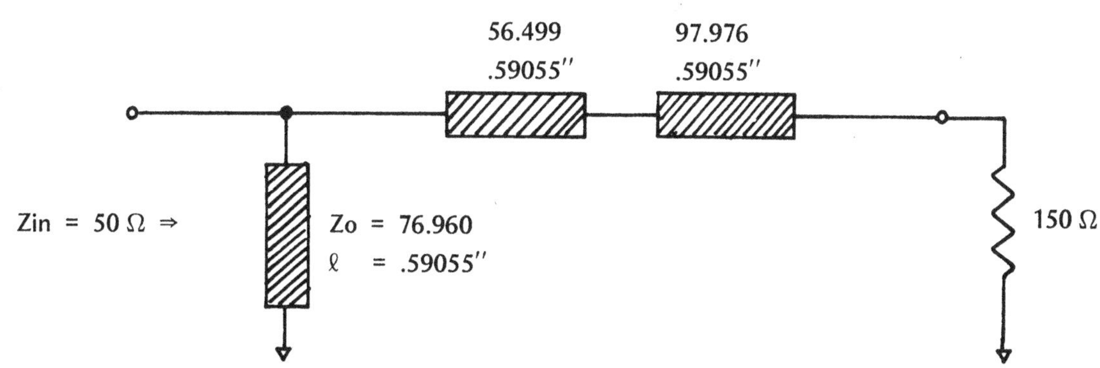

The network will first be analyzed at 3.5 GHz using the normal keyboard entry.

Store the analysis frequency (3.5) into Reg (A)

store the 150 ohm termination impedance into Reg (1) and (2), 150 into Reg (1), and 0 into Reg (2)

connect the 97 ohm transmission line, 97.976, [ENTER ↑], .59055, [f][A]

connect the 56 ohm transmission line, 56.499, [ENTER ↑], .59055, [f][A]

connect the 76 ohm shorted stub, 76.960, [ENTER ↑], .59055, [f][E] (parallel connection), [D]

convert to reflection coefficient, [f][B].

Z= 150.0000000 0.000000000°	TERMINATION IMPEDANCE
Z= 77.01080146 -19.54915141°	AFTER CONNECTION OF 97 OHM TRANSMISSION LINE
Z= 35.46634838 -3.516929720°	AFTER CONNECTION OF 56 OHM TRANSMISSION LINE
Z= 35.00848734 9.859269838°	AFTER CONNECTION OF 76 OHM SHORTED STUBS INPUT IMPEDANCE

Γ= 0.196292630 154.8089762°	INPUT REFLECTION COEFFICIENT
VSWR= 1.488467912	INPUT VSWR
IL= -0.170646220	NETWORK INSERTION LOSS

The network topology will now be programmed to automatically calculate the insertion loss from 1.5 to 8.5 GHz. The frequency and element values are stored as follows:

- Reg (A), start frequency of 1.5
- Reg (B), stop frequency of 8.5
- Reg (C), step frequency of 0.5
- Reg (3), 97.976 impedance of first transmission line
- Reg (4), 56.499 impedance of second transmission line
- Reg (5), 76.960 impedance of shorted stub
- Reg (6), .59055 commensurate line length

[GTO] [1], switch to [W/PRGM], and program the network as follows:

LBL 1

Steps	Function	
150 STO 1 0 STO 2	TERMINATION IMPEDANCE	
RCL 3 RCL 6 [f] [A]	97.976 OHM TRANSMISSION LINE	USER PROGRAM
RCL 4 RCL 6 [f] [A]	56.499 OHM TRANSMISSION LINE	
RCL 5 RCL 6 f E [D]	76.960 OHM SHORTED STUB	

GSB b Convert to reflection coefficient, part of program. SINGLE STEP

[g] [x^2] CHS 1 + [f] [LOG] 10 X	CALCULATE INSERTION LOSS FROM REFLECTION COEFFICIENT.	USER PROGRAM

RCL A
R/S — Recall frequency, stop with frequency in "x", insertion loss (dB) in "y".

The user should execute [R/S] to begin calculations at the next frequency. The program may be halted at any point, a new start frequency stored in Reg (A), and restarted. Execute [h] [RTN] [R/S] to restart.

Program output for this example is as follows:

FREQ	LOSS
1.50	-1.402478282
2.00	-0.173811009
2.50	-0.006654066
3.00	-0.114657269
3.50	-0.170646220
4.00	-0.125054538
4.50	-0.040555832
5.00	-0.000006186
5.50	-0.040552959
6.00	-0.125051351
6.50	-0.170646233
7.00	-0.114661508
7.50	-0.006656467
8.00	-0.173789815
8.50	-1.402394424

The reader is referred to the example of Program MP-61 for the TI-59 calculator for an example of a lumped element network. This example is similar to the distributed element network, and is programmed in the same manner. Note that in those cases which require a single inductor or capacitor, the unwanted element is assigned a value such that it will not contribute to the overall impedance. For example, a capacitor is programmed by using the series inductor-capacitor element with a 0 nh series inductor.

HP-41C COMPATIBILITY

The HP-41C Programmable Calculator represents a significant improvement over the HP-67/97 calculator, particularly in terms of available user memory and speed of execution. The HP-67/97 Microwave Pac, as described in this book, is compatible with the HP-41C by virtue of the software built into the HP-41C card reader. Thus, all of the analysis and design routines for the HP-67/97 may be used directly with the HP-41C, provided that the user has the card reader and at least one memory module. The HP-41C card reader can either read an HP-67/97 program card and convert it to HP-41C instructions, or interpret each HP-67/97 program step which is an HP-67/97 instruction. The HP-41C user who wishes to use the Microwave Pac must have the card reader and at least one memory module as a minimal system configuration. The user may elect to load HP-67/97 cards or program the cards directly from Appendix A. If HP-67/97 cards are used, it is suggested that each be given a global name, such as MP21, which will make it much easier to find and delete the program to make room for another program. All of the two-card programs can be contained in program memory. For example, MP541 and MP542 (MP-54.1 and MP-54.2) can be loaded together, MP541 executed, then [GTO] "MP542" used to position the calculator at the second program. Users who are not familiar with the HP-67/97, and who wish to program the HP-41C from Appendix A, should refer to section 3 of HP-82104A Card Reader Owner's Handbook for the corresponding storage registers and functions.

The HP-41C system was used to print the listings found in the HP-67/97 section and also the examples found in the last chapter of this book. The listing for the program used to print Two-Port A [XEQ] [A] and Two-Port B [XEQ] [B] is given below for those users with an HP-41C Printer.

```
♦LBL "PT"
♦LBL A
1
STO Z
"A"
ASTO X
GTO 00
♦LBL B
11
STO Z
"B"
ASTO X
♦LBL 00
" TWO-PORT "
ARCL X
SF 12
PRA
"S-PARAMETERS"
PRA
CF 12
"S11= "
XEQ 01
"S12= "
XEQ 01
"S21= "
XEQ 01
"S22= "
XEQ 01
RTN
♦LBL 01
ACA
RCL IND Z
ACX
PRBUF
RDN
CLA
5
SKPCHR
RDN
4
ST+ T
RDN
RCL IND Z
ACX
PRBUF
RDN
4
ST- T
RDN
ISG Z
.END.
```

```
USER KEYS.
 11 "PT"
                    CAT 1
LBL'PT
.END.       140 BYTES
```

Note to HP-41C users: Some users encountered problems with the data input flag. This flag is 3 for the HP-67/97, and 22 for the HP-41C. All other flags should be the same.

HP-41C COMPATIBILITY

[illegible]

EXAMPLES

The examples in this chapter are intended to familiarize the Microwave Pac user with the methods which can be used to analyze a wide variety of networks — low frequency as well as microwave. In each case, a step-by-step analysis procedure will be outlined; however, specific program operation will not be covered: these instructions can be found in the applicable section for the desired type of calculator. The HP-41C calculator is used for the following examples. Output from other calculators may differ slightly due to the internal algorithms.

Example 1. 60 Hz Notch Filter.

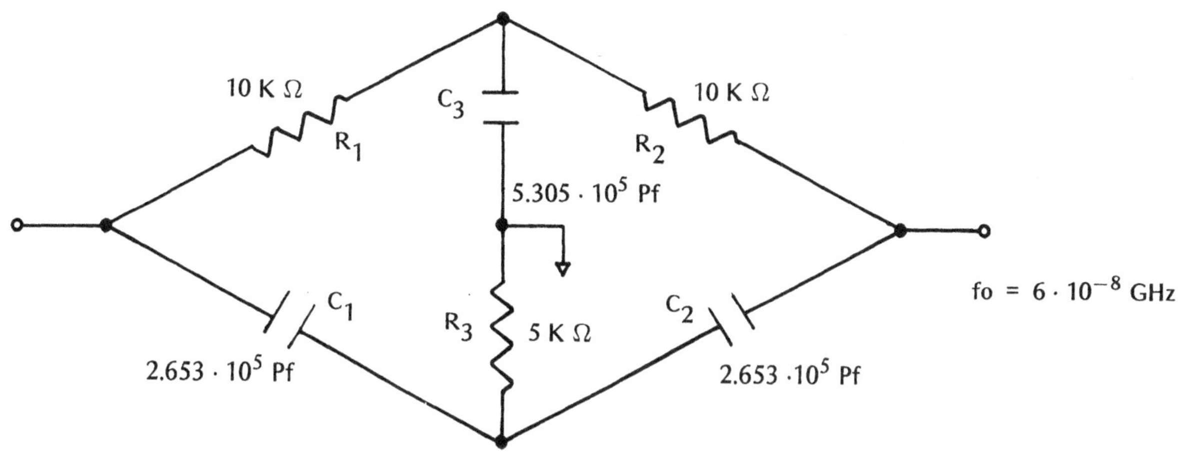

Figure 1. 60 Hz Notch Filter.

The 60 Hz notch filter shown in Figure 1 will be analyzed. Note that the element values are in ohms and picofarads, while the frequency must be in GHz to be compatible with the units of the Microwave Pac. The Microwave Pac uses a two-port analysis technique where the two-port elements are interconnected using the appropriate interconnection. The determination of the specific type of interconnection, and the order in which to interconnect the elements, constitute what is, perhaps, the most difficult — and important — task for the user. This problem is not unique to this Microwave Pac, as most larger CAD programs also use two-port analysis. The user who masters these techniques should have no trouble with either calculator or computer analysis; in fact, the interconnections, and, in many cases, the order of interconnection, are exactly the same.

Whenever there is any doubt as to the correct interconnection, the network should be redrawn as interconnected two-ports. The type of interconnections should then become obvious. The user who is not familiar with two-port interconnections should study the figures which accompany the programs of MP-20.

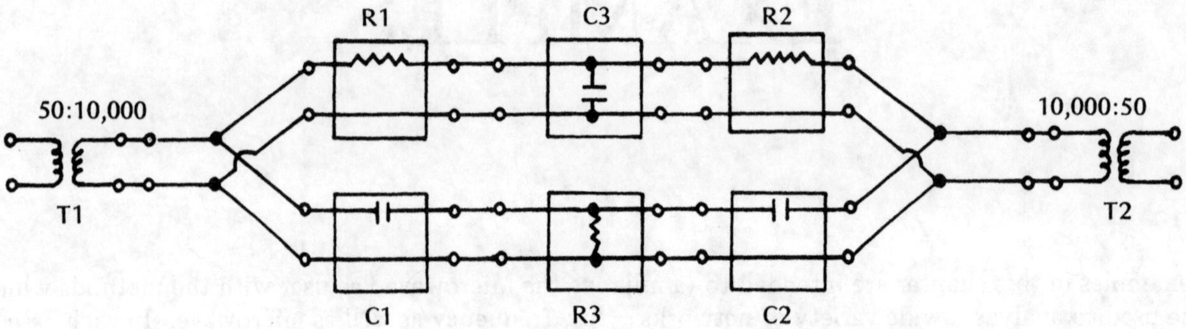

Figure 2. Two-Port Representation of Notch Filter.

Figure 2 shows the notch filter drawn in this manner.

It is clear from this drawing that R1, C3, and R2 can be cascaded to form a new two-port, while the cascade of C1, R3, and C2 forms another two-port. The new two-ports are connected in parallel to form the complete notch filter. A 60 Hz notch filter would probably be used with an Operational Amplifier, in which case the operating impedance would not be 50 ohms. Transformers T1 and T2 are used to transform the 50 ohm reference impedance to 10,000 ohms — note that each port of the notch filter is terminated with 10,000 ohms. The overall S-parameters calculated in this manner are, in fact, the same S-parameters which would be calculated with a 10,000 ohm reference impedance. The cascade of T1, the notch filter, and T2 completes the analysis. The dependent sources of MP-41 can be used to model the Operational Amplifier if a more complete analysis is desired.

The Microwave Pac will be used to analyze this notch filter at 50, 60, and 70 Hz. The order of interconnection will be that described above. The 50 Hz ($5.0 \cdot 10^{-8}$ GHz) analysis frequency should be stored in the frequency register ((00) for the TI-59, (A) for the HP-67/97, and (20) for the HP-41C). Load Program MP-31, which contains the series resistor, enter the 10 K value of R1, and calculate Two-Port B. Since this is the first calculation, there is no Two-Port A which can be interconnected with Two-Port B; therefore, interchange Two-Ports A and B to place Two-Port B into A. Load MP-32 and calculate C3 as Two-Port B, or use the "USE PA" option of Program MP-31 for the same calculation. R1 and C3 should be cascaded using Program MP-21 to form Two-Port A. Reload Program MP-31, calculate Two-Port B for R2, and cascade this element to form Two-Port A for the branch R1, C3, and R2. Two-Port A must now be saved on a data card in preparation for the calculation of the C1, R3, and C2 branch. This portion of the network is analyzed in exactly the same fashion; MP-31 used for Two-Port B of C1, Two-Port A and B interchanged, and MP-32 used to calculate Two-Port B for R3. C1 and R3 are cascaded to form Two-Port A, MP-31 used to calculate Two-Port B for C2, then the cascade program used to form Two-Port A for C1, R3, and C2.

At this point, the lower branch of C1, R3, and C2 is saved as Two-Port A, and the upper branch of R1, C3, and R2, is saved on a data card. This data should now be reloaded as Two-Port B in preparation for the parallel connection. The parallel connection of Program MP-22 can now be used to parallel connect the two branches to form the overall S-parameters of the notch filter in a 50 ohm system. The user should note that it does not matter which branch is Two-Port A or B for the parallel connection — the results are exactly the same. The 50 ohm S-parameters must now be referenced to 10,000 ohms through the use of the ideal transformers. Two-Port B for the 50-to-10,000 ohm transformer is created using MP-41. The S-matrix for the notch filter is now Two-Port A, and the transformer is Two-Port B. Two-Ports A and B must be interchanged before the cascade connection, otherwise Port 1 of the transformer will be connected to Port 2 of the filter, which is not the desired configuration. The output 10,000-to-50 ohm transformer is created as Two-Port B, and cascaded with Two-Port A to form the overall S-parameters of the notch filter in a 10,000 ohm system. The analysis at other frequencies is exactly the same, with the desired analysis frequency stored in the frequency register. Output from the 50, 60, and 70 Hz analysis is shown below.

50 Hz	60 Hz	70 Hz
TWO-PORT A	TWO-PORT A	TWO-PORT A
S-PARAMETERS	S-PARAMETERS	S-PARAMETERS
S11= 0.433700146	S11= 0.447236256	S11= 0.462848709
-110.0524230	-116.5683156	-121.7170203
S12= 0.052345272	S12= 0.000033921	S12= 0.042876947
-90.80816390	-301.2615926	-284.3333472
S21= 0.052345272	S21= 0.000033921	S21= 0.042876947
-90.80816390	-301.2615926	-284.3333472
S22= 0.433700144	S22= 0.447236256	S22= 0.462848710
-110.0524235	-116.5683155	-121.7170203

The transducer gain (in the 10,000 ohm system) can easily be calculated from the relationship: G = 20 Log / S_{21} / dB. The result of this calculation is

FREQ (HZ)	GAIN (dB)
50	-25.62245078
60	-89.39062706
70	-27.35552292

which shows the desired notch at 60 Hz.

The above procedure may seem confusing at first; however, after several examples the user will find that, in fact, analysis using the Microwave Pac is easy and straightforward. There are several points concerning the operation of the Microwave Pac which, if always remembered, will help reduce mistakes. Only two two-ports can be held in the calculator at any time, hence the necessity to save the two-port on a data card in the preceding example. The interconnection programs connect Two-Port A and Two-Port B, then place the results in Two-Port A. All circuit elements are created as Two-Port B, automatically set up for an interconnection; however, the first element must be placed in Two-Port A (use the Exchange A and B of MP-01), the second element created as Two-Port B, then interconnected. The following examples will further illustrate these techniques.

Example 2. Parallel-Coupled Microwave Filter.

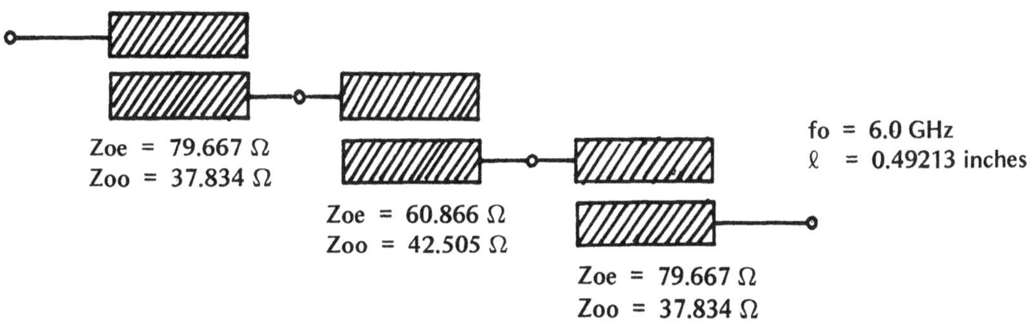

Figure 3. 6.0 GHz Parallel-Coupled Microwave Filter.

A parallel-coupled microwave filter has been designed for 5 percent bandwidth centered around 6.0 GHz. The filter should have 0.01 dB ripple in the passband and a minimum of 20 dB rejection at 8.0 GHz. The design should be verified through analysis, since an error in the initial calculations will not be discovered until the fabrication and test are complete, which can be expensive and time-consuming.

The filter is a cascade of three coupled line sections. The signal enters the input of each coupled line pair and exits the isolated port, or what would be the isolated port if the other ports were terminated in the characteristic impedance of the coupler. The other ports are terminated with an open, in this case, $\Gamma_2 = \Gamma_3 = 1 \angle 0°$.

Store the analysis frequency of 5.85 GHz (lower passband edge) in the frequency register and load the coupled line program, MP-42. Run label [C], "ISO", with the even and odd mode impedance of 79.667 and 37.834 ohms for the first section. The center frequency of each pair is the filter center frequency, 6.0 GHz. The S-matrix for the first pair is now in Two-Port B; exchange Two-Port A and B since no interconnection can be made. Again load MP-42, enter the data for the second pair, and create Two-Port B. The first and second sections can now be cascaded using MP-21 to create Two-Port A, which represents the first two coupled sections. Load MP-42, calculate Two-Port B for the final pair, and cascade using MP-21 to obtain the overall S-parameters for the filter at 5.85 GHz. This output is shown below.

5.85 GHz	6.15 GHz	8.00 GHz
TWO-PORT A	TWO-PORT A	TWO-PORT A
S-PARAMETERS	S-PARAMETERS	S-PARAMETERS
S11= 0.046920046	S11= 0.046920046	S11= 0.998727849
30.61827095	-30.61827283	127.8062825
S12= 0.998898651	S12= 0.998898650	S12= 0.050425041
-239.3817258	-300.6182742	-502.1937176
S21= 0.998898651	S21= 0.998898650	S21= 0.050425041
-239.3817258	-300.6182742	-502.1937176
S22= 0.046920046	S22= 0.046920045	S22= 0.998727849
30.61827143	-30.61827091	127.8062825

FREQ (GHz)	GAIN (dB)
5.85	-0.009571468
6.15	-0.009571476
8.00	-25.94707480

Repeat the above analysis at 6.15 GHz, the upper passband edge, and at the rejection frequency of 8.0 GHz. This output, as well as the transducer gain computed from $|S_{21}|$, are also shown. The analysis confirms that the design is indeed correct, with approximately 0.01 dB ripple at each bandedge, and greater than 20 dB rejection at 8.0 GHz.

The user should note that the symmetry of this network could have been used to simplify the analysis. The first and last coupled line pairs are identical, except one is the image of the other. That is, Port 1 of the first pair is the same as Port 2 of the last pair (in this case, the S-matrix is symmetrical; however, this is not always true). The image, "IMG", routine of Program MP-41 could have been used to save the image of the first pair on a data card, then reloaded after the cascade of the first and second pair. When used with larger networks, this technique can have a significant impact on the analysis time. For example, a six section filter would have symmetry about the center, thus the first three sections would be cascaded as Two-Port A, an image created in Two-Port B, and cascaded to form the complete network.

Another technique which may be used to analyze this type of structure is to make use of two-port equivalent circuits for coupled transmission lines which can be found in the literature. Figure 4 shows the equivalence for the type used in this parallel-coupled filter.

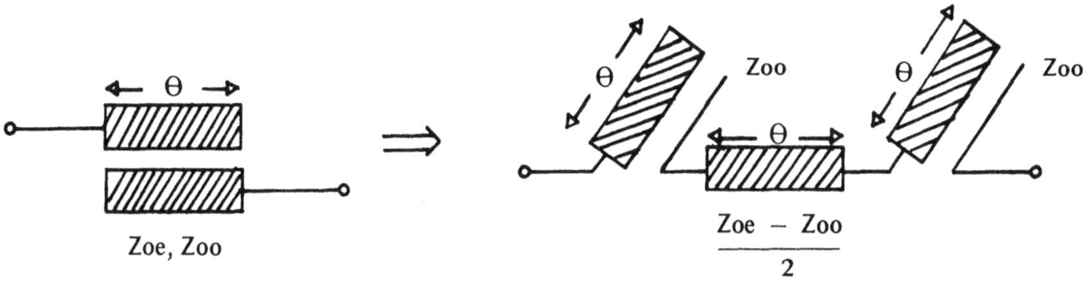

Figure 4. Two-Port Equivalent for Symmetrical Coupled Lines.

This equivalent circuit does not have any advantage when using two-port analysis, as it has nine elements instead of three (although symmetry may be used to reduce the number of elements); however, the Ladder Analysis Program, MP-60, may be used for the analysis. The entire filter may be programmed, and the analysis can be made over a frequency band without user interaction. The gain response is the most important parameter for the filter, and MP-60 can only be used to calculate the input impedance, or S_{11}; however, for a lossless two-port, $|S_{21}|^2 = 1 - |S_{11}|^2$. Thus, the gain response can also be calculated.

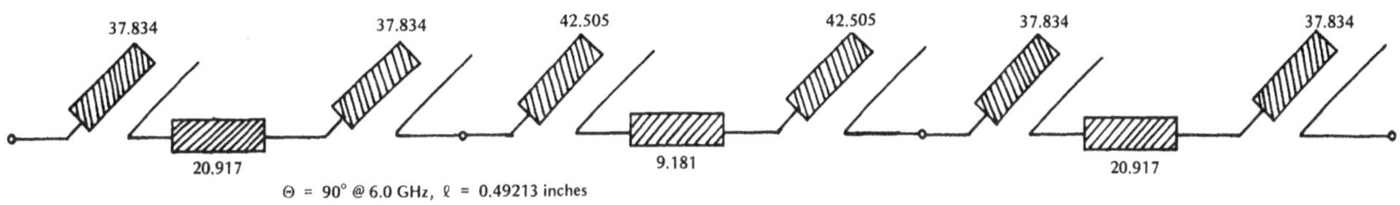

Figure 5. Two-Port Equivalent for Parallel-Coupled Filter.

Figure 5 shows the equivalent two-port model for the filter. Each of the four characteristic impedances, as well as the common line length, should be stored in a data register. Go to the appropriate label (depending upon the type of calculator), switch to program, and begin the topology programming. The first portion of the network should be the load impedance of 50 $\angle 0°$ ohms, followed by the series connected open stubs and cascade transmission lines. The last portion of the program should be the gain calculation from input reflection coefficient (this function is provided with label $\boxed{C'}$ for the TI-59), and either a print statement if a printer is available, or "STOP" to display the calculated values. Switch to normal mode; store the start, stop, and step frequencies; and begin execution. The results of the analysis are shown below.

FREQ (GHz)	GAIN (dB)		
		3.00	-34.59863685
5.75	-0.198437017	4.00	-25.94598584
5.85	-0.009540471	5.00	-13.00143209
5.95	-0.006111518	6.00	-0.010016756
6.05	-0.006098078	7.00	-13.00313423
6.15	-0.009591004	8.00	-25.94696252
6.25	-0.198809770	9.00	-34.59943829

Note that two frequency bands have been used: a small step size through the passband, and a larger step to observe the rejection characteristics. These results are in excellent agreement with the previous two-port analysis, although the element values were rounded to three decimal places. This program may be saved on a program card and used later with element values from another design.

Example 3. Microwave Power Divider.

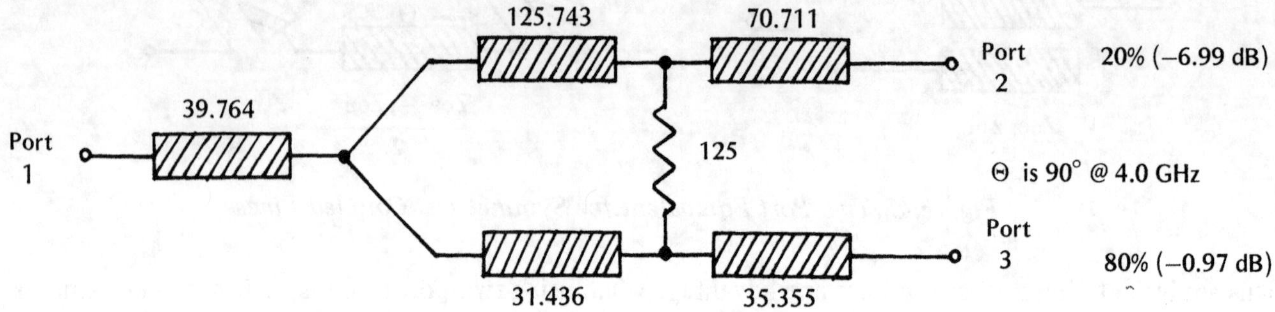

Figure 6. 4-to-1 Power Divider.

Figure 6 shows a 4-to-1 microwave power divider. This network will be analyzed to determine the power split at Ports 2 and 3 from Port 1, and also the isolation between Ports 2 and 3. The center frequency of the divider is 4.0 GHz; however, the analysis frequency will be at 3.5 GHz to determine the degradation in power split and isolation with frequency. This network is actually a three-port; however, a 50 ohm termination resistor can be placed at the third port, resulting in a two-port which can be analyzed using the Microwave Pac.

The power split from Port 1 to Ports 2 and 3 will be analyzed first. Figure 7 shows the network which results when Port 3 is terminated with the 50 ohm reference impedance, and redrawn to emphasize the parallel connection.

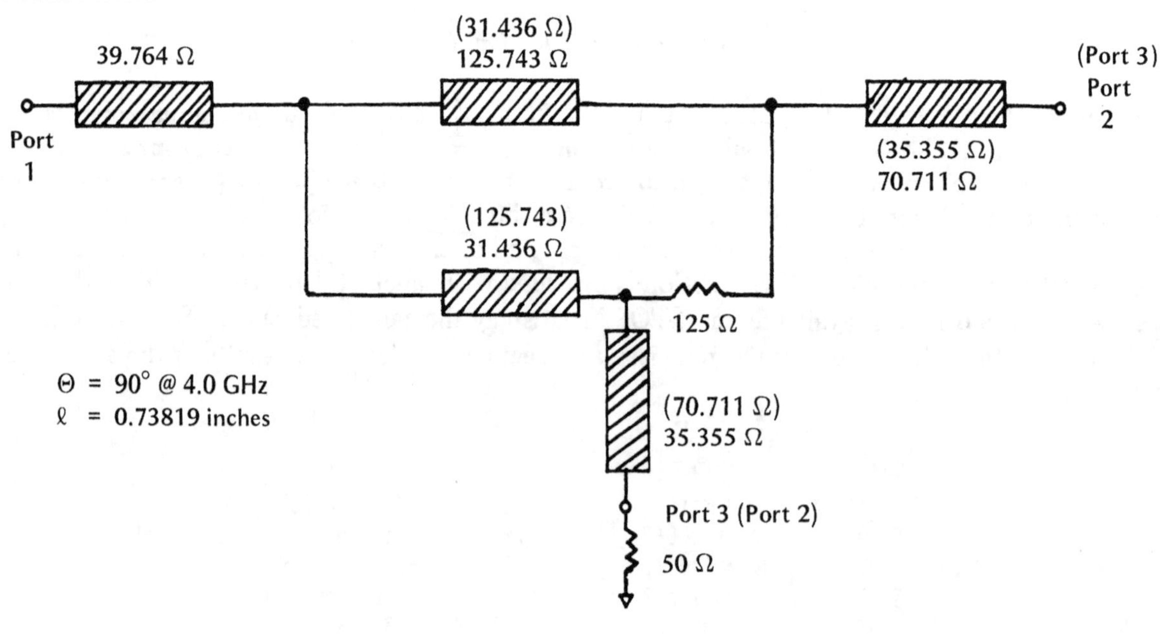

Figure 7. Two-Port Circuit from Port 1 to Ports 2 and 3.

Note that this network is symmetrical about the center line, which means that the equivalent topology will be the same from Port 1 to 2, as from Port 1 to 3. Element values and port numbers shown in parantheses will be used for the analysis from Port 1 to Port 3. Some thought should be given to the sequence of analysis, in order to minimize the number of two-ports which must be saved and reloaded from data cards. The 35.355 ohm transmission line and 50 ohm terminating resistor will have to be calculated as a one-port, converted to a two-port, then cascaded, since there is no single element which represents a shunt transmission line with a termination resistance. This operation will require that either the S-parameters for the 31.436 ohm line be saved, or the shunt branch calculated, exchanged with Two-Port A, and restored following the calculation for the 31.436 ohm line. The parallel branches of the 125.743 ohm line and the 31.436 ohm line, one-port, and 125 ohm resistor must also be analyzed in a particular order to avoid another save and load operation. The lower branch should be analyzed first, so that Two-Ports A and B may be used for the cascade of the three elements. The upper branch, which consists of the 125.743 ohm line, can be calculated as Two-Port B and paralleled with Two-Port A. The input 39.764 ohm line is next calculated as Two-Port B, exchanged with A (the elements must be in order), and cascaded. The 70.711 ohm output line is computed and cascaded with Two-Port A, which results in the S-matrix for the power divider from Port 1 to Port 2.

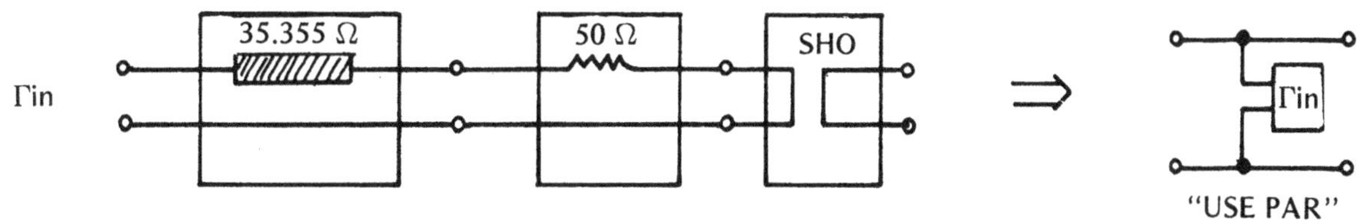

Figure 8. One-Port Branch Connected in Parallel.

Figure 8 shows the technique which should be used for any parallel (or series) branch which consists of several elements. The elements are connected together to form the network in the normal manner, which results in a one-port network. That is, the input reflection coefficient for the desired branch has been calculated. Then, the "USE PAR" (or "USE SER") routine of Program MP-26 is used to calculate the S-matrix for the parallel connected one-port. The user should always terminate the network with a short or open if there is the slightest doubt that the termination will be the desired one. In this case, the series 50 ohm resistor and short can be replaced with a 50 ohm parallel resistor and an open for the termination with identical results. However, should the short not be included, the actual termination will be 100 ohms, since the normal S-parameter calculations are made with the 50 ohm reference impedance at each port. The reader who is not comfortable with this concept should spend the time to understand it. Most larger CAD programs use the same technique, and a complete amplifier model usually includes at least one bias network which must be analyzed in this fashion.

The Port 1 to Port 2 analysis will follow the procedure outlined above. Store the analysis frequency of 3.5 GHz into the frequency register. The one-port network will be calculated first to avoid a save operation. Load the transmission line program, MP-33, enter the 35.355 ohm impedance and length, then calculate Two-Port B. It is not necessary to include the series 50 ohm resistor and short in this case, because the S-parameters are calculated with a 50 ohm termination resistor; however, Two-Ports A and B must be exchanged, since the one-port program, MP-26, expects the input reflection coefficient to be S_{11} of Two-Port A. Load MP-26 and create Two-Port B, then exchange Two-Ports A and B in preparation for the 31.436 ohm line calculation. Load MP-33, calculate Two-Port B, again exchange A and B, then cascade. Load MP-31, calculate Two-Port B for the 125 ohm resistor, and cascade to form the lower parallel branch. Next, load MP-33, calculate Two-Port B for the 125.743 ohm line, then parallel with Two-Port A. Load MP-33, calculate the S-parameters for the 39.764 ohm input line, exchange Two-Ports A and B to obtain the proper order, then cascade. Reload MP-33, calculate the 70.711 ohm output line, and cascade to form the Port 1 to Port 2 S-parameters shown below. The Port 1 to Port 3 S-parameters are also shown below, and are calculated in exactly the same manner, except the element values in parentheses from Figure 7 are used.

Port 1 - Port 2

```
TWO-PORT A
S-PARAMETERS
S11= 0.053942686
    -15.83997294
S12= 0.442189040
    -235.0698532
S21= 0.442189040
    -235.0698532
S22= 0.128492194
    -68.87936809
```

Port 1 - Port 3

```
TWO-PORT A
S-PARAMETERS
S11= 0.053942684
    -15.83997186
S12= 0.893627346
    -234.8940263
S21= 0.893627346
    -234.8940263
S22= 0.064186752
     87.29231897
```

This data shows that, at 3.5 GHz, approximately 19.6 percent of the power (—7.09 dB) is split to Port 2, and 79.9 percent (—0.98 dB) is split to Port 3. The remaining 0.5 percent is lost due to mismatch and the isolation resistor. The worst-case VSWR is 1.29:1 at Port 2, which indicates that the power divider performance is acceptable if the Port 2 to Port 3 isolation is above the desired value.

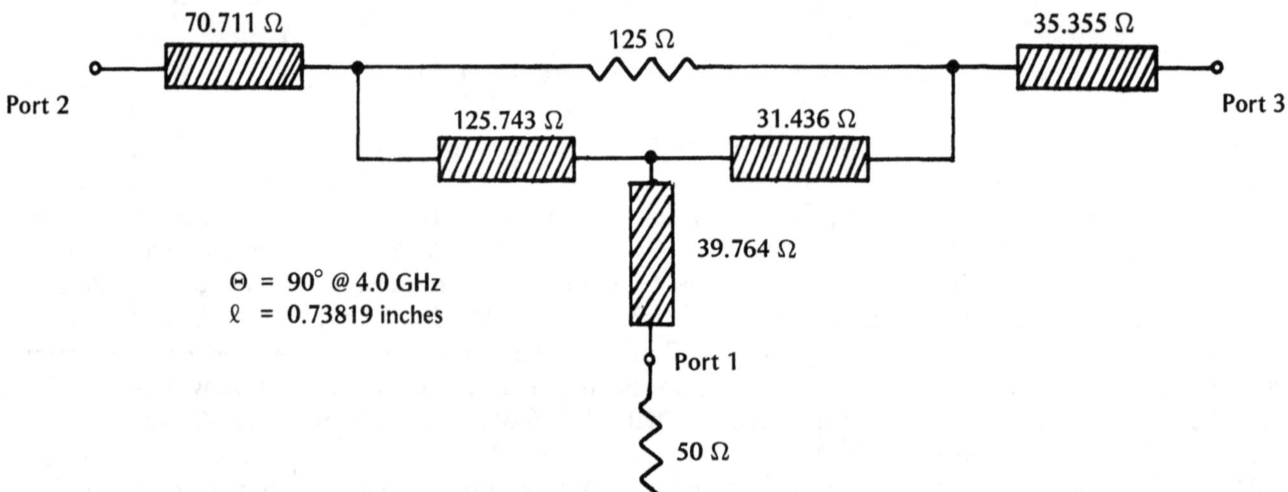

Figure 9. Equivalent Circuit from Port 2 to Port 3.

Figure 9 shows the equivalent network from Port 2 to Port 3 with Port 1 terminated, which will be used to calculate the isolation. The analysis procedure is almost identical to that outlined above, and is left as an exercise for the reader. This output is shown below and indicates there is approximately 29.3 dB isolation between Ports 2 and 3.

Port 2 - Port 3

```
TWO-PORT A
S-PARAMETERS
S11= 0.128492195
    -68.87936826
S12= 0.034244450
    -253.1047605
S21= 0.034244450
    -253.1047605
S22= 0.064186749
     87.29231843
```

The reader should note that the input reflection coefficient at each port has been calculated twice using different circuit models. These calculations should agree, and serve as a check for errors which may occur in either the analysis or the equivalent circuit model.

Example 4. Transmission Line Transformer.

A transmission line transformer described by Ruthroff[1] is shown in Figure 10. Note that the schematic

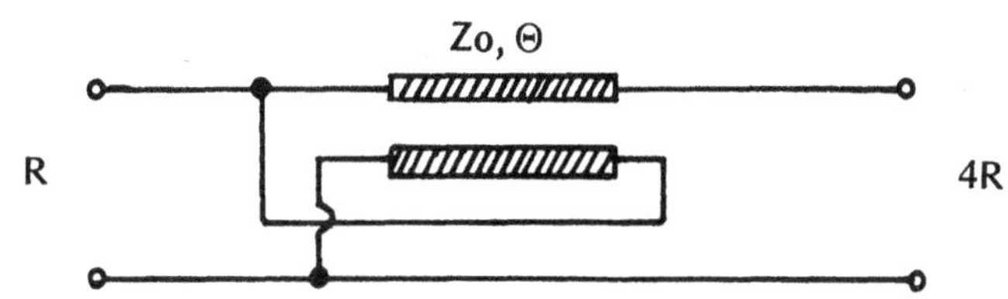

Figure 10. Transmission Line Transformer.

diagram represents the two conductors of a transmission line, not a pair of coupled lines. This is an excellent example of a network which must be redrawn in terms of two-ports to determine the appropriate interconnections. In this case, as shown in Figure 11, a dummy two-port has been used to demonstrate that this network is an interconnection of two two-ports.

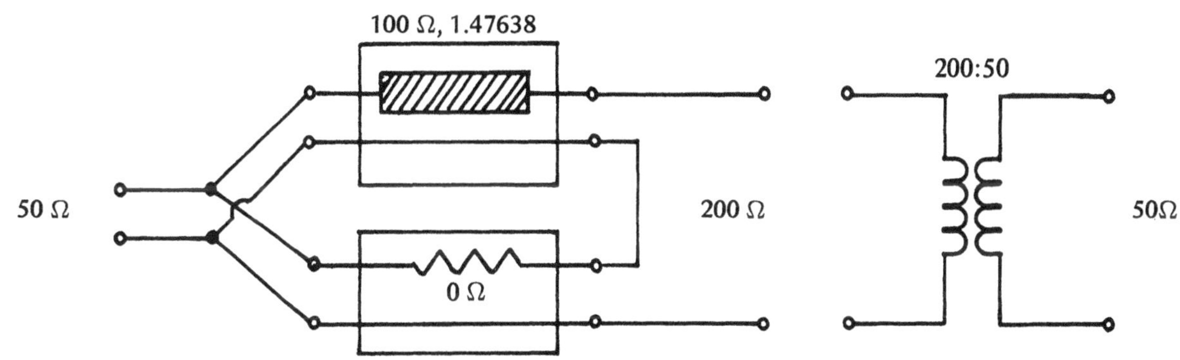

Figure 11. Two-Port Representation of Transmission Line Transformer.

The dummy two-port should be ideal; that is, it should have $S_{11} = S_{22} = 0$ and $S_{12} = S_{21} = 1$. The user may enter these parameters as Two-Port A or B, use MP-31 to calculate a 0 ohm series resistor, or MP-33 to calculate a 0 length, 50 ohm transmission line. Inspection of Figure 11 shows that the input (Port 1) of each two-port is connected in parallel, while the output (Port 2) is connected in series. Thus, the "PAR-SER" connection of MP-24 will be required. This transformer provides a 4-to-1 impedance transformation, or 200 ohms using the 50 ohm reference impedance of the Microwave Pac. The S-parameters for this network do not provide much useful information, since one port is 200 ohms — although the output reflection coefficient may be converted to impedance. A much more useful set of S-parameters results when the output port is referenced to 200 ohms using an ideal transformer as shown in Figure 11. The input and output VSWR, as well as the insertion loss, may easily be determined from the renormalized S-parameters.

[1] C.L. Ruthroff, "Some Broad-Band Transformers", *Proc. IEEE*, Vol. 47, August 1959, pp. 1337-1342.

The characteristic impedance of the transmission line should be twice the value of R, where R is the impedance at the input port. The length should be less than one-eighth wavelength at the upper frequency. Assume the upper frequency to be 1.0 GHz; therefore, a 100 ohm, 1.47638 inch transmission line will be required. Store the analysis frequency of 0.1 GHz into the frequency register, then store the dummy two-port into A. Store zero into all registers except $|S_{12}|$ and $|S_{21}|$, where a "1" should be entered ($S_{11} = S_{22} = 0 \angle 0°$, $S_{12} = S_{21} = 1 \angle 0°$). Load MP-33 and calculate Two-Port B for the transmission line. The "PAR-SER" connection of MP-24 is executed to obtain the S-parameters with all ports normalized to 50 ohms. Load MP-41 and calculate Two-Port B for the 200-to-50 ohm transformer, then cascade to normalize Port 2 of the network to 200 ohms. The output is shown below for both normalizations at an analysis frequency of 0.1, 0.5, and 1.0 GHz.

50 OHM NORMALIZATION

0.1 GHz

```
TWO-PORT A
S-PARAMETERS
S11=  0.600493379
     -184.4995445
S12=  0.799629727
     -362.2506973
S21=  0.799629727
     -362.2506973
S22=  0.600493379
     -0.001850200
```

0.5 GHz

```
TWO-PORT A
S-PARAMETERS
S11=  0.648346874
     -224.6116464
S12=  0.761345081
     -383.2160851
S21=  0.761345081
     -383.2160851
S22=  0.648346874
     -1.820523800
```

1.0 GHz

```
TWO-PORT A
S-PARAMETERS
S11=  0.612273917
     -202.4446726
S12=  0.790645717
     -371.3374345
S21=  0.790645717
     -371.3374345
S22=  0.612273916
     -0.230196700
```

50 - 200 OHM NORMALIZATION

0.1 GHz

```
TWO-PORT A
S-PARAMETERS
S11=  0.000771859
      177.7482265
S12=  0.999999701
     -362.2517394
S21=  0.999999701
     -362.2517394
S22=  0.000771857
     -2.251769325
```

0.5 GHz

```
TWO-PORT A
S-PARAMETERS
S11=  0.019779251
      168.5288849
S12=  0.999804370
     -371.4711059
S21=  0.999804370
     -371.4711059
S22=  0.019779248
     -11.47110723
```

1.0 GHz

```
TWO-PORT A
S-PARAMETERS
S11=  0.085472767
      155.6255445
S12=  0.996340507
     -384.3744564
S21=  0.996340507
     -384.3744564
S22=  0.085472766
     -24.37445763
```

This analysis shows that the worst-case performance is at the upper cutoff frequency of 1.0 GHz. The VSWR at both ports is 1.19 = 1, and the insertion loss from a 50 ohm source to a 200 ohm load is 0.03 dB. Practical transmission line transformers do, in fact, exhibit bandwidths in excess of a decade. The normal construction technique uses a twisted wire pair for the transmission line, where the conductor diameter, insulation material, and thickness, along with the number of twists per unit length, determine the characteristic impedance and phase velocity. The transmission lines are usually wound on a ferrite core to extend the low frequency response. Transformers with other impedance ratios can be obtained from interconnections of more than one transmission line. These transformers can also be analyzed using the Microwave Pac.

The user should note that, in this example, Two-Port A contained the dummy two-port, while Two-Port B contained the S-parameters for the 100 ohm transmission line when the parallel-series connection was executed. This is not the two-port representation of Figure 11! However, the Microwave Pac, as well as many CAD programs, calculates this connection by converting to another parameter set, adding the resulting matrices, then converting back to the S-parameter set. Because of the matrix addition, it does not matter which two-port is A or B — as long as it is a valid connection. The same is true of all interconnections, with the exception of the cascade connection. This fact can often be used to save program steps, as in this case where the A and B interchange was omitted following the calculation of the 100 ohm transmission line.

Example 5. Microwave Switch.

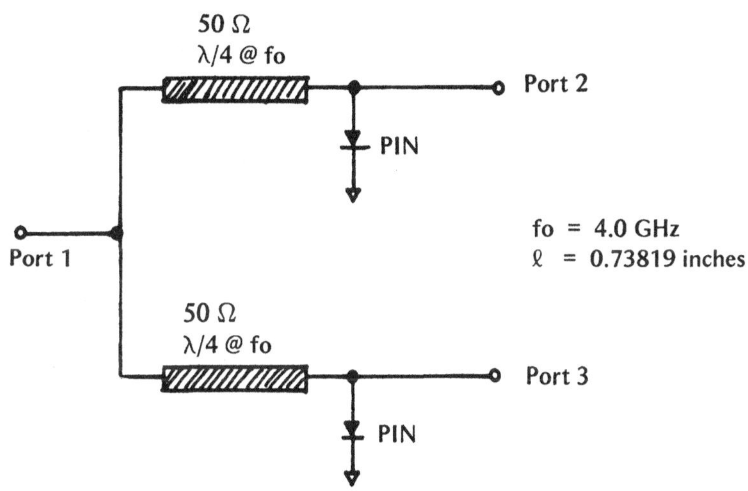

Figure 12. Microwave Switch.

The microwave switch shown in Figure 12 is similar to the power divider of Example 3; however, a useful analysis technique can be illustrated through this example. Operation of this switch depends on the RF characteristics of the PIN diodes. An ideal forward biased PIN diode appears as a short, while a reversed bias diode appears as an open. If one diode is forward biased and the other reverse biased in Figure 12, the short will reflect through the quarter-wavelength line and appear as an open at the junction of the two lines. Thus, no power will be transmitted through the shorted branch; however, it will also not affect the input impedance, since it reflects as an open. The RF signal can travel from Port 1 through the branch with the reverse biased (or "OFF") diode, with complete isolation to the "ON" branch. This network tends to be narrow band because of the required quarter-wavelength, and practical PIN diodes are not opens and shorts.

Figure 13. (a) Forward and (b) Reverse Biased PIN Diode Models.

A first order approximation of an actual PIN diode is shown in Figure 13. A more complete diode model would include bonding inductance, additional losses, and package parasitics if a packaged diode is used. The circuit model of Figure 12 should also include D.C. blocking capacitors as well as the bias networks; however, the simple models will be used for this analysis. The parameters of interest for the switch are the VSWR for the thru path, insertion loss for the thru path, isolation from the input to the off port, and isolation between Ports 2 and 3.

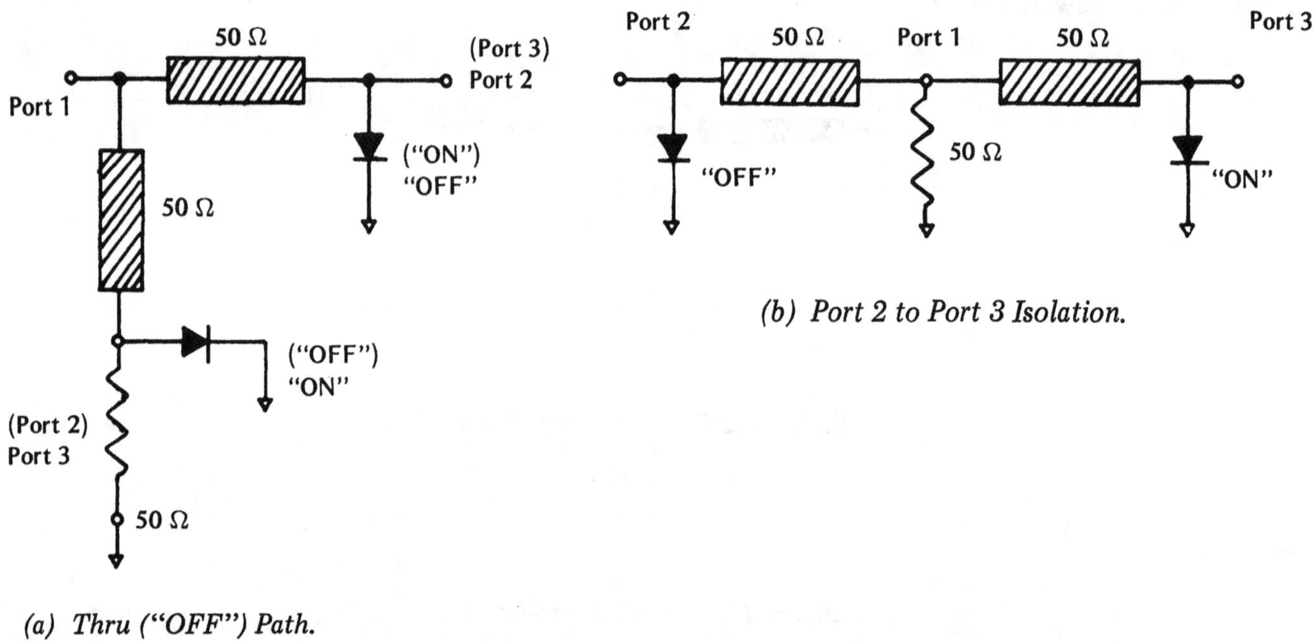

Figure 14. Thru Path, Off Path, and Isolated Path.

Figure 14 shows the circuit models which will be used to analyze the three cases. As with Example 3, the symmetry of the network allows the topology for the two cases of Figure 14(a) to be the same; however, the models for the "ON" and "OFF" diodes will differ. At this point, an observation can be made which will significantly reduce the analysis time. All three models use a transmission line cascaded with a shunt connected diode. If the transmission line cascaded with an "OFF" diode (the thru path in Figure 14(a)) is labeled Two-Port I, and the "ON" diode combination Two-Port II, the equivalent networks of Figure 15 result.

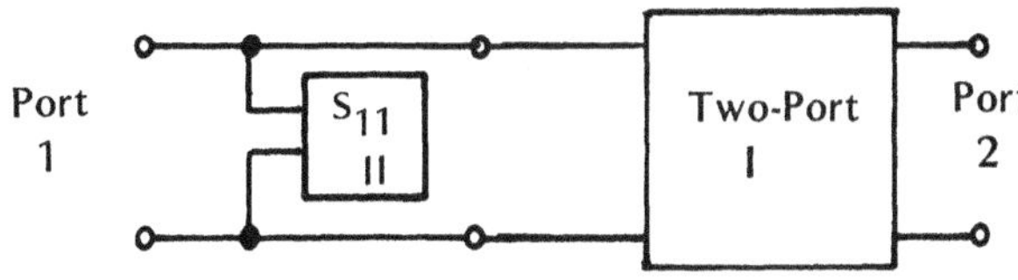

(a) Thru Path.

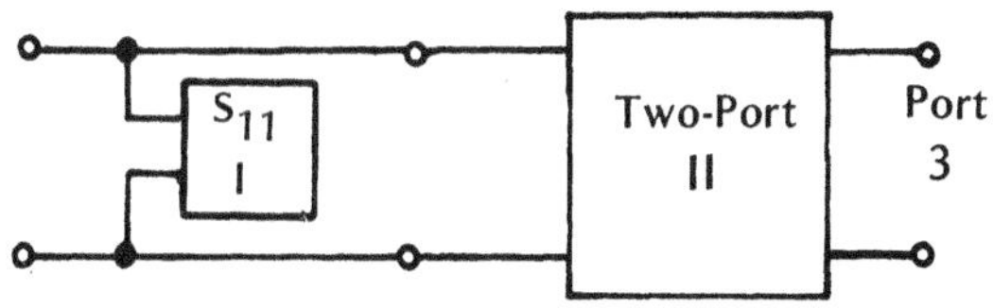

(b) Off Path.

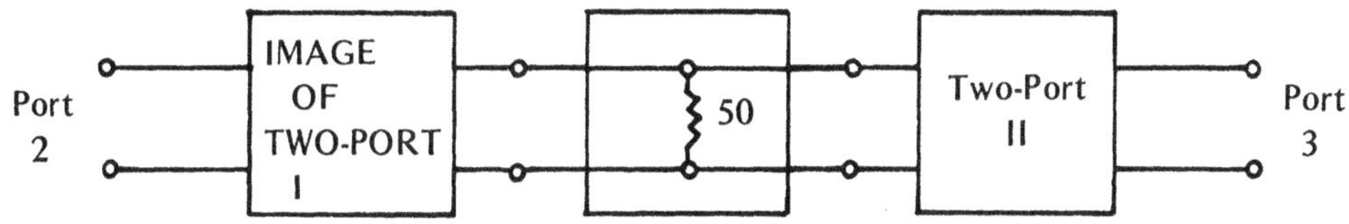

(c) Isolated Path.

Figure 15. Two-Port Representation of Microwave Switch.

Thus the analysis will only require that Two-Port I and Two-Port II be calculated at the analysis frequency and interconnected as shown. Note that S_{11} of Two-Ports I and II is used for the required one-port "USE PAR" since the 50 ohm termination is already present (refer to Example 3).

The analysis will be performed at 3.0, 4.0, and 5.0 GHz. Store the analysis frequency of 3.0 GHz into the frequency register, load MP-33, calculate Two-Port B for the 50 ohm line, and exchange A and B. Load MP-31, enter the 5 K ohm and 0.1 pf data for the "OFF" diode model, and calculate Two-Port B (do not forget to execute [E′] or [f] [E] to connect the series R-C in shunt). Cascade Two-Ports A and B to form Two-Port I, then save on a data card. Calculate Two-Port II in the same manner using the "ON" model, and also save on a data card. Load Two-Port II into Two-Port A, then use the "USE PAR" connection of MP-26 to create the first two-port of the thru path in Two-Port B. Exchange A and B, load Two-Port I in B, and cascade to compute the S-parameters for the thru path. Repeat the above procedure exchanging Two-Ports I and II to calculate the S-parameters for Port 1 to Port 3, the off path. Now, load Two-Port I into A, then use the "IMG" routine of MP-41 to create an image in Two-Port B. Exchange A and B to place the image in Two-Port A, calculate Two-Port B for the shunt 50Ω resistor, cascade, load Two-Port II into Two-Port B, and cascade again, which results in the S-parameters from Port 2 to Port 3.

The results of this analysis are shown below and indicate that the worst-case VSWR occurs at each band edge for the input port. The VSWR changes from 1.01:1 at the center frequency, to 1.52:1. The insertion loss in the thru path varies from 0.13 dB at the center frequency, to 0.33 dB at each end frequency.

3.0 GHz

PORT 1 - PORT 2 "THRU"	PORT 1 - PORT 3 "OFF"	PORT 2 - PORT 3 "ISOLATION"
TWO-PORT A S-PARAMETERS	TWO-PORT A S-PARAMETERS	TWO-PORT A S-PARAMETERS
S11= 0.204820899	S11= 0.204820899	S11= 0.194698632
104.0825457	104.0825457	-31.15699312
S12= 0.962930275	S12= 0.020601049	S12= 0.020435944
-55.94012348	-77.72508367	-145.4747890
S21= 0.962930275	S21= 0.020601049	S21= 0.020435944
-55.94012348	-77.72508367	-145.4747890
S22= 0.194698631	S22= 0.961213371	S22= 0.961213371
-31.15699310	179.9867634	179.9867634

(Continued)

4.0 GHz

TWO-PORT A S-PARAMETERS	TWO-PORT A S-PARAMETERS	TWO-PORT A S-PARAMETERS
S11= 0.004875427	S11= 0.004875427	S11= 0.004685325
175.4969203	175.4969203	-4.873766791
S12= 0.985347916	S12= 0.019512543	S12= 0.019320547
-90.02294643	-89.97796349	-180.0229489
S21= 0.985347916	S21= 0.019512543	S21= 0.019320547
-90.02294643	-89.97796349	-180.0229489
S22= 0.004685325	S22= 0.961166913	S22= 0.961166913
-4.873767553	-179.9999911	-179.9999911

5.0 GHz

TWO-PORT A S-PARAMETERS	TWO-PORT A S-PARAMETERS	TWO-PORT A S-PARAMETERS
S11= 0.204393286	S11= 0.204393286	S11= 0.194238483
-103.9000686	-103.9000686	30.84451007
S12= 0.962971060	S12= 0.020615253	S12= 0.020436846
-124.1167281	-102.2511813	-214.5820682
S21= 0.962971060	S21= 0.020615253	S21= 0.020436846
-124.1167281	-102.2511813	-214.5820682
S22= 0.194238483	S22= 0.961213706	S22= 0.961213706
30.84451004	-179.9867566	-179.9867566

The off path and isolation between output ports is almost identical (as expected), and varies from 34.2 dB to 33.7 dB. The insertion loss and isolation are strong functions of the diode "ON" and "OFF" characteristics. These effects can easily be investigated, since the transmission line diode combination can be saved on a data card, and used to minimize the analysis time. More isolation can be obtained by using additional diodes separated by transmission lines. The more complicated structure can also be analyzed as a combination of Two-Ports I and II.

Example 6. Active Isolator.

The mapping programs found in the MP-50 section provide the user with an extremely powerful design tool. In many cases the data, which is in an easy-to-use format, can be obtained faster than when using larger CAD programs. The MP-50 programs enable the user to investigate the effects of a feedback impedance on any of the S-parameters. The output is a mapping of the desired S-parameter as a function of all possible feedback impedances, which is easy to interpret, and in some cases will provide information not available from a point-by-point analysis.

The design of an active isolator will be used to demonstrate the concept of feedback mapping. A commercially available 0.5 micron gate-length GaAsFET with the following S-parameters will be used at 18 GHz for the design:

$$S_{11} = 0.62 \angle -42^\circ \qquad S_{12} = 0.18 \angle 169^\circ$$

$$S_{21} = 2.46 \angle 116^\circ \qquad S_{22} = 0.55 \angle -89^\circ$$

The first step should be to characterize the device using MP-11. K = 0.51556, so the device is unstable and the maximum available gain is not defined; however, the maximum stable gain is 11.35663 dB. The potential instability means that the device cannot be matched at the input and output to provide low VSWR. Thus, the device must be stabilized through the use of lossy elements to allow a match at the input and output. This procedure, however, will not tend to improve the amplifier isolation. Recall from the examples in the MP-10 section that a decrease in the value of the reverse transmission coefficient, S_{12}, increases the stability factor — and will also improve the isolation. Therefore, the design approach will be to minimize S_{12}, then match into the input and output to provide an amplifier with low VSWR and acceptable isolation.

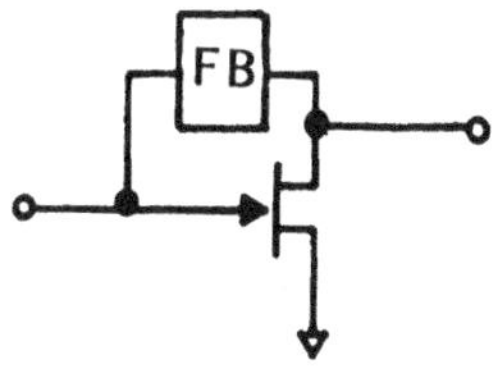

Figure 16. FET with Shunt Feedback.

Figure 16 shows the FET with shunt feedback, which is the configuration to be investigated. A trial network could be used for the feedback, then analyzed to determine the response, and another investigated for better results. This approach may or may not work, depending on the user's insight and a certain amount of luck. Even when using a CAD program with optimization, the results may be only a local, and not global, minimum. A far better approach is to use the mappings of MP-50 to provide a graphical representation of S_{12} for all values of feedback impedance.

Load MP-51, store the device S-parameters in the corresponding data registers, then calculate the three-port parameters for the shunt feedback configuration as outlined in the program instructions. Next, load MP-54.1 and calculate the mapping constants shown below for the S_{12} mapping.

TI-59	HP-67/97	
Reg (10)	Reg (18)	R18= -0.225481135
Reg (11)	Reg (19)	R19= 0.199921022
Reg (12)	Reg (21)	R21= 1.362717262
Reg (13)	Reg (22)	R22= -1.605078091
Reg (08)	Reg (23)	R23= 0.048788242
Reg (09)	Reg (24)	R24= -0.165575402

(NOTE: These registers correspond to (08) through (13) for the TI-59, with the value in register (13) the negative of the above output.) Next, load MP-54.1 and compute the center and radius for the contours of constant resistance and reactance shown.

```
       R=  0.000000000
  CENTER=  0.172613749
          -73.58192890
  RADIUS=  0.301347237
```

```
       R=  99.9999995
  CENTER=  0.092033540
          -137.5967849
  RADIUS=  0.145318609
```

(Continued)

```
     R=  15.00000000
CENTER=  0.138951258
         -82.76028372
RADIUS=  0.259546040

     R=  50.00000000
CENTER=  0.100322493
         -107.3856148
RADIUS=  0.196081083

     R=  175.0000000
CENTER=  0.104446052
         -160.3653412
RADIUS=  0.104671694

     R=  400.0000001
CENTER=  0.134150915
         -178.5424233
RADIUS=  0.056913798

     X=  250.0000000
CENTER=  0.300005905
         -162.6686749
RADIUS=  0.165342449

     X=  150.0000001
CENTER=  0.517677095
         -148.9834013
RADIUS=  0.402405824

     X=  100.0000000
CENTER=  1.520798551
         -137.4111040
RADIUS=  1.421356192

     X=  80.30000000
CENTER=  608.9636776
         -131.5761237
RADIUS=  608.8721346

     X=  50.00000000
CENTER=  0.850410166
         58.93983607
RADIUS=  0.927689681

     X=  25.00000000
CENTER=  0.444336313
         68.85384332
RADIUS=  0.507950257

     X=  0.000000000
CENTER=  0.301144308
         79.41498955
RADIUS=  0.349718001

     X= -25.00000000
CENTER=  0.233866866
         89.94832614
RADIUS=  0.266652673

     X= -80.00000000
CENTER=  0.176109424
         110.0936628
RADIUS=  0.175136045

     X= -250.0000001
CENTER=  0.155132935
         140.8544777
RADIUS=  0.084983810
```

Each set of outputs for a particular value of resistance or reactance defines the contour of constant resistance (reactance) for all values of reactance (resistance) on the S_{12} plane. For example, the R = 0 ohms contour defines all possible values of S_{12} which result from purely reactive feedback impedances.

The contours are constructed by using the phase and magnitude scale of the plane for the desired parameter. Figure 17 shows the S_{12} plane with the shunt feedback mapping. A polar plot is usually used for the transmission coefficients, since magnitude and angle are the important parameters, while the input and output parameters are plotted on a Smith Chart to provide additional impedance information. The maximum magnitude for the S_{12} plane is 0.5, as there is no positive real feedback impedance which will cause a greater value. The R = 0 ohm contour should always be constructed first, since the mapping will be contained within this circle (in most cases). Locate the center at $0.17 \angle -73.6°$ using the magnitude and angle scale of the polar graph, then construct a circle of radius 0.3 around this point. Construct the remaining resistance contours in a similar fashion. Note that the circles become smaller as the resistance increases, and they meet at a single point exactly as the normal Smith Chart mapping. This point is the infinite resistance point. Contours of constant reactance are constructed in exactly the same way; however, only that portion which falls within the R = 0 contour is drawn. The mapping outside the R = 0 contour represents those values of negative real feedback impedance. Note that the constant reactance contours also meet at the infinite resistance point.

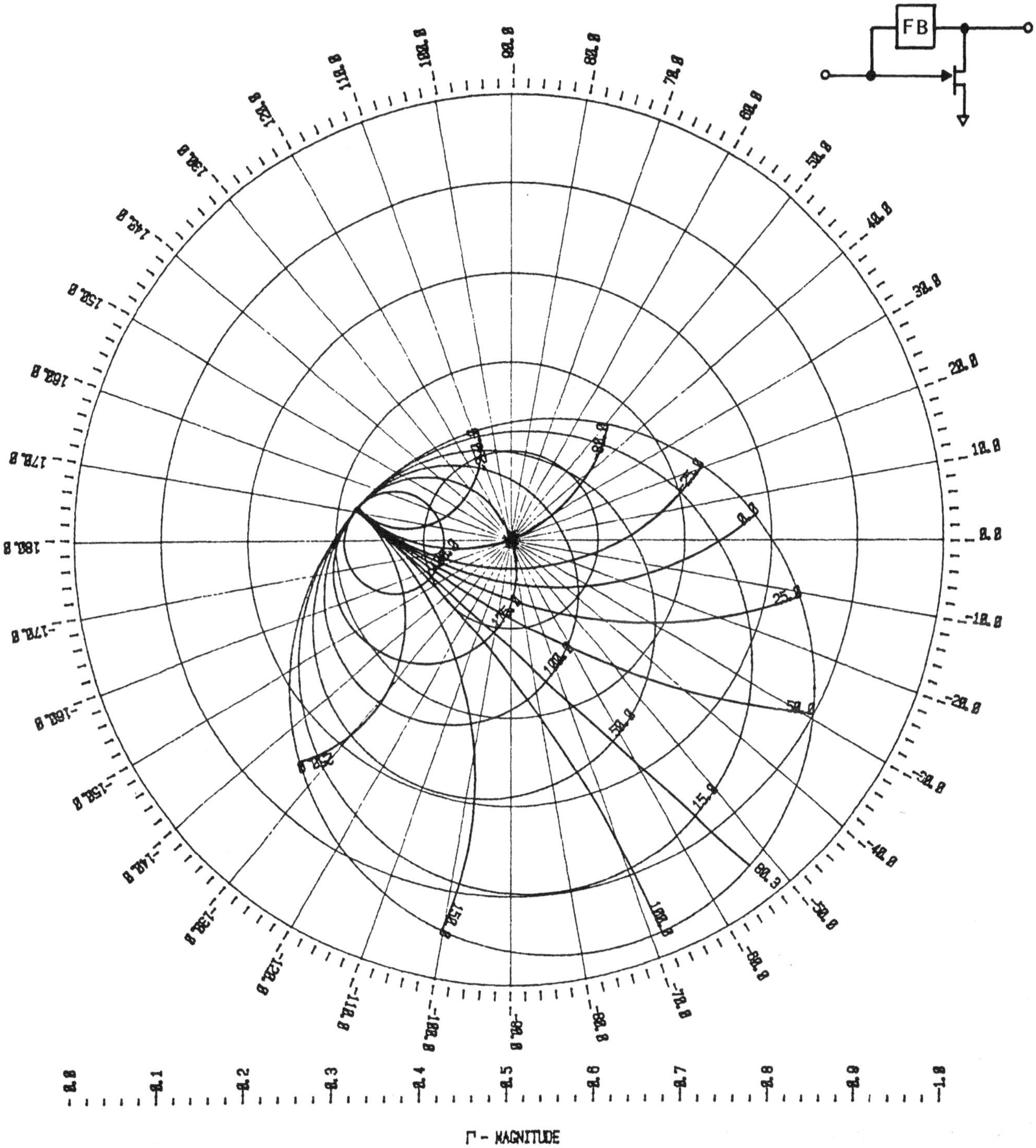

Figure 17. S_{12} *Mapping for Shunt Feedback.*

The value of S_{12} which results from a particular value of feedback impedance is read directly from the mapping on the S_{12} plane. For example, the point on the S_{12} plane where the R = 50 ohm contour intersects the x = +j50 ohm contour is approximately 0.22 $\angle -45°$ as read from the scale of the polar graph. Thus, the value of S_{12} which will result from a feedback impedance of Z = 50 + j50 ohms is 0.22 $\angle -45°$. The Z = ∞ ohm point provides a check for the accuracy of the mapping. This point corresponds to an open circuit, or no feedback, and should result in the device S_{12}, which in fact it does. The Z = 0 ohm point can be used for the series feedback case to check the accuracy.

Since S_{12} should be minimized, the feedback impedance which corresponds to S_{12} = 0 should be located. This point is approximately Z = 175 − j80, or a 175 ohm resistor in series with a 0.11052 pf capacitor, as shown in Figure 18.

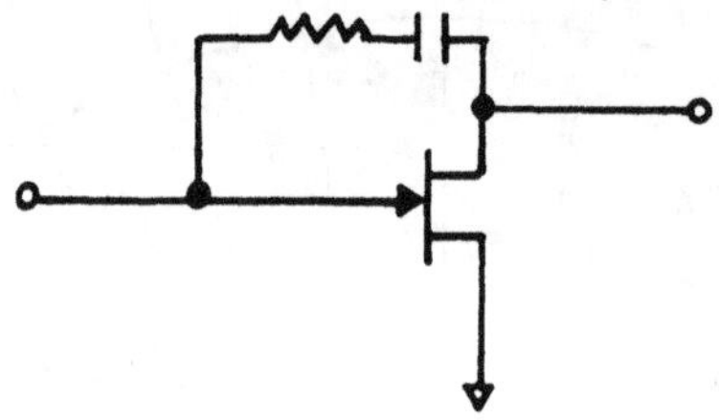

Figure 18. Feedback Impedance which Minimizes S_{12}.

Reload the device parameters into Two-Port A, then use MP-31 to create Two-Port B for the feedback impedance, and parallel connect using MP-22 (NOTE: save the three-port parameters on a data card for later use). The results of this analysis are shown below, and indicate that approximately 60 dB isolation can be achieved using this network!

```
TWO-PORT A
S-PARAMETERS
S11=  0.166659328
     -78.65312650
S12=  0.001001806
     -263.1144673
S21=  1.452234890
     -241.2152833
S22=  0.362485574
     -130.2707913
```

An analysis using MP-11 indicates that the stability factor is now approximately 290, and the maximum available gain is 3.98 dB. If more gain is desired, all four S-parameters should be mapped, and the results of shunt feedback investigated for each. S_{11}, S_{22}, and S_{21} should remain as large as possible, while S_{12} is reduced to an acceptable value. The above example shows a reduction in S_{21} from 7.8 dB to 3.2 dB, while the maximum gain is reduced from 11.4 dB to 4.0 dB. The gain reduction in excess of the decrease in S_{21} is due to lower values of S_{11} and S_{22}. Recall that the value of S_{21} is calculated in a 50 ohm system and includes the input and output mismatch. The objective of this example, however, is to design an active isolator, and any gain is better than the insertion loss associated with a ferrite device.

The above parameters for the device with shunt feedback indicate that, along with the reduction of S_{12}, S_{11} is reduced from 0.62 to 0.17. It may be possible to find a value of feedback impedance which will reduce S_{12} below − 25 dB (or some other specification), and minimize S_{11} to the point where an input matching network is not required. Although S_{11} will change slightly due to the load impedance (S'_{11}), a small value of S_{12} will minimize this interaction. In fact, if $S_{12} = 0$, the input would be completely isolated from the output. This possibility is easily investigated by generating a mapping of S_{11}, and observing the effect of feedback on S_{11} and S_{12} simultaneously.

Reload the three-port parameters saved above, and use MP-54.1 to calculate the mapping constants for the S_{11} mapping.

```
R18=  0.399702604
R19= -0.730163303

R21=  1.862717262
R22= -1.605078091
R23=  0.061047166
R24=  0.315302326
```

Then use MP-54.2 to compute the contours of constant resistance and reactance as shown.

```
      R=  0.000000000
 CENTER=  0.321157774
          79.04227147
 RADIUS=  0.832406524

      R=  15.00000000
 CENTER=  0.243667993
          61.44021742
 RADIUS=  0.716939763

      R=  50.00000000
 CENTER=  0.209518304
          16.71013956
 RADIUS=  0.541631555

      R=  99.99999995
 CENTER=  0.275250469
         -13.17879336
 RADIUS=  0.401411208

      R=  150.0000000
 CENTER=  0.336022663
         -23.71868648
 RADIUS=  0.318362463

      R=  200.0000000
 CENTER=  0.380571447
         -28.71927781
 RADIUS=  0.264474293

      X=  250.0000000
 CENTER=  0.883292785
         -12.79075153
 RADIUS=  0.456722730

      X=  150.0000001
 CENTER=  1.440691176
          4.733367741
 RADIUS=  1.111558999

      X=  100.0000000
 CENTER=  4.172373723
          20.63507179
 RADIUS=  3.926188864

      X=  25.00000000
 CENTER=  1.333403823
         -125.2733816
 RADIUS=  1.403102653

      X=  0.000000000
 CENTER=  0.960012400
         -113.7482974
 RADIUS=  0.966020289

      X= -25.00000000
 CENTER=  0.790579465
         -103.5590069
 RADIUS=  0.736570300
```

(Continued)

```
      X=  80.30000000               X= -50.00000000
 CENTER=  1,682.082547         CENTER=  0.703341611
          28.67704204                   -95.00334745
 RADIUS=  1,681.877497         RADIUS=  0.595198281

      X=  50.00000000               X= -200.0000000
 CENTER=  2.429122528          CENTER=  0.589519930
         -137.3640869                   -68.28703630
 RADIUS=  2.562541967          RADIUS=  0.276630937
```

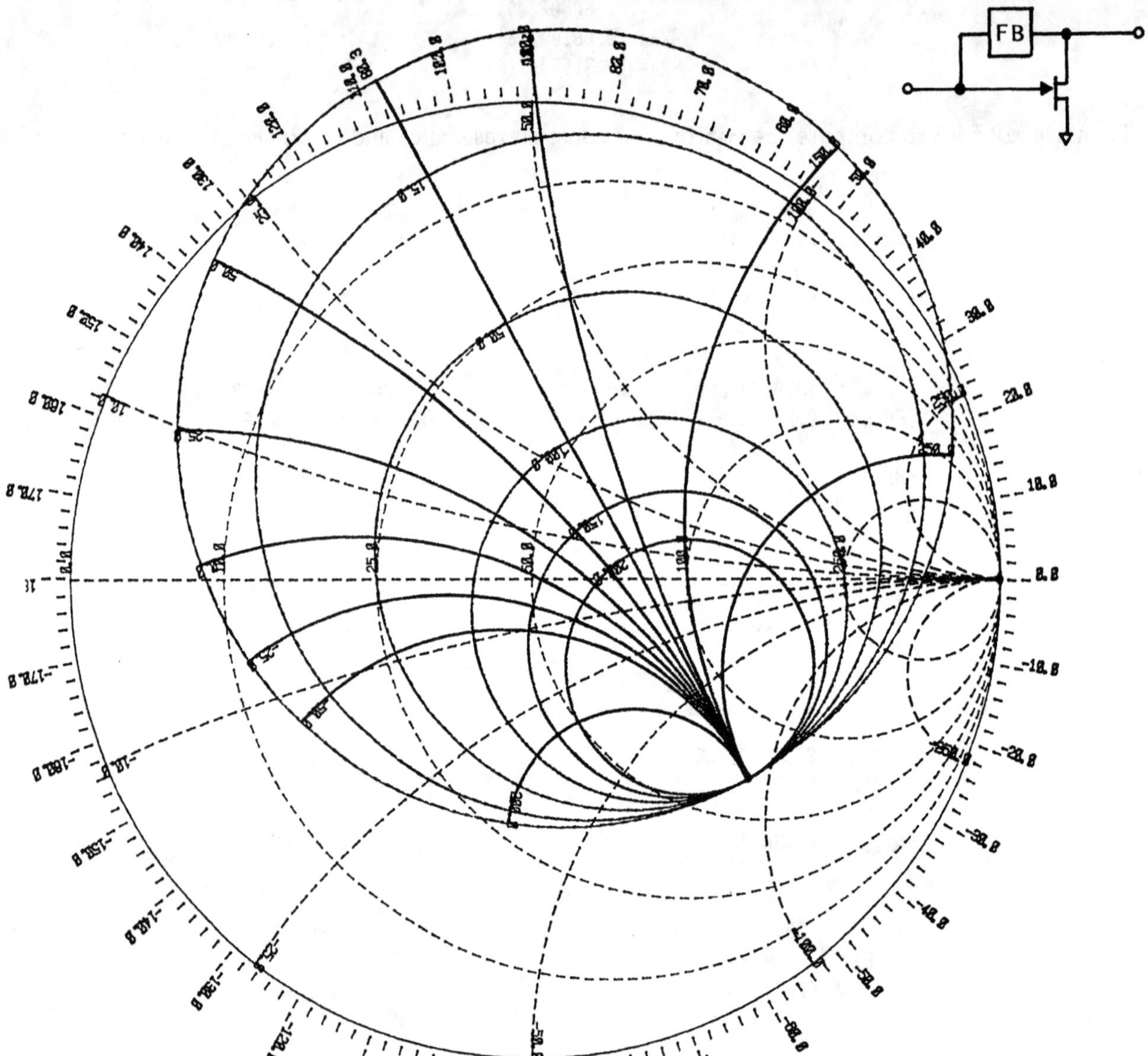

Figure 19. S_{11} Mapping for Shunt Feedback.

Construct the S_{11} mapping as shown in Figure 19. Note that a Smith Chart has been used for the S_{11} plane, which provides VSWR and impedance information as well as magnitude and angle. The task is to find a value of feedback impedance which will minimize S_{11} and provide a value of S_{12} below −25 dB. Construct a circle of radius 0.05 (−26 dB) around the center of the S_{12} plane. Any feedback impedance within this region will result in the desired value for S_{12}. The optimum value for S_{12} is $Z = 175 - j80$; however, when this same impedance is located on the S_{11} mapping, the input reflection coefficient is found to be $0.18 \angle -78°$, which is in good agreement with the calculated value above. The feedback impedance which is found to minimize S_{11} is approximately 140 + j0 ohms, which results in $S_{12} = 0.07 \angle -50°$. This point is not acceptable because it falls outside the specified 0.05 circle. An impedance which does satisfy both requirements is 150 − j25. The S_{11} mapping indicates that this impedance will result in $S_{11} = 0.07 \angle -88°$ (VSWR of 1.15:1), and the S_{12} mapping shows a value of $S_{12} = 0.04 \angle -56°$, or −28 dB isolation.

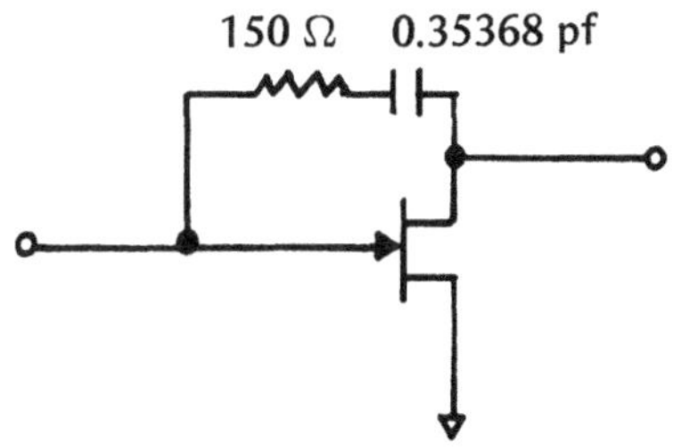

Figure 20. Feedback Impedance which Minimizes S_{11} and S_{12}.

Figure 20 shows the resulting circuit values. Analyze this network as in the previous example. The results shown below are in excellent agreement with the predictions from the mappings.

```
TWO-PORT A
S-PARAMETERS
S11=  0.055405806
     -91.12127705
S12=  0.040116373
     -52.12426370
S21=  1.312216719
     -233.5961434
S22=  0.303129077
     -140.4456975
```

If more accuracy is required, the mappings should be constructed using small step sizes for the resistance and reactance contours, and a larger scale for the S_{11} and S_{12} planes — perhaps 0.2 full scale for S_{12}, and 0.5 full scale for S_{11}. The final step in the design is to match S_{22} into 50 ohms, as shown in Figure 21, and analyze the complete circuit.

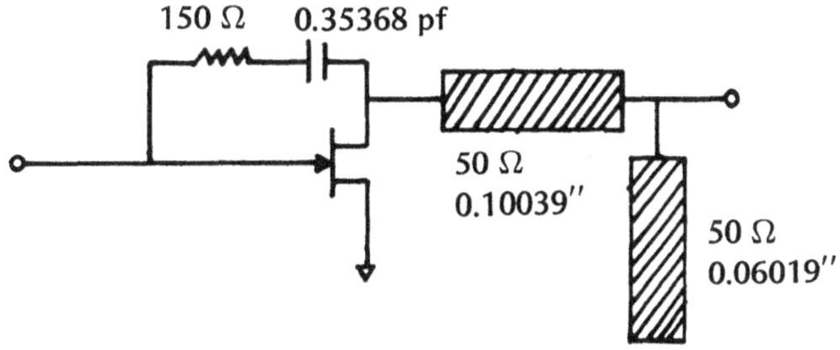

Figure 21. Active Isolator.

```
TWO-PORT A
S-PARAMETERS
S11= 0.067822273
    -103.2616236
S12= 0.042094732
    -125.0614195
S21= 1.376929346
    -306.5332992
S22= 0.010517334
    -158.5889240
```

The final results show that the active isolator has 27.5 dB isolation and 2.78 dB gain, along with an input and output VSWR of 1.15:1 and 1.02:1, respectively.

The network obtained through the use of the mappings is, most likely, much simpler than the network which would have resulted without the use of mappings. In fact, the entire input matching network is eliminated. The designer using another technique might have noticed the decrease in S_{11} and S_{22}; however, depending on the approach, the optimum feedback impedance may not be found. Certainly, if discrete values are being used, it is possible to skip the optimum completely and reach an incorrect conclusion. The mappings, however, provide the designer with information which completely describes the effect of a feedback impedance on any of the parameters, in a format which can readily be used.

Example 7. Amplifier Design.

The final example illustrates the analysis techniques which may be applied when using the Microwave Pac for the design of active networks. A narrow bandwidth amplifier will be designed at 18.0 GHz using a 0.5 micron gate-length GaAsFET. The S-parameters for this device are shown below.

$$S_{11} = 0.62 \angle{-42°} \qquad S_{12} = 0.18 \angle{169°}$$
$$S_{21} = 2.46 \angle{116°} \qquad S_{22} = 0.55 \angle{-89°}$$

The device should first be characterized using MP-11 to determine the stability factor and maximum available gain, as shown below.

```
FREQ= 18.00000
   K= 0.51556
 MSG= 11.35263
```

The results show that the device is potentially unstable, which will require that either the device be stabilized using a lossy element or the load and source impedances be selected to ensure that they do not fall within the regions of instability on the source and load planes. MP-12 can be used to compute the locations of the stability circles on the input and output planes.

```
INPUT PLANE                      OUTPUT PLANE
  FREQ= 18.00000                   FREQ= 18.00000
     K= 0.51556                       K= 0.51556
CENTER= 2.50466                  CENTER= 3.40958
        22.45173                         64.17897
RADIUS= 1.83797                  RADIUS= 2.78460
  CODE= 0.00000                    CODE= 0.00000
       (STABLE OUTSIDE)                 (STABLE OUTSIDE)
```

The stability circles constructed on the source and load planes are shown in Figures 22 and 23, respectively.

The specifications for this amplifier will be assumed to be 25 dB gain, input and output VSWR less than 2.0:1, and unconditional stability. The maximum stable gain is 11.4 dB, which will require three stages using this device. The gain of each stage should be 8.3 dB; however, due to circuit losses at this frequency, the design goal will be 9.3 dB per stage. The design can be accomplished through the use of lossy elements which may be used to stabilize the device and provide a maximum available gain of 9.3 dB with an input and output match. However, any lossy elements at the input will increase the noise figure, and lossy elements in the output circuit will reduce the output power. Therefore, the approach will be to use reactive mismatch to obtain the necessary gain, and use a balanced amplifier configuration which will reduce the input and output VSWR and stabilize the amplifier.

The design procedure will be to select a load and source reflection coefficient which results in a transducer gain of 9.3 dB and which is not within the unstable region on the load and source planes. The matching networks will then be synthesized to provide the desired impedance, and the overall network analyzed to confirm that the design is correct. How should these reflection coefficients be selected? Is there one set of reflection coefficients which may be better than another? The Microwave Pac includes some features which easily enable the designer to determine the answers to these questions. The constant gain circles of MP-13 can be constructed on the input or output planes, a reflection coefficient selected which falls on the desired gain circle, then the appropriate S'_{11} or S'_{22} calculation of MP-13 used to determine the corresponding reflection coefficient on the other plane.

Figure 22 shows the 8.3, 9.3, and 10.3 dB constant gain circles constructed on the input plane using the output from MP-13 shown below.

```
   INPUT  PLANE
   FREQ=  18.0000

   GAIN=  8.3000
 CENTER=  0.5312
          22.4517
 RADIUS=  0.6753

   GAIN=  9.3000
 CENTER=  0.6339
          22.4517
 RADIUS=  0.6450

   GAIN=  10.3000
 CENTER=  0.7489
          22.4517
 RADIUS=  0.6294
```

Although the gain circles can be constructed on either plane, the output impedance is usually more sensitive to changes in source impedance than vice versa; in fact, if S_{12} is zero, the input impedance which results from all possible load impedances is a single point, S_{11}. As will be shown, the ability to select a source impedance which results in a particular output impedance can be used to advantage.

The three constant gain contours can be used to determine desirable regions for the source reflection coefficient. In this case, the desired gain contour passes close to the origin of the source plane. This is a good candidate for the source because no input matching network will be required. In addition, the distance between the gain contours is a maximum at this point, which indicates that the gain sensitivity as a function of source impedance will be minimized. For example, the source impedance can vary from approximately 35 ohms to 65 ohms with a plus and minus 1 dB change in gain. This point is also far removed from the unstable region. Recall that the gain contours are computed with the constraint that the opposite plane is matched. Therefore, the output reflection coefficient is not a design variable; it must be the conjugate of S'_{22}, which corresponds to the source reflection coefficient. In this case, Γl is S^*_{22}, since Γs is zero.

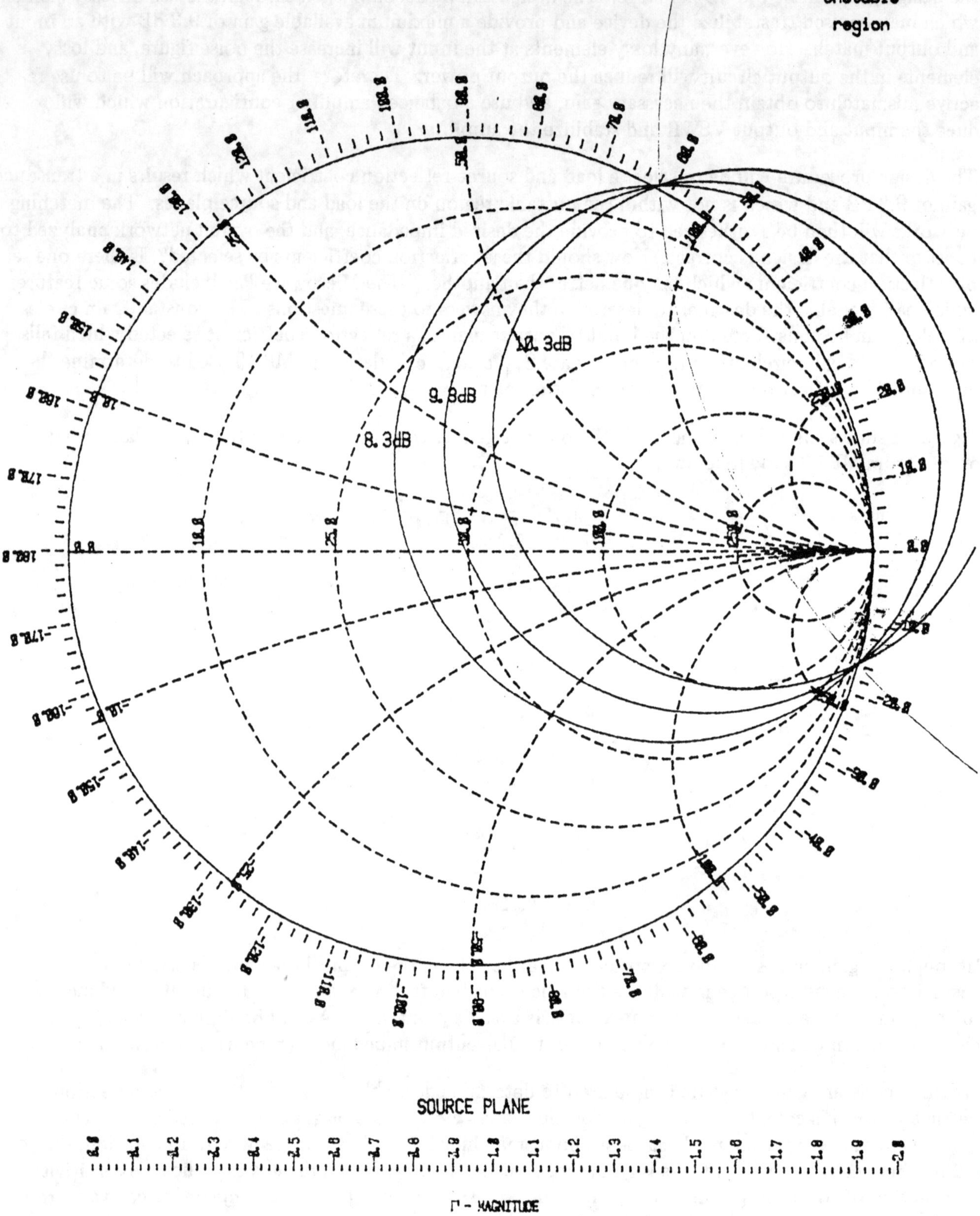

Figure 22. FET Source Plane.

Most designs, however, will not reduce to this simple case, and S'_{22} will have to be calculated for each source reflection coefficient. It may turn out that the calculated Γl is not desirable (high impedance region or undesired matching network), in which case another Γs will have to be selected and another Γl calculated. This is not a particularly time-consuming task using the Microwave Pac; however, this approach provides no insight which can be used for this selection process. The source and load mappings of MP-55, which do provide the necessary insight, can be used to select the most desirable set of source and load impedances.

The load and source mappings are, essentially, a mapping of one port with all possible impedances presented to the other port. The source mapping in Figure 23 shows the values of S_{22} which result from all possible source impedances, or, in fact, a mapping of S'_{22}. For the procedure used to calculate the data shown below for the source mapping of Figure 23, the reader is referred to the MP-55 program heading in that section which corresponds to the appropriate calculator.

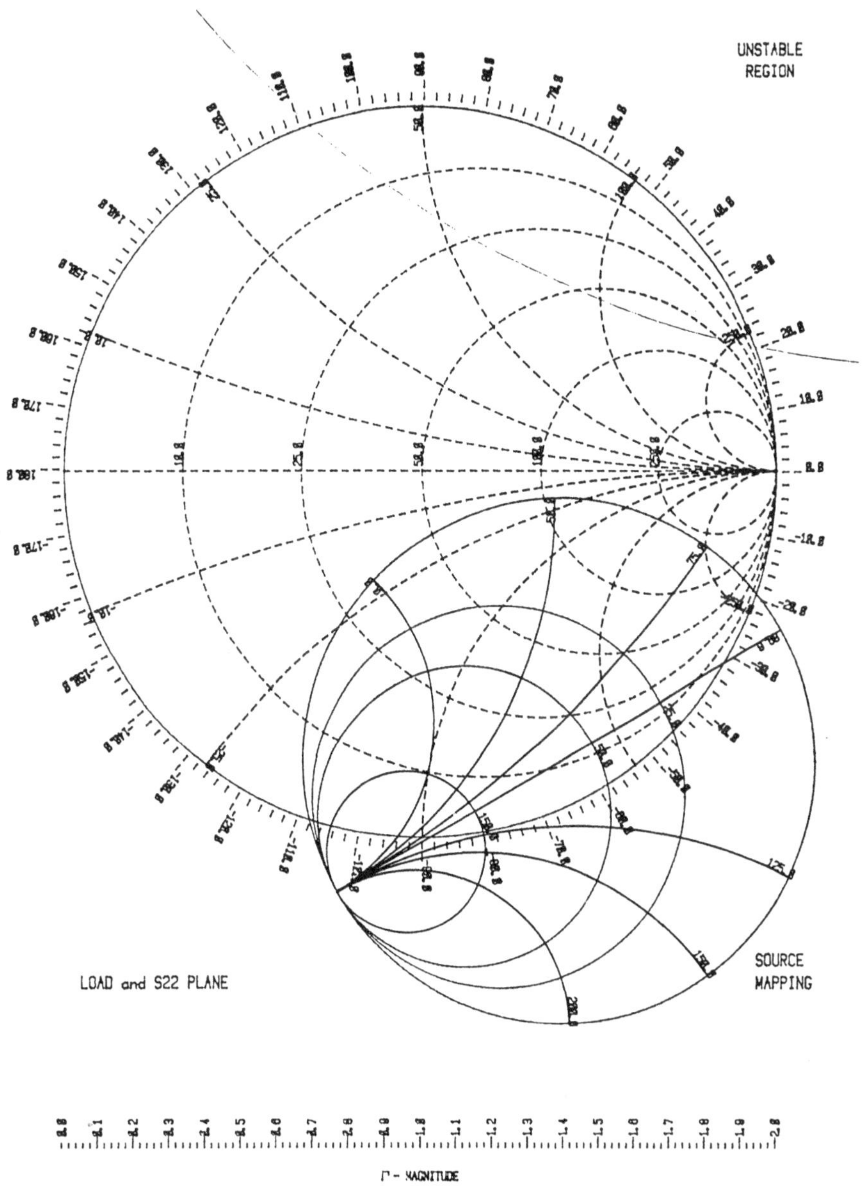

Figure 23. FET Load and S_{22} Plane.

The interpretation of this mapping is exactly the same as in the previous example; however, in this case the values of impedance correspond to the source impedance presented to the device. Thus, to determine S'_{22} for $\Gamma s = 0 \angle 0°$ or 50 ohms, locate the intersection of the R = 50 ohm and x = j0 ohm contours and read the value of S'_{22}, $0.55 \angle -89°$, from the scale of the S_{22} plane. This point is S_{22} of the device,

```
R18= -0.623834207
R19= -0.358079388

R21=  1.329875667
R22= -1.792441576
R23=  0.383616470
R24= -0.792806132
```

```
      R=  0.00000              R=  50.00000
 CENTER=  0.88074         CENTER=  0.95356
         -64.17897                -83.02105
 RADIUS=  0.71930         RADIUS=  0.41057

      R=  25.00000             R=  150.00000
 CENTER=  0.91580         CENTER=  1.04204
         -76.54060                -92.67396
 RADIUS=  0.52276         RADIUS=  0.22092

      X=  0.00000              X=  125.00000
 CENTER=  0.85400         CENTER=  2.36337
         -126.32559               -79.44790
 RADIUS=  0.53367         RADIUS=  1.35194

      X=  50.00000             X=  150.00000
 CENTER=  0.84755         CENTER=  1.84436
         -172.95391               -85.20622
 RADIUS=  1.20713         RADIUS=  0.79216

      X=  75.00000             X=  200.00000
 CENTER=  2.51678         CENTER=  1.52689
          137.94049               -90.91953
 RADIUS=  3.27100         RADIUS=  0.43332
```

and should always be used as a check for possible errors during the construction of the mappings. The selection of Γs and Γl can now be obtained directly from the two mappings. An impedance is selected which falls on the desired gain contour on the source plane, then the impedance is located on the source mapping, and Γl immediately determined. Two points are shown on the source mapping which result from values of Γs which fall on the 9.3 dB gain contour. These values correspond to 50 +j0 and 25 + j50 ohms. The line which connects these two points crosses the R = 50 ohm contour of the S_{22} plane, which will require a single series element to match the output into 50 ohms. This point is approximately 30 + j40 ohms, or $\Gamma s = 0.49 \angle 90.5°$. The corresponding S'_{22} from the mapping is $0.47 \angle -61°$, or

$\Gamma l = 0.47 \angle 61^\circ$. The transducer gain calculation of MP-14 can be used as a quick check to ensure that these reflection coefficients are correct before the matching networks are synthesized. The result of this calculation is 9.3483 dB, which is very accurate, considering that the values were determined graphically.

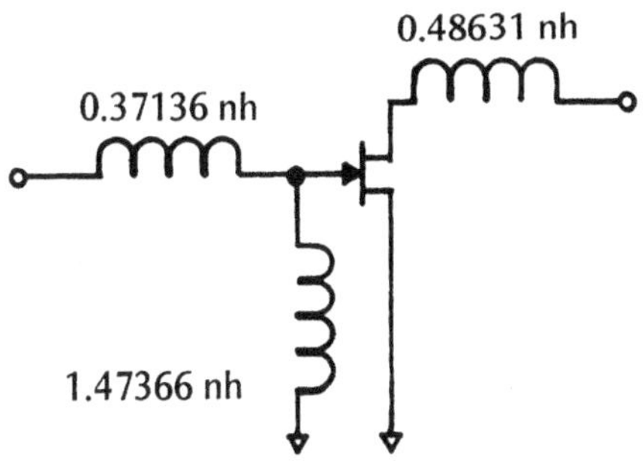

Figure 24. Matching Networks for 9.3 dB Gain Stage.

The required matching networks shown in Figure 24 are simple and realizable at this frequency using the inductance of short lengths of bonding wire. The structure used for the gate will provide the DC ground required for self bias. The output network will require a high impedance quarter-wavelength transmission line and blocking capacitor to provide drain bias. These design values can also be rapidly verified using the Ladder Analysis Program, MP-60. The results of this analysis show that $\Gamma s = 0.5041 \angle 88.9808^\circ$, and $\Gamma l = 0.4819 \angle 61.1890^\circ$. The transducer gain corresponding to these values is 9.4135 dB, which is acceptable.

Before proceeding, several comments should be made about the above design. First, another approach which also results in design flexibility is to increase the gain on the input plane, and then to use mismatch circles on the output plane, which allows the designer to select a desirable value of load impedance. For example, the 10.3 dB gain contour is used to select Γs, then the source mapping is used to determine S'_{22}, and a load impedance is selected which falls on the 1 dB mismatch circle of MP-14.

The reader should also observe that, when using the load and source mappings, the stability circles are not necessary. The stability circle is the boundary for those values of Γs or Γl which result in a reflection coefficient of 1 on the opposite plane. Values selected from the unstable region result in corresponding reflection coefficients greater than one, while those selected from the stable region are less than one. The load and source mappings define these regions very clearly without the need to determine a stable and unstable region.

The amplifier stage of Figure 24 can now be analyzed using the two-port analysis programs, which results in the two-port parameters shown below.

```
 TWO-PORT A
S-PARAMETERS
S11= 0.868506869
    -2.511172218
S12= 0.216274331
     143.4693505
S21= 2.955749193
     90.46935049
S22= 0.020706393
    -21.20604906
```

The insertion gain with a 50 ohm load and source is 9.4134 dB, which agrees very well with the above results from the Ladder Analysis Program and the transducer gain calculation. The stability factor calculated from this set of S-parameters is 0.51556, or exactly the same as the device without the matching networks. Thus, this single gain stage does not meet the design specification of unconditional stability, and has an unacceptable input VSWR. However, a balanced amplifier should meet these specifications.

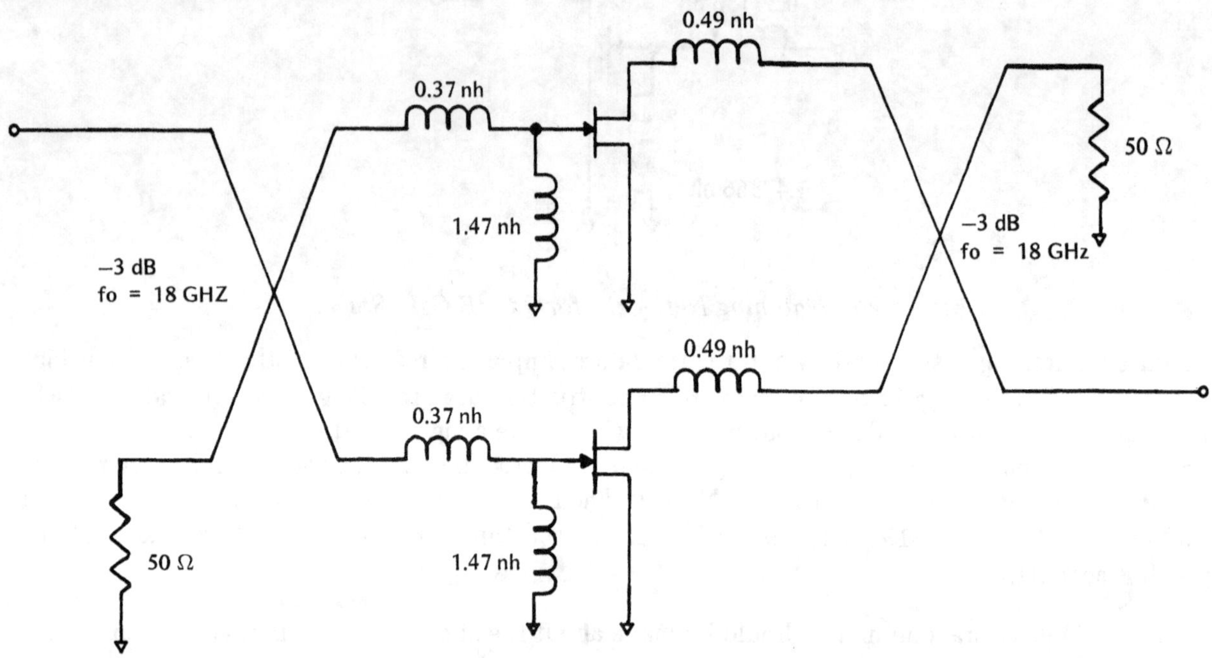

Figure 25. 18 GHz Balanced Amplifier.

Figure 25 shows the balanced amplifier configuration. The characteristics of the directional couplers used to combine the two gain stages result in a significant improvement in the performance of the balanced configuration over that of an individual gain stage. The 3 dB couplers at the input and output of Figure 25 are drawn in a manner slightly different from the normal representation. The structure is shown crossed over so that the coupled and direct ports are on the same side. Couplers which are used to physically realize the balanced amplifier are crossed over as well, which allows the gain stages to be connected as shown without undesirable lengths of transmission line. The RF signal enters the input port and is split equally between the direct port at the bottom and the coupled port at the top. The phase difference between the two ports is such that equal reflection coefficients at the two ports will cancel at the RF input port, and be absorbed by the terminating resistor at the isolated port. Thus, equal reflection coefficients due to the input VSWR of the single amplifier stages will result in a low input VSWR at the RF input port. The value of the input reflection coefficient is one-half the difference between the reflection coefficients at the coupled and direct ports for a 3 dB coupler at the center frequency. The coupler is symmetrical, so that the reflection coefficient at the coupled and direct port is also one-half the difference of two reflections, in this case the RF input and termination. The termination has zero reflection coefficient, therefore, the input to each of the single amplifier stages is one-half the input reflection coefficient.

In terms of the mappings, the entire RF input plane will map into a region with a maximum magnitude of 0.5 at the input to the amplifier stages. Since the unstable region of the stability circles will be outside this mapping, the balanced amplifier will be unconditionally stable.

The phase of the signals through the two sides of the amplifier is such that they will recombine at the output port shown. The output VSWR is similar to the input, and is also a function of the reflections at the two input ports of the coupler. The resulting two-port parameters can easily be calculated using the hybrid connection of MP-23. Use MP-41 to create an image of Two-Port A in Two-Port B, then use MP-23 to calculate the two-port parameters for the two identical gain stages connected with two 18 GHz, 3 dB couplers.

```
  TWO-PORT A
S-PARAMETERS
S11=  0.00206
     -2.51117
S12=  0.21627
      53.46935
S21=  2.95574
      0.46935
S22=  0.00005
     -21.20605
```

Thus, the balanced amplifier of Figure 25 has an input and output VSWR less than 1.01:1, gain of 9.41 dB, and provides unconditional stability, which meets all the design specifications.

The balanced amplifier configuration is a common design technique which is often used for designs — such as low noise amplifiers — where the device must be mismatched, but the overall amplifier stage must provide low VSWR. This technique also results in a 3 dB increase in output power because of the power combination at the output coupler, and is commonly used for this purpose. The equal power split and exact phase difference from the two ports of the coupler only occur at the center frequency. The performance of this configuration is degraded at other frequencies, as well as by differences between the two amplifier stages. The hybrid connection provided by the Microwave Pac can easily be used to investigate these effects.

APPENDIX A: TI-59 Program Listings

DATA INPUT AND OUTPUT — MP-01

1 ◁				▷ 2
LIST A	LIST B			PRT
ENTER A	ENTER B	A ⇌ B	LOAD B	SAVE A

```
000  76 LBL
001  11  A
002  01  1
003  05  5
004  42 STO
005  01  01
006  61 GTO
007  33 X²
008  76 LBL
009  12  B
010  02  2
011  03  3
012  42 STO
013  01  01
014  76 LBL
015  33 X²
016  01  1
017  01  1
018  91 R/S
019  72 ST*
020  01  01
021  32 X:T
022  69 OP
023  31  31
024  72 ST*
025  01  01
026  03  3
027  44 SUM
028  01  01
029  01  1
030  02  2
031  91 R/S
032  72 ST*
033  01  01
034  32 X:T
035  69 OP
036  31  31
037  72 ST*
038  01  01
039  03  3
040  44 SUM
041  01  01
042  02  2
043  01  1
044  91 R/S
045  72 ST*
046  01  01
047  69 OP
048  31  31
049  32 X:T
050  72 ST*
051  01  01
052  03  3
053  44 SUM
054  01  01
055  02  2
056  02  2
057  91 R/S
058  72 ST*
059  01  01
060  69 OP
061  31  31
062  32 X:T
063  72 ST*
064  01  01
065  09  9
066  09  9
067  92 RTN
068  76 LBL
069  13  C
070  02  2
071  02  2
072  42 STO
073  01  01
074  01  1
075  04  4
076  42 STO
077  02  02
078  08  8
079  42 STO
080  03  03
081  76 LBL
082  28 LOG
083  73 RC*
084  01  01
085  63 EX*
086  02  02
087  72 ST*
088  01  01
089  69 OP
090  21  21
091  69 OP
092  22  22
093  97 DSZ
094  03  03
095  28 LOG
096  09  9
097  09  9
098  92 RTN
099  76 LBL
100  14  D
101  03  3
102  00  0
103  42 STO
104  01  01
105  02  2
106  02  2
107  42 STO
108  02  02
109  03  3
110  22 INV
111  96 WRT
112  71 SBR
113  18 C'
114  09  9
115  09  9
116  92 RTN
117  76 LBL
118  15  E
119  01  1
120  04  4
121  42 STO
122  01  01
123  03  3
124  00  0
125  42 STO
126  02  02
127  71 SBR
128  18 C'
129  03  3
130  96 WRT
131  09  9
132  09  9
133  92 RTN
134  76 LBL
135  18 C'
136  08  8
137  42 STO
138  03  03
139  76 LBL
140  34 √X
141  73 RC*
142  01  01
143  72 ST*
144  02  02
145  69 OP
146  21  21
147  69 OP
148  22  22
149  97 DSZ
```

```
150  03  03
151  34  ΓX
152  92  RTN
153  76  LBL
154  10  E'
155  86  STF
156  00  00
157  03  3
158  06  6
159  00  0
160  02  2
161  00  0
162  02  2
163  42  STO
164  05  05
165  85  +
166  01  1
167  95  =
168  42  STO
169  06  06
170  03  3
171  06  6
172  00  0
173  03  3
174  00  0
175  02  2
176  42  STO
177  07  07
178  85  +
179  01  1
180  95  =
181  42  STO
182  08  08
183  03  3
184  07  7
185  04  4
186  03  3
187  03  3
188  02  2
189  02  2
190  00  0
191  03  3
192  03  3
193  42  STO
194  09  09
195  03  3
196  02  2
197  03  3
198  05  5
199  03  3
200  07  7
201  00  0
202  00  0
203  01  1
204  03  3
205  42  STO
206  10  10
207  09  9
208  09  9
209  92  RTN
210  76  LBL
211  16  A'
212  04  4
213  42  STO
214  03  03
215  00  0
216  42  STO
217  04  04
218  01  1
219  04  4
220  42  STO
221  01  01
222  01  1
223  05  5
224  42  STO
225  02  02
226  87  IFF
227  00  00
228  19  D'
229  61  GTO
230  35  1/X
231  76  LBL
232  17  B'
233  04  4
234  42  STO
235  03  03
236  01  1
237  42  STO
238  04  04
239  02  2
240  02  2
241  42  STO
242  01  01
243  02  2
244  03  3
245  42  STO
246  02  02
247  87  IFF
248  00  00
249  19  D'
250  76  LBL
251  35  1/X
252  73  RC*
253  02  02
254  32  X:T
255  73  RC*
256  01  01
257  91  R/S
258  02  2
259  44  SUM
260  01  01
261  44  SUM
262  02  02
263  97  DSZ
264  03  03
265  35  1/X
266  09  9
267  09  9
268  92  RTN
269  76  LBL
270  19  D'
271  22  INV
272  58  FIX
273  69  OP
274  00  00
275  43  RCL
276  09  09
277  69  OP
278  02  02
279  43  RCL
280  10  10
281  85  +
282  43  RCL
283  04  04
284  95  =
285  69  OP
286  03  03
287  69  OP
288  05  05
289  43  RCL
290  05  05
291  69  OP
292  04  04
293  73  RC*
294  01  01
295  69  OP
296  06  06
297  73  RC*
298  02  02
299  99  PRT
```

```
300  43 RCL
301  06  06
302  69 OP
303  04  04
304  02  2
305  44 SUM
306  01  01
307  44 SUM
308  02  02
309  73 RC*
310  01  01
311  69 OP
312  06  06
313  73 RC*
314  02  02
315  99 PRT
316  43 RCL
317  07  07
318  69 OP
319  04  04
320  02  2
321  44 SUM
322  01  01
323  44 SUM
324  02  02
325  73 RC*
326  01  01
327  69 OP
328  06  06
329  73 RC*
330  02  02
331  99 PRT
332  43 RCL
333  08  08
334  69 OP
335  04  04
336  02  2
337  44 SUM
338  01  01
339  44 SUM
340  02  02
341  73 RC*
342  01  01
343  69 OP
344  06  06
345  73 RC*
346  02  02
347  99 PRT
348  98 ADV
349  22 INV
350  86 STF
351  00  00
352  92 RTN
```

1 ◀				▶ 2
PARAMETER NORMALIZATION				MP-02
RCL Z	RCL Ymmho	RCL H	RCL G	PRT
ENT Z	ENT Ymmho	ENT H	ENG G	

```
000  76 LBL
001  11  A
002  05  5
003  00  0
004  35 1/X
005  42 STO
006  01  01
007  61 GTO
008  00  00
009  17  17
010  76 LBL
011  12  B
012  93  .
013  00  0
014  05  5
015  42 STO
016  01  01
017  01  1
018  01  1
019  91 R/S
020  42 STO
021  15  15
022  32 X:T
023  42 STO
024  14  14
025  01  1
026  02  2
027  91 R/S
028  42 STO
029  17  17
030  32 X:T
031  42 STO
032  16  16
033  02  2
034  01  1
035  91 R/S
036  42 STO
037  19  19
038  32 X:T
039  42 STO
040  18  18
041  02  2
042  02  2
043  91 R/S
044  42 STO
045  21  21
046  32 X:T
047  42 STO
048  20  20
049  43 RCL
050  01  01
051  49 PRD
052  14  14
053  49 PRD
054  16  16
055  49 PRD
056  18  18
057  49 PRD
058  20  20
059  09  9
060  09  9
061  92 RTN
062  76 LBL
063  13  C
064  05  5
065  00  0
066  42 STO
067  01  01
068  61 GTO
069  00  00
070  78  78
071  76 LBL
072  14  D
073  05  5
074  00  0
075  35 1/X
076  42 STO
077  01  01
078  01  1
079  01  1
080  91 R/S
081  42 STO
082  15  15
083  32 X:T
084  42 STO
085  14  14
086  01  1
087  02  2
088  91 R/S
089  42 STO
090  17  17
091  32 X:T
092  42 STO
093  16  16
094  02  2
095  01  1
096  91 R/S
097  42 STO
098  19  19
099  32 X:T
100  42 STO
101  18  18
102  02  2
103  02  2
104  91 R/S
105  42 STO
106  21  21
107  32 X:T
108  42 STO
109  20  20
110  43 RCL
111  01  01
112  22 INV
113  49 PRD
114  14  14
115  49 PRD
116  20  20
117  09  9
118  09  9
119  92 RTN
120  76 LBL
121  16 A'
122  05  5
123  00  0
124  42 STO
125  01  01
126  04  4
127  06  6
128  42 STO
129  02  02
130  04  4
131  06  6
132  00  0
133  00  0
134  00  0
135  00  0
136  42 STO
137  03  03
138  61 GTO
139  01  01
140  59  59
141  76 LBL
142  17 B'
143  02  2
144  00  0
145  42 STO
146  01  01
147  04  4
148  05  5
149  42 STO
```

Step	Code	Key
150	02	02
151	04	4
152	05	5
153	00	0
154	00	0
155	00	0
156	00	0
157	42	STO
158	03	03
159	06	6
160	42	STO
161	04	04
162	01	1
163	04	4
164	42	STO
165	05	05
166	08	8
167	42	STO
168	00	00
169	73	RC*
170	05	05
171	72	ST*
172	04	04
173	69	OP
174	24	24
175	69	OP
176	25	25
177	97	DSZ
178	00	00
179	01	01
180	69	69
181	43	RCL
182	01	01
183	49	PRD
184	06	06
185	49	PRD
186	08	08
187	49	PRD
188	10	10
189	49	PRD
190	12	12
191	87	IFF
192	01	01
193	24	CE
194	01	1
195	01	1
196	91	R/S
197	43	RCL
198	07	07
199	32	X:T
200	43	RCL
201	06	06
202	91	R/S
203	01	1
204	02	2
205	91	R/S
206	43	RCL
207	09	09
208	32	X:T
209	43	RCL
210	08	08
211	91	R/S
212	02	2
213	01	1
214	91	R/S
215	43	RCL
216	11	11
217	32	X:T
218	43	RCL
219	10	10
220	91	R/S
221	02	2
222	02	2
223	91	R/S
224	43	RCL
225	13	13
226	32	X:T
227	43	RCL
228	12	12
229	91	R/S
230	09	9
231	08	8
232	92	RTN
233	76	LBL
234	23	LNX
235	43	RCL
236	02	02
237	69	OP
238	01	01
239	02	2
240	00	0
241	03	3
242	03	3
243	01	1
244	03	3
245	03	3
246	05	5
247	01	1
248	03	3
249	69	OP
250	02	02
251	03	3
252	00	0
253	01	1
254	07	7
255	03	3
256	07	7
257	01	1
258	07	7
259	03	3
260	05	5
261	69	OP
262	03	03
263	03	3
264	06	6
265	00	0
266	00	0
267	00	0
268	00	0
269	00	0
270	00	0
271	00	0
272	00	0
273	69	OP
274	04	04
275	69	OP
276	05	05
277	92	RTN
278	76	LBL
279	24	CE
280	71	SBR
281	23	LNX
282	43	RCL
283	03	03
284	85	+
285	02	2
286	00	0
287	02	2
288	95	=
289	69	OP
290	04	04
291	43	RCL
292	06	06
293	69	OP
294	06	06
295	43	RCL
296	07	07
297	99	PRT
298	43	RCL
299	03	03

```
300  85  +
301  02  2
302  00  0
303  03  3
304  95  =
305  69  OP
306  04  04
307  43  RCL
308  08  08
309  69  OP
310  06  06
311  43  RCL
312  09  09
313  99  PRT
314  43  RCL
315  03  03
316  85  +
317  03  3
318  00  0
319  02  2
320  95  =
321  69  OP
322  04  04
323  43  RCL
324  10  10
325  69  OP
326  06  06
327  43  RCL
328  11  11
329  99  PRT
330  43  RCL
331  03  03
332  85  +
333  03  3
334  00  0
335  03  3
336  95  =
337  69  OP
338  04  04
339  43  RCL
340  12  12
341  69  OP
342  06  06
343  43  RCL
344  13  13
345  99  PRT
346  09  9
347  09  9
348  22  INV
349  86  STF
350  01  01
351  92  RTN
352  76  LBL
353  18  C'
354  05  5
355  00  0
356  42  STO
357  01  01
358  02  2
359  03  3
360  42  STO
361  02  02
362  02  2
363  03  3
364  00  0
365  00  0
366  00  0
367  00  0
368  42  STO
369  03  03
370  61  GTO
371  03  03
372  92  92
373  76  LBL
374  19  D'
375  05  5
376  00  0
377  35  1/X
378  42  STO
379  01  01
380  02  2
381  02  2
382  42  STO
383  02  02
384  02  2
385  02  2
386  00  0
387  00  0
388  00  0
389  00  0
390  42  STO
391  03  03
392  06  6
393  42  STO
394  04  04
395  01  1
396  04  4
397  42  STO
398  05  05
399  08  8
400  42  STO
401  00  00
402  73  RC*
403  05  05
404  72  ST*
405  04  04
406  69  OP
407  24  24
408  69  OP
409  25  25
410  97  DSZ
411  00  00
412  04  04
413  02  02
414  43  RCL
415  01  01
416  49  PRD
417  06  06
418  22  INV
419  49  PRD
420  12  12
421  61  GTO
422  01  01
423  91  91
424  76  LBL
425  10  E'
426  86  STF
427  01  01
428  92  RTN
429  00  0
430  00  0
431  00  0
432  00  0
433  00  0
434  00  0
435  00  0
436  00  0
437  00  0
438  00  0
439  00  0
440  00  0
441  00  0
442  00  0
443  00  0
444  00  0
445  00  0
446  00  0
447  00  0
448  00  0
449  00  0
```

1 ◁				▷ 2
MISCELLANEOUS RELATIONS				MP-03
Z ⇒ Γ	Ymmho ⇒ Γ	RFL LOSS ⇌ Γ	MW ⇒ dBm	Rp, Xp ⇒ Rs, Xs
Γ ⇒ Z	Γ ⇒ Ymmho	RTN LOSS ⇌ Γ	VSWR ⇌ Γ	Rs, Xs ⇌ Rp, Xp

```
000  76  LBL
001  85   +
002  86  STF
003  00   00
004  76  LBL
005  75   -
006  42  STO
007  01   01
008  73  RC*
009  01   01
010  87  IFF
011  00   00
012  00   00
013  15   15
014  94  +/-
015  32  X⇌T
016  42  STO
017  02   02
018  69  OP
019  21   21
020  73  RC*
021  01   01
022  37  P/R
023  42  STO
024  04   04
025  73  RC*
026  02   02
027  32  X⇌T
028  42  STO
029  03   03
030  69  OP
031  22   22
032  73  RC*
033  02   02
034  37  P/R
035  44  SUM
036  04   04
037  32  X⇌T
038  44  SUM
039  03   03
040  43  RCL
041  03   03
042  32  X⇌T
043  43  RCL
044  04   04
045  22  INV
046  37  P/R
047  32  X⇌T
048  22  INV
049  86  STF
050  00   00
051  92  RTN
052  76  LBL
053  65   ×
054  42  STO
055  01   01
056  73  RC*
057  01   01
058  42  STO
059  03   03
060  69  OP
061  21   21
062  73  RC*
063  01   01
064  42  STO
065  04   04
066  32  X⇌T
067  42  STO
068  01   01
069  73  RC*
070  01   01
071  49  PRD
072  03   03
073  69  OP
074  21   21
075  73  RC*
076  01   01
077  44  SUM
078  04   04
079  43  RCL
080  04   04
081  32  X⇌T
082  43  RCL
083  03   03
084  92  RTN
085  76  LBL
086  24  CE
087  05   5
088  32  X⇌T
089  07   7
090  71  SBR
091  75   -
092  42  STO
093  09   09
094  05   5
095  32  X⇌T
096  42  STO
097  10   10
098  07   7
099  71  SBR
100  85   +
101  35  1/X
102  42  STO
103  11   11
104  32  X⇌T
105  94  +/-
106  42  STO
107  12   12
108  09   9
109  32  X⇌T
110  01   1
111  01   1
112  71  SBR
113  65   ×
114  42  STO
115  05   05
116  32  X⇌T
117  42  STO
118  06   06
119  32  X⇌T
120  92  RTN
121  76  LBL
122  11   A
123  42  STO
124  08   08
125  32  X⇌T
126  94  +/-
127  42  STO
128  07   07
129  01   1
130  42  STO
131  05   05
132  00   0
133  42  STO
134  06   06
135  71  SBR
136  24  CE
137  65   ×
138  05   5
139  00   0
140  95   =
141  92  RTN
142  76  LBL
143  16   A'
144  42  STO
145  06   06
146  32  X⇌T
147  55   ÷
148  05   5
149  00   0
```

```
150  95  =
151  42  STO
152  05   05
153  01   1
154  42  STO
155  07   07
156  00   0
157  42  STO
158  08   08
159  61  GTO
160  24  CE
161  76  LBL
162  12   B
163  42  STO
164  08   08
165  32  X:T
166  42  STO
167  07   07
168  01   1
169  42  STO
170  05   05
171  00   0
172  42  STO
173  06   06
174  71  SBR
175  24  CE
176  65   ×
177  02   2
178  00   0
179  95   =
180  92  RTN
181  76  LBL
182  17  B'
183  94  +/-
184  42  STO
185  06   06
186  32  X:T
187  55   ÷
188  02   2
189  00   0
190  95   =
191  35  1/X
192  42  STO
193  05   05
194  01   1
195  42  STO
196  07   07
197  00   0
198  42  STO
199  08   08
200  61  GTO
201  24  CE
202  76  LBL
203  13   C
204  32  X:T
205  00   0
206  22  INV
207  77   GE
208  02   02
209  18   18
210  32  X:T
211  55   ÷
212  02   2
213  00   0
214  95   =
215  22  INV
216  28  LOG
217  92  RTN
218  32  X:T
219  28  LOG
220  65   ×
221  02   2
222  00   0
223  95   =
224  92  RTN
225  76  LBL
226  18  C'
227  32  X:T
228  00   0
229  22  INV
230  77   GE
231  02   02
232  46   46
233  32  X:T
234  55   ÷
235  01   1
236  00   0
237  95   =
238  22  INV
239  28  LOG
240  94  +/-
241  85   +
242  01   1
243  95   =
244  34  ΓX
245  92  RTN
246  32  X:T
247  33  X²
248  94  +/-
249  85   +
250  01   1
251  95   =
252  28  LOG
253  65   ×
254  01   1
255  00   0
256  95   =
257  92  RTN
258  76  LBL
259  14   D
260  32  X:T
261  01   1
262  22  INV
263  77   GE
264  02   02
265  82   82
266  32  X:T
267  42  STO
268  01   01
269  01   1
270  85   +
271  43  RCL
272  01   01
273  95   =
274  55   ÷
275  53   (
276  01   1
277  75   -
278  43  RCL
279  01   01
280  95   =
281  92  RTN
282  32  X:T
283  42  STO
284  01   01
285  75   -
286  01   1
287  95   =
288  55   ÷
289  53   (
290  43  RCL
291  01   01
292  85   +
293  01   1
294  95   =
295  92  RTN
296  76  LBL
297  19  D'
298  28  LOG
299  65   ×
```

```
300  01  1
301  00  0
302  95  =
303  92  RTN
304  76  LBL
305  15  E
306  22  INV
307  37  P/R
308  32  X⇄T
309  35  1/X
310  32  X⇄T
311  37  P/R
312  35  1/X
313  32  X⇄T
314  35  1/X
315  92  RTN
316  76  LBL
317  10  E'
318  35  1/X
319  32  X⇄T
320  35  1/X
321  32  X⇄T
322  22  INV
323  37  P/R
324  32  X⇄T
325  35  1/X
326  32  X⇄T
327  37  P/R
328  32  X⇄T
329  92  RTN
330  00  0
331  00  0
332  00  0
333  00  0
334  00  0
335  00  0
```

1 ◀				▶ 2
TWO-PORT CHARACTERIZATION				MP-11
MSG			PRT	PRT FREQ
ANALYZE	Ga max	Γml	Γms	K

```
000  76 LBL
001  75  -
002  42 STO
003  01  01
004  73 RC*
005  01  01
006  94 +/-
007  32 X:T
008  42 STO
009  02  02
010  69 OP
011  21  21
012  73 RC*
013  01  01
014  37 P/R
015  42 STO
016  04  04
017  73 RC*
018  02  02
019  32 X:T
020  42 STO
021  03  03
022  69 OP
023  22  22
024  73 RC*
025  02  02
026  37 P/R
027  44 SUM
028  04  04
029  32 X:T
030  44 SUM
031  03  03
032  43 RCL
033  03  03
034  32 X:T
035  43 RCL
036  04  04
037  22 INV
038  37 P/R
039  32 X:T
040  22 INV
041  86 STF
042  00  00
043  92 RTN
044  76 LBL
045  65  ×
046  42 STO
047  01  01
048  73 RC*
049  01  01
050  42 STO
051  03  03
052  69 OP
053  21  21
054  73 RC*
055  01  01
056  42 STO
057  04  04
058  32 X:T
059  42 STO
060  01  01
061  73 RC*
062  01  01
063  49 PRD
064  03  03
065  69 OP
066  21  21
067  73 RC*
068  01  01
069  44 SUM
070  04  04
071  43 RCL
072  04  04
073  32 X:T
074  43 RCL
075  03  03
076  22 INV
077  86 STF
078  00  00
079  92 RTN
080  76 LBL
081  11  A
082  01  1
083  04  4
084  32 X:T
085  02  2
086  00  0
087  71 SBR
088  65  ×
089  42 STO
090  05  05
091  01  1
092  06  6
093  32 X:T
094  42 STO
095  06  06
096  01  1
097  08  8
098  71 SBR
099  65  ×
100  42 STO
101  07  07
102  05  5
103  32 X:T
104  42 STO
105  08  08
106  07  7
107  71 SBR
108  75  -
109  42 STO
110  05  05
111  32 X:T
112  42 STO
113  06  06
114  01  1
115  94 +/-
116  49 PRD
117  21  21
118  05  5
119  32 X:T
120  02  2
121  00  0
122  71 SBR
123  65  ×
124  42 STO
125  09  09
126  01  1
127  04  4
128  32 X:T
129  42 STO
130  10  10
131  09  9
132  71 SBR
133  75  -
134  42 STO
135  11  11
136  32 X:T
137  94 +/-
138  42 STO
139  12  12
140  01  1
141  94 +/-
142  49 PRD
143  21  21
144  49 PRD
145  15  15
146  05  5
147  32 X:T
148  01  1
149  04  4
```

Step	Code	Key
150	71	SBR
151	65	×
152	42	STO
153	09	09
154	02	2
155	00	0
156	32	X⇌T
157	42	STO
158	10	10
159	09	9
160	71	SBR
161	75	-
162	42	STO
163	09	09
164	32	X⇌T
165	94	+/-
166	42	STO
167	10	10
168	01	1
169	94	+/-
170	49	PRD
171	15	15
172	01	1
173	85	+
174	43	RCL
175	05	05
176	33	X²
177	75	-
178	43	RCL
179	14	14
180	33	X²
181	75	-
182	43	RCL
183	20	20
184	33	X²
185	95	=
186	55	÷
187	53	(
188	02	2
189	65	×
190	43	RCL
191	07	07
192	54	)
193	95	=
194	42	STO
195	13	13
196	01	1
197	85	+
198	43	RCL
199	14	14
200	33	X²
201	75	-
202	43	RCL
203	20	20
204	33	X²
205	75	-
206	43	RCL
207	05	05
208	33	X²
209	95	=
210	42	STO
211	06	06
212	01	1
213	85	+
214	43	RCL
215	20	20
216	33	X²
217	75	-
218	43	RCL
219	14	14
220	33	X²
221	75	-
222	43	RCL
223	05	05
224	33	X²
225	95	=
226	42	STO
227	05	05
228	01	1
229	42	STO
230	07	07
231	43	RCL
232	06	06
233	32	X⇌T
234	00	0
235	77	GE
236	02	02
237	42	42
238	01	1
239	94	+/-
240	42	STO
241	07	07
242	43	RCL
243	13	13
244	92	RTN
245	76	LBL
246	12	B
247	02	2
248	02	2
249	01	1
250	03	3
251	32	X⇌T
252	43	RCL
253	13	13
254	33	X²
255	75	-
256	01	1
257	95	=
258	34	√X
259	65	×
260	43	RCL
261	07	07
262	85	+
263	43	RCL
264	13	13
265	95	=
266	65	×
267	43	RCL
268	18	18
269	95	=
270	55	÷
271	43	RCL
272	16	16
273	95	=
274	28	LOG
275	65	×
276	01	1
277	00	0
278	95	=
279	87	IFF
280	01	01
281	22	INV
282	92	RTN
283	76	LBL
284	14	D
285	03	3
286	00	0
287	03	3
288	06	6
289	42	STO
290	08	08
291	43	RCL
292	12	12
293	32	X⇌T
294	53	(
295	43	RCL
296	06	06
297	33	X²
298	75	-
299	53	(

```
300  04  4
301  65  ×
302  43  RCL
303  11  11
304  33  X²
305  95  =
306  34  ΓX
307  65  ×
308  43  RCL
309  07  07
310  85  +
311  43  RCL
312  06  06
313  95  =
314  55  ÷
315  53  (
316  02  2
317  65  ×
318  43  RCL
319  11  11
320  33  X²
321  95  =
322  65  ×
323  43  RCL
324  11  11
325  95  =
326  87  IFF
327  01  01
328  23  LNX
329  92  RTN
330  76  LBL
331  13  C
332  03  3
333  00  0
334  02  2
335  07  7
336  42  STO
337  08  08
338  53  (
339  43  RCL
340  05  05
341  33  X²
342  75  -
343  53  (
344  04  4
345  65  ×
346  43  RCL
347  09  09
348  33  X²
349  95  =
350  34  ΓX
351  42  STO
352  01  01
353  00  0
354  32  X:T
355  43  RCL
356  05  05
357  22  INV
358  77  GE
359  03  03
360  65  65
361  01  1
362  94  +/-
363  49  PRD
364  01  01
365  43  RCL
366  10  10
367  32  X:T
368  43  RCL
369  01  01
370  85  +
371  43  RCL
372  05  05
373  95  =
374  55  ÷
375  53  (
376  02  2
377  65  ×
378  43  RCL
379  09  09
380  33  X²
381  95  =
382  65  ×
383  43  RCL
384  09  09
385  95  =
386  87  IFF
387  01  01
388  23  LNX
389  92  RTN
390  76  LBL
391  15  E
392  02  2
393  06  6
394  32  X:T
395  43  RCL
396  13  13
397  87  IFF
398  01  01
399  22  INV
400  92  RTN
401  76  LBL
402  16  A'
403  03  3
404  00  0
405  03  3
406  06  6
407  02  2
408  02  2
409  32  X:T
410  43  RCL
411  18  18
412  55  ÷
413  43  RCL
414  16  16
415  95  =
416  28  LOG
417  65  ×
418  01  1
419  00  0
420  95  =
421  87  IFF
422  01  01
423  22  INV
424  92  RTN
425  76  LBL
426  10  E'
427  02  2
428  01  1
429  03  3
430  05  5
431  01  1
432  07  7
433  03  3
434  04  4
435  32  X:T
436  43  RCL
437  00  00
438  76  LBL
439  22  INV
440  32  X:T
441  69  OP
442  04  04
443  32  X:T
444  69  OP
445  06  06
446  22  INV
447  86  STF
448  01  01
449  92  RTN
```

```
76 LBL
19 D'
86 STF
01  01
92 RTN
76 LBL
23 LNX
48 EXC
08  08
69 OP
04  04
43 RCL
08  08
69 OP
06  06
25 CLR
69 OP
00  00
43 RCL
08  08
32 X:T
69 OP
06  06
32 X:T
22 INV
86 STF
01  01
92 RTN
00  0
00  0
```

1 ◀				▶ 2
STABILITY			MP-12	
		0 = outside 1 = inside	10 = ABS	
ANALYZE	INPUT	OUTPUT	PRT	

Step	Code	Key
000	76	LBL
001	75	-
002	42	STO
003	01	01
004	73	RC*
005	01	01
006	94	+/-
007	32	X:T
008	42	STO
009	02	02
010	69	OP
011	21	21
012	73	RC*
013	01	01
014	37	P/R
015	42	STO
016	04	04
017	73	RC*
018	02	02
019	32	X:T
020	42	STO
021	03	03
022	69	OP
023	22	22
024	73	RC*
025	02	02
026	37	P/R
027	44	SUM
028	04	04
029	32	X:T
030	44	SUM
031	03	03
032	43	RCL
033	03	03
034	32	X:T
035	43	RCL
036	04	04
037	22	INV
038	37	P/R
039	32	X:T
040	22	INV
041	86	STF
042	00	00
043	92	RTN
044	76	LBL
045	65	×
046	42	STO
047	01	01
048	73	RC*
049	01	01
050	42	STO
051	03	03
052	69	OP
053	21	21
054	73	RC*
055	01	01
056	42	STO
057	04	04
058	32	X:T
059	42	STO
060	01	01
061	73	RC*
062	01	01
063	49	PRD
064	03	03
065	69	OP
066	21	21
067	73	RC*
068	01	01
069	44	SUM
070	04	04
071	43	RCL
072	04	04
073	32	X:T
074	43	RCL
075	03	03
076	22	INV
077	86	STF
078	00	00
079	92	RTN
080	76	LBL
081	11	A
082	01	1
083	04	4
084	32	X:T
085	02	2
086	00	0
087	71	SBR
088	65	×
089	42	STO
090	05	05
091	01	1
092	06	6
093	32	X:T
094	42	STO
095	06	06
096	01	1
097	08	8
098	71	SBR
099	65	×
100	42	STO
101	07	07
102	05	5
103	32	X:T
104	42	STO
105	08	08
106	07	7
107	71	SBR
108	75	-
109	42	STO
110	08	08
111	32	X:T
112	42	STO
113	09	09
114	01	1
115	94	+/-
116	49	PRD
117	21	21
118	08	8
119	32	X:T
120	02	2
121	00	0
122	71	SBR
123	65	×
124	42	STO
125	05	05
126	01	1
127	04	4
128	32	X:T
129	42	STO
130	06	06
131	05	5
132	71	SBR
133	75	-
134	42	STO
135	05	05
136	32	X:T
137	94	+/-
138	42	STO
139	06	06
140	01	1
141	94	+/-
142	49	PRD
143	21	21
144	49	PRD
145	15	15
146	08	8
147	32	X:T
148	01	1
149	04	4

```
150  71 SBR
151  65  ×
152  42 STO
153  10  10
154  02  2
155  00  0
156  32 X⇄T
157  42 STO
158  11  11
159  01  1
160  00  0
161  71 SBR
162  75  -
163  42 STO
164  10  10
165  32 X⇄T
166  94 +/-
167  42 STO
168  11  11
169  01  1
170  94 +/-
171  49 PRD
172  15  15
173  01  1
174  85  +
175  43 RCL
176  08  08
177  33 X²
178  75  -
179  43 RCL
180  14  14
181  33 X²
182  75  -
183  43 RCL
184  20  20
185  33 X²
186  95  =
187  55  ÷
188  53  (
189  02  2
190  65  ×
191  43 RCL
192  07  07
193  95  =
194  42 STO
195  13  13
196  43 RCL
197  20  20
198  33 X²
199  75  -
200  43 RCL
201  08  08
202  33 X²
203  95  =
204  35 1/X
205  42 STO
206  01  01
207  65  ×
208  43 RCL
209  07  07
210  95  =
211  50 I×I
212  42 STO
213  12  12
214  43 RCL
215  10  10
216  65  ×
217  43 RCL
218  01  01
219  95  =
220  32 X⇄T
221  43 RCL
222  11  11
223  37 P/R
224  22 INV
225  37 P/R
226  42 STO
227  11  11
228  32 X⇄T
229  42 STO
230  10  10
231  43 RCL
232  14  14
233  33 X²
234  75  -
235  43 RCL
236  08  08
237  33 X²
238  95  =
239  35 1/X
240  42 STO
241  01  01
242  65  ×
243  43 RCL
244  07  07
245  95  =
246  50 I×I
247  42 STO
248  07  07
249  43 RCL
250  05  05
251  65  ×
252  43 RCL
253  01  01
254  95  =
255  32 X⇄T
256  43 RCL
257  06  06
258  37 P/R
259  22 INV
260  37 P/R
261  42 STO
262  06  06
263  32 X⇄T
264  42 STO
265  05  05
266  00  0
267  42 STO
268  02  02
269  42 STO
270  08  08
271  42 STO
272  09  09
273  43 RCL
274  05  05
275  75  -
276  43 RCL
277  07  07
278  95  =
279  32 X⇄T
280  00  0
281  22 INV
282  77  GE
283  02  02
284  88  88
285  01  1
286  42 STO
287  08  08
288  32 X⇄T
289  50 I×I
290  32 X⇄T
291  01  1
292  77  GE
293  02  02
294  97  97
295  44 SUM
296  02  02
297  43 RCL
298  10  10
299  75  -
```

```
300  43  RCL
301  12   12
302  95   =
303  32  X:T
304  00   0
305  22  INV
306  77   GE
307  03   03
308  12   12
309  01   1
310  42  STO
311  09   09
312  32  X:T
313  50  I×I
314  32  X:T
315  01   1
316  77   GE
317  03   03
318  21   21
319  44  SUM
320  02   02
321  32  X:T
322  43  RCL
323  14   14
324  77   GE
325  03   03
326  33   33
327  01   1
328  44  SUM
329  02   02
330  61  GTO
331  03   03
332  37   37
333  01   1
334  94  +/-
335  44  SUM
336  09   09
337  43  RCL
338  20   20
339  77   GE
340  03   03
341  48   48
342  01   1
343  44  SUM
344  02   02
345  61  GTO
346  03   03
347  52   52
348  01   1
349  94  +/-
350  44  SUM
351  08   08
352  43  RCL
353  08   08
354  50  I×I
355  42  STO
356  08   08
357  43  RCL
358  09   09
359  50  I×I
360  42  STO
361  09   09
362  04   4
363  32  X:T
364  43  RCL
365  02   02
366  22  INV
367  67   EQ
368  03   03
369  76   76
370  01   1
371  00   0
372  42  STO
373  08   08
374  42  STO
375  09   09
376  43  RCL
377  13   13
378  92  RTN
379  76  LBL
380  12   B
381  02   2
382  04   4
383  03   3
384  01   1
385  69  OP
386  02   02
387  43  RCL
388  05   05
389  42  STO
390  01   01
391  43  RCL
392  06   06
393  42  STO
394  02   02
395  43  RCL
396  07   07
397  42  STO
398  03   03
399  43  RCL
400  08   08
401  42  STO
402  04   04
403  61  GTO
404  04   04
405  32   32
406  76  LBL
407  13   C
408  03   3
409  02   2
410  04   4
411  01   1
412  03   3
413  07   7
414  69  OP
415  02   02
416  43  RCL
417  10   10
418  42  STO
419  01   01
420  43  RCL
421  11   11
422  42  STO
423  02   02
424  43  RCL
425  12   12
426  42  STO
427  03   03
428  43  RCL
429  09   09
430  42  STO
431  04   04
432  87  IFF
433  01   01
434  04   04
435  48   48
436  43  RCL
437  02   02
438  32  X:T
439  43  RCL
440  01   01
441  91  R/S
442  43  RCL
443  04   04
444  32  X:T
445  43  RCL
446  03   03
447  92  RTN
448  69  OP
449  05   05
```

```
450   43 RCL
451   01  01
452   99 PRT
453   43 RCL
454   02  02
455   99 PRT
456   43 RCL
457   03  03
458   99 PRT
459   43 RCL
460   04  04
461   99 PRT
462   22 INV
463   86 STF
464   01  01
465   92 RTN
466   76 LBL
467   14  D
468   69 OP
469   00  00
470   86 STF
471   01  01
472   92 RTN
473   00  0
474   00  0
475   00  0
476   00  0
477   00  0
478   00  0
479   00  0
```

1 ◀				▶ 2
CONST. GAIN, S'_{11} & S'_{22}				MP-13
			PRT ⇒	SET FLAG 2
INPUT GAIN	OUTPUT GAIN	S'_{11}	S'_{22}	

```
76 LBL
75  -
42 STO
01  01
73 RC*
01  01
94 +/-
32 X:T
42 STO
02  02
69 OP
21  21
73 RC*
01  01
37 P/R
42 STO
04  04
73 RC*
02  02
32 X:T
42 STO
03  03
69 OP
22  22
73 RC*
02  02
37 P/R
44 SUM
04  04
32 X:T
44 SUM
03  03
43 RCL
03  03
32 X:T
43 RCL
04  04
22 INV
37 P/R
32 X:T
22 INV
86 STF
00  00
92 RTN
76 LBL
65  ×
42 STO
01  01
73 RC*
01  01
42 STO
03  03
69 OP
21  21
73 RC*
01  01
42 STO
04  04
32 X:T
42 STO
01  01
73 RC*
01  01
49 PRD
03  03
69 OP
21  21
73 RC*
01  01
44 SUM
04  04
43 RCL
04  04
32 X:T
43 RCL
03  03
22 INV
86 STF
00  00
92 RTN
76 LBL
32 X:T
43 RCL
14  14
32 X:T
43 RCL
20  20
42 STO
14  14
32 X:T
42 STO
20  20
43 RCL
15  15
32 X:T
43 RCL
21  21
42 STO
15  15
32 X:T
42 STO
21  21
92 RTN
76 LBL
11  A
86 STF
01  01
76 LBL
12  B
01  1
04  4
32 X:T
02  2
00  0
71 SBR
65  ×
42 STO
05  05
01  1
06  6
32 X:T
42 STO
06  06
01  1
08  8
71 SBR
65  ×
42 STO
07  07
05  5
32 X:T
42 STO
08  08
07  7
71 SBR
75  -
42 STO
05  05
32 X:T
42 STO
06  06
01  1
85  +
43 RCL
05  05
33 X²
75  -
43 RCL
14  14
33 X²
```

```
150  75  -
151  43  RCL
152  20   20
153  33  X²
154  95   =
155  55   ÷
156  53   (
157  02   2
158  65   ×
159  43  RCL
160  07   07
161  95   =
162  42  STO
163  13   13
164  87  IFF
165  01   01
166  01   01
167  70   70
168  71  SBR
169  32  X⇌T
170  01   1
171  94  +/-
172  49  PRD
173  21   21
174  05   5
175  32  X⇌T
176  02   2
177  00   0
178  71  SBR
179  65   ×
180  42  STO
181  09   09
182  01   1
183  04   4
184  32  X⇌T
185  42  STO
186  10   10
187  09   9
188  71  SBR
189  75   -
190  42  STO
191  09   09
192  32  X⇌T
193  94  +/-
194  42  STO
195  10   10
196  01   1
197  94  +/-
198  49  PRD
199  21   21
200  43  RCL
201  14   14
202  33  X²
203  75   -
204  43  RCL
205  05   05
206  33  X²
207  95   =
208  42  STO
209  11   11
210  02   2
211  04   4
212  87  IFF
213  01   01
214  02   02
215  20   20
216  71  SBR
217  32  X⇌T
218  03   3
219  02   2
220  69  OP
221  04   04
222  22  INV
223  86  STF
224  01   01
225  02   2
226  01   1
227  91  R/S
228  42  STO
229  01   01
230  55   ÷
231  01   1
232  00   0
233  95   =
234  22  INV
235  28  LOG
236  55   ÷
237  43  RCL
238  18   18
239  33  X²
240  95   =
241  42  STO
242  08   08
243  65   ×
244  43  RCL
245  11   11
246  85   +
247  01   1
248  95   =
249  35  1/X
250  65   ×
251  43  RCL
252  08   08
253  95   =
254  65   ×
255  43  RCL
256  09   09
257  95   =
258  42  STO
259  02   02
260  43  RCL
261  10   10
262  42  STO
263  03   03
264  43  RCL
265  08   08
266  33  X²
267  65   ×
268  43  RCL
269  07   07
270  33  X²
271  75   -
272  53   (
273  02   2
274  65   ×
275  43  RCL
276  13   13
277  65   ×
278  43  RCL
279  07   07
280  65   ×
281  43  RCL
282  08   08
283  54   )
284  85   +
285  01   1
286  95   =
287  34  √X
288  55   ÷
289  53   (
290  43  RCL
291  11   11
292  65   ×
293  43  RCL
294  08   08
295  85   +
296  01   1
297  95   =
298  42  STO
299  04   04
```

```
300  87  IFF
301  02   02
302  03   03
303  19   19
304  43  RCL
305  03   03
306  32  X⇌T
307  43  RCL
308  02   02
309  91  R/S
310  43  RCL
311  01   01
312  32  X⇌T
313  43  RCL
314  04   04
315  91  R/S
316  61  GTO
317  02   02
318  25   25
319  43  RCL
320  01   01
321  69  OP
322  06   06
323  43  RCL
324  02   02
325  99  PRT
326  43  RCL
327  03   03
328  99  PRT
329  43  RCL
330  04   04
331  99  PRT
332  61  GTO
333  02   02
334  25   25
335  76  LBL
336  13   C
337  86  STF
338  01   01
339  76  LBL
340  14   D
341  42  STO
342  13   13
343  32  X⇌T
344  42  STO
345  12   12
346  87  IFF
347  01   01
348  03   03
349  52   52
350  71  SBR
351  32  X⇌T
352  01   1
353  06   6
354  32  X⇌T
355  01   1
356  08   8
357  71  SBR
358  65   ×
359  42  STO
360  10   10
361  01   1
362  02   2
363  32  X⇌T
364  42  STO
365  11   11
366  01   1
367  00   0
368  71  SBR
369  65   ×
370  42  STO
371  10   10
372  02   2
373  00   0
374  32  X⇌T
375  42  STO
376  11   11
377  01   1
378  02   2
379  71  SBR
380  65   ×
381  42  STO
382  08   08
383  32  X⇌T
384  42  STO
385  09   09
386  01   1
387  42  STO
388  06   06
389  00   0
390  42  STO
391  07   07
392  06   6
393  32  X⇌T
394  08   8
395  71  SBR
396  75   -
397  42  STO
398  06   06
399  32  X⇌T
400  42  STO
401  07   07
402  43  RCL
403  10   10
404  94  +/-
405  55   ÷
406  43  RCL
407  06   06
408  95   =
409  42  STO
410  10   10
411  43  RCL
412  11   11
413  75   -
414  43  RCL
415  07   07
416  95   =
417  42  STO
418  11   11
419  01   1
420  04   4
421  32  X⇌T
422  01   1
423  00   0
424  71  SBR
425  75   -
426  42  STO
427  02   02
428  32  X⇌T
429  42  STO
430  03   03
431  02   2
432  00   0
433  02   2
434  87  IFF
435  01   01
436  04   04
437  43   43
438  71  SBR
439  32  X⇌T
440  03   3
441  00   0
442  03   3
443  69  OP
444  04   04
445  22  INV
446  86  STF
447  01   01
448  87  IFF
449  02   02
```

```
450  04  04
451  64  64
452  43  RCL
453  03  03
454  32  X:T
455  43  RCL
456  02  02
457  91  R/S
458  43  RCL
459  13  13
460  32  X:T
461  43  RCL
462  12  12
463  92  RTN
464  43  RCL
465  02  02
466  69  OP
467  06  06
468  43  RCL
469  03  03
470  99  PRT
471  43  RCL
472  12  12
473  99  PRT
474  43  RCL
475  13  13
476  99  PRT
477  92  RTN
478  00  0
479  00  0
```

1 ◄				► 2
GT AND MISMATCH				MP-14
			PRINT ⇒	SET FLAG 1
GT	MISMATCH CIRCLES	MISMATCH LOSS		

```
76 LBL
75 -
42 STO
01 01
73 RC*
01 01
94 +/-
32 X:T
42 STO
02 02
69 OP
21 21
73 RC*
01 01
37 P/R
42 STO
04 04
73 RC*
02 02
32 X:T
42 STO
03 03
69 OP
22 22
73 RC*
02 02
37 P/R
44 SUM
04 04
32 X:T
44 SUM
03 03
43 RCL
03 03
32 X:T
43 RCL
04 04
22 INV
37 P/R
32 X:T
22 INV
86 STF
00 00
92 RTN
76 LBL
65 ×
42 STO
01 01
73 RC*
01 01
```

```
42 STO
03 03
69 OP
21 21
73 RC*
01 01
42 STO
04 04
32 X:T
42 STO
01 01
73 RC*
01 01
49 PRD
03 03
69 OP
21 21
73 RC*
01 01
44 SUM
04 04
43 RCL
04 04
32 X:T
43 RCL
03 03
22 INV
86 STF
00 00
92 RTN
76 LBL
22 INV
94 +/-
32 X:T
37 P/R
32 X:T
85 +
01 1
95 =
32 X:T
22 INV
37 P/R
32 X:T
92 RTN
76 LBL
11 A
01 1
01 1
91 R/S
42 STO
```

```
06 06
32 X:T
42 STO
05 05
02 2
02 2
91 R/S
42 STO
08 08
01 1
04 4
32 X:T
42 STO
07 07
05 5
71 SBR
65 ×
71 SBR
22 INV
42 STO
12 12
07 7
32 X:T
42 STO
13 13
02 2
00 0
71 SBR
65 ×
71 SBR
22 INV
42 STO
10 10
01 1
00 0
32 X:T
42 STO
11 11
01 1
02 2
71 SBR
65 ×
42 STO
12 12
01 1
06 6
32 X:T
42 STO
13 13
01 1
```

```
150  08   8
151  71  SBR
152  65   X
153  42  STO
154  10   10
155  07   7
156  32  X:T
157  42  STO
158  11   11
159  01   1
160  00   0
161  71  SBR
162  65   X
163  42  STO
164  10   10
165  05   5
166  32  X:T
167  42  STO
168  11   11
169  01   1
170  00   0
171  71  SBR
172  65   X
173  42  STO
174  10   10
175  01   1
176  02   2
177  32  X:T
178  42  STO
179  11   11
180  01   1
181  00   0
182  71  SBR
183  75   -
184  33  X²
185  42  STO
186  10   10
187  53   (
188  01   1
189  75   -
190  43  RCL
191  05   05
192  33  X²
193  54   )
194  65   X
195  53   (
196  01   1
197  75   -
198  43  RCL
199  07   07
200  33  X²
201  95   =
202  65   X
203  43  RCL
204  18   18
205  33  X²
206  95   =
207  55   ÷
208  43  RCL
209  10   10
210  95   =
211  28  LOG
212  65   X
213  01   1
214  00   0
215  95   =
216  42  STO
217  09   09
218  87  IFF
219  01   01
220  02   02
221  23   23
222  92  RTN
223  02   2
224  02   2
225  03   3
226  07   7
227  69  OP
228  04   04
229  43  RCL
230  09   09
231  69  OP
232  06   06
233  05   5
234  02   2
235  03   3
236  06   6
237  69  OP
238  04   04
239  43  RCL
240  05   05
241  69  OP
242  06   06
243  43  RCL
244  06   06
245  99  PRT
246  05   5
247  02   2
248  02   2
249  07   7
250  69  OP
251  04   04
252  43  RCL
253  07   07
254  69  OP
255  06   06
256  43  RCL
257  08   08
258  99  PRT
259  92  RTN
260  76  LBL
261  12   B
262  01   1
263  01   1
264  91  R/S
265  42  STO
266  06   06
267  32  X:T
268  42  STO
269  05   05
270  02   2
271  01   1
272  91  R/S
273  42  STO
274  09   09
275  94  +/-
276  55   ÷
277  01   1
278  00   0
279  95   =
280  22  INV
281  28  LOG
282  42  STO
283  10   10
284  01   1
285  75   -
286  53   (
287  43  RCL
288  05   05
289  33  X²
290  65   X
291  53   (
292  01   1
293  75   -
294  43  RCL
295  10   10
296  95   =
297  35  1/X
298  42  STO
299  11   11
```

```
300  43 RCL
301  05  05
302  65  ×
303  43 RCL
304  10  10
305  65  ×
306  43 RCL
307  11  11
308  95  =
309  42 STO
310  07  07
311  43 RCL
312  06  06
313  94 +/-
314  42 STO
315  08  08
316  01  1
317  75  -
318  43 RCL
319  10  10
320  95  =
321  34 √X
322  65  ×
323  53  (
324  01  1
325  75  -
326  43 RCL
327  05  05
328  33 X^2
329  95  =
330  65  ×
331  43 RCL
332  11  11
333  95  =
334  42 STO
335  12  12
336  87 IFF
337  01  01
338  03  03
339  55  55
340  43 RCL
341  08  08
342  32 X<>T
343  43 RCL
344  07  07
345  91 R/S
346  43 RCL
347  09  09
348  32 X<>T
349  43 RCL
350  12  12
351  91 R/S
352  61 GTO
353  02  02
354  70  70
355  02  2
356  02  2
357  69 OP
358  04  04
359  43 RCL
360  09  09
361  69 OP
362  06  06
363  01  1
364  05  5
365  03  3
366  07  7
367  03  3
368  05  5
369  69 OP
370  04  04
371  43 RCL
372  07  07
373  69 OP
374  06  06
375  43 RCL
376  08  08
377  99 PRT
378  03  3
379  05  5
380  01  1
381  03  3
382  01  1
383  06  6
384  69 OP
385  04  04
386  43 RCL
387  12  12
388  69 OP
389  06  06
390  05  5
391  02  2
392  69 OP
393  04  04
394  43 RCL
395  05  05
396  69 OP
397  06  06
398  43 RCL
399  06  06
400  99 PRT
401  61 GTO
402  02  02
403  70  70
404  76 LBL
405  13  C
406  01  1
407  91 R/S
408  42 STO
409  06  06
410  32 X<>T
411  42 STO
412  05  05
413  02  2
414  91 R/S
415  42 STO
416  08  08
417  32 X<>T
418  42 STO
419  07  07
420  05  5
421  32 X<>T
422  07  7
423  71 SBR
424  65  ×
425  42 STO
426  09  09
427  32 X<>T
428  42 STO
429  10  10
430  01  1
431  42 STO
432  11  11
433  00  0
434  42 STO
435  12  12
436  01  1
437  01  1
438  32 X<>T
439  09  9
440  71 SBR
441  75  -
442  33 X^2
443  35 1/X
444  42 STO
445  13  13
446  01  1
447  75  -
448  43 RCL
449  07  07
```

```
450  33 X²
451  95  =
452  65  ×
453  53  (
454  01  1
455  75  -
456  43 RCL
457  05  05
458  33 X²
459  95  =
460  65  ×
461  43 RCL
462  13  13
463  95  =
464  28 LOG
465  65  ×
466  01  1
467  00  0
468  95  =
469  42 STO
470  09  09
471  87 IFF
472  01  01
473  02  02
474  23  23
475  92 RTN
476  00  0
477  00  0
478  00  0
479  00  0
```

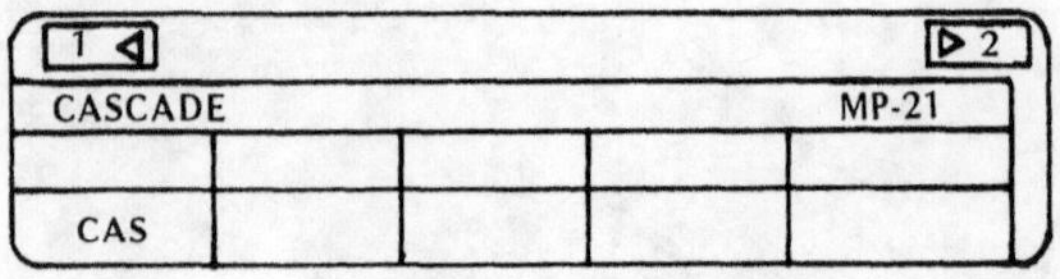


```
000  76 LBL
001  85  +
002  86 STF
003  00  00
004  76 LBL
005  75  -
006  42 STO
007  01  01
008  73 RC*
009  01  01
010  87 IFF
011  00  00
012  00  00
013  15  15
014  94 +/-
015  32 X:T
016  42 STO
017  02  02
018  69 OP
019  21  21
020  73 RC*
021  01  01
022  37 P/R
023  42 STO
024  04  04
025  73 RC*
026  02  02
027  32 X:T
028  42 STO
029  03  03
030  69 OP
031  22  22
032  73 RC*
033  02  02
034  37 P/R
035  44 SUM
036  04  04
037  32 X:T
038  44 SUM
039  03  03
040  43 RCL
041  03  03
042  32 X:T
043  43 RCL
044  04  04
045  22 INV
046  37 P/R
047  32 X:T
048  22 INV
049  86 STF
050  00  00
051  92 RTN
052  76 LBL
053  65  ×
054  86 STF
055  00  00
056  76 LBL
057  55  ÷
058  42 STO
059  01  01
060  73 RC*
061  01  01
062  87 IFF
063  00  00
064  00  00
065  67  67
066  35 1/X
067  42 STO
068  03  03
069  69 OP
070  21  21
071  73 RC*
072  01  01
073  87 IFF
074  00  00
075  00  00
076  78  78
077  94 +/-
078  42 STO
079  04  04
080  32 X:T
081  42 STO
082  01  01
083  73 RC*
084  01  01
085  49 PRD
086  03  03
087  69 OP
088  21  21
089  73 RC*
090  01  01
091  44 SUM
092  04  04
093  43 RCL
094  04  04
095  32 X:T
096  43 RCL
097  03  03
098  22 INV
099  86 STF
100  00  00
101  92 RTN
102  76 LBL
103  11  A
104  01  1
105  42 STO
106  05  05
107  00  0
108  42 STO
109  06  06
110  02  2
111  00  0
112  32 X:T
113  02  2
114  02  2
115  71 SBR
116  65  ×
117  42 STO
118  07  07
119  05  5
120  32 X:T
121  42 STO
122  08  08
123  07  7
124  71 SBR
125  75  -
126  35 1/X
127  42 STO
128  05  05
129  01  1
130  06  6
131  32 X:T
132  94 +/-
133  42 STO
134  06  06
135  01  1
136  08  8
137  71 SBR
138  65  ×
139  42 STO
140  07  07
141  07  7
142  32 X:T
143  42 STO
144  08  08
145  02  2
146  02  2
147  71 SBR
148  65  ×
149  42 STO
```

```
07 07
05 5
32 X:T
42 STO
08 08
07 7
71 SBR
65 ×
42 STO
07 07
07 7
32 X:T
42 STO
08 08
01 1
04 4
71 SBR
85 +
42 STO
14 14
01 1
06 6
32 X:T
42 STO
15 15
02 2
04 4
71 SBR
65 ×
42 STO
07 07
07 7
32 X:T
42 STO
08 08
05 5
71 SBR
65 ×
42 STO
16 16
01 1
08 8
32 X:T
42 STO
17 17
02 2
06 6
71 SBR
65 ×
42 STO
07 07
07 7
32 X:T
42 STO
08 08
05 5
71 SBR
65 ×
42 STO
18 18
02 2
04 4
32 X:T
42 STO
19 19
02 2
06 6
71 SBR
65 ×
42 STO
07 07
07 7
32 X:T
42 STO
08 08
02 2
00 0
71 SBR
65 ×
42 STO
07 07
07 7
32 X:T
42 STO
08 08
05 5
71 SBR
65 ×
42 STO
07 07
02 2
08 8
32 X:T
42 STO
08 08
07 7
71 SBR
85 +
42 STO
20 20
32 X:T
42 STO
21 21
09 9
09 9
92 RTN
00 0
00 0
00 0
00 0
00 0
00 0
00 0
00 0
00 0
00 0
00 0
00 0
00 0
00 0
00 0
00 0
00 0
00 0
00 0
00 0
00 0
00 0
00 0
00 0
00 0
00 0
00 0
00 0
00 0
00 0
00 0
00 0
00 0
00 0
00 0
00 0
00 0
00 0
00 0
00 0
00 0
00 0
00 0
00 0
```

1 ◁				▷ 2
SERIES AND PARALLEL				MP-22
SER	PAR			

```
000  76  LBL
001  85  +
002  86  STF
003  00  00
004  76  LBL
005  75  -
006  42  STO
007  01  01
008  73  RC*
009  01  01
010  87  IFF
011  00  00
012  00  00
013  15  15
014  94  +/-
015  32  X:T
016  42  STO
017  02  02
018  69  OP
019  21  21
020  73  RC*
021  01  01
022  37  P/R
023  42  STO
024  04  04
025  73  RC*
026  02  02
027  32  X:T
028  42  STO
029  03  03
030  69  OP
031  22  22
032  73  RC*
033  02  02
034  37  P/R
035  44  SUM
036  04  04
037  32  X:T
038  44  SUM
039  03  03
040  43  RCL
041  03  03
042  32  X:T
043  43  RCL
044  04  04
045  22  INV
046  37  P/R
047  32  X:T
048  22  INV
049  86  STF
050  00  00
051  92  RTN
052  76  LBL
053  65  ×
054  42  STO
055  01  01
056  73  RC*
057  01  01
058  42  STO
059  03  03
060  69  OP
061  21  21
062  73  RC*
063  01  01
064  42  STO
065  04  04
066  32  X:T
067  42  STO
068  01  01
069  73  RC*
070  01  01
071  49  PRD
072  03  03
073  69  OP
074  21  21
075  73  RC*
076  01  01
077  44  SUM
078  04  04
079  43  RCL
080  04  04
081  32  X:T
082  43  RCL
083  03  03
084  92  RTN
085  76  LBL
086  32  X:T
087  01  1
088  42  STO
089  05  05
090  00  0
091  42  STO
092  06  06
093  05  5
094  32  X:T
095  01  1
096  04  4
097  71  SBR
098  85  +
099  42  STO
100  07  07
101  05  5
102  32  X:T
103  42  STO
104  08  08
105  02  2
106  00  0
107  71  SBR
108  85  +
109  42  STO
110  09  09
111  01  1
112  06  6
113  32  X:T
114  42  STO
115  10  10
116  01  1
117  08  8
118  71  SBR
119  65  ×
120  42  STO
121  11  11
122  07  7
123  32  X:T
124  42  STO
125  12  12
126  09  9
127  71  SBR
128  65  ×
129  42  STO
130  30  30
131  03  3
132  00  0
133  32  X:T
134  42  STO
135  31  31
136  01  1
137  01  1
138  71  SBR
139  75  -
140  35  1/X
141  65  ×
142  02  2
143  94  +/-
144  95  =
145  42  STO
146  11  11
147  01  1
148  01  1
149  32  X:T
```

```
150  94  +/-
151  42  STO
152  12   12
153  09   9
154  71  SBR
155  65   x
156  42  STO
157  32   32
158  03   3
159  02   2
160  32  X:T
161  42  STO
162  33   33
163  05   5
164  71  SBR
165  85   +
166  94  +/-
167  42  STO
168  14   14
169  01   1
170  01   1
171  32  X:T
172  42  STO
173  15   15
174  01   1
175  06   6
176  71  SBR
177  65   x
178  42  STO
179  16   16
180  01   1
181  01   1
182  32  X:T
183  42  STO
184  17   17
185  01   1
186  08   8
187  71  SBR
188  65   x
189  42  STO
190  18   18
191  01   1
192  01   1
193  32  X:T
194  42  STO
195  19   19
196  07   7
197  71  SBR
198  65   x
199  42  STO
200  32   32
201  03   3
202  02   2
203  32  X:T
204  42  STO
205  33   33
206  05   5
207  71  SBR
208  85   +
209  94  +/-
210  42  STO
211  20   20
212  32  X:T
213  42  STO
214  21   21
215  92  RTN
216  76  LBL
217  22  INV
218  02   2
219  02   2
220  42  STO
221  01   01
222  01   1
223  04   4
224  42  STO
225  02   02
226  08   8
227  42  STO
228  03   03
229  73  RC*
230  01   01
231  63  EX*
232  02   02
233  72  ST*
234  01   01
235  69  OP
236  21   21
237  69  OP
238  22   22
239  97  DSZ
240  03   03
241  02   02
242  29   29
243  92  RTN
244  76  LBL
245  23  LNX
246  04   4
247  42  STO
248  07   07
249  01   1
250  04   4
251  42  STO
252  05   05
253  02   2
254  02   2
255  42  STO
256  06   06
257  43  RCL
258  05   05
259  32  X:T
260  43  RCL
261  06   06
262  71  SBR
263  85   +
264  72  ST*
265  05   05
266  69  OP
267  25   25
268  32  X:T
269  72  ST*
270  05   05
271  69  OP
272  25   25
273  69  OP
274  26   26
275  69  OP
276  26   26
277  97  DSZ
278  07   07
279  02   02
280  57   57
281  92  RTN
282  76  LBL
283  24  CE
284  04   4
285  42  STO
286  03   03
287  01   1
288  04   4
289  42  STO
290  01   01
291  73  RC*
292  01   01
293  32  X:T
294  69  OP
295  21   21
296  73  RC*
297  01   01
298  37  P/R
299  22  INV
```

```
300  37 P/R
301  72 ST*
302  01  01
303  69 OP
304  31  31
305  32 X:T
306  72 ST*
307  01  01
308  69 OP
309  21  21
310  69 OP
311  21  21
312  97 DSZ
313  03  03
314  02  02
315  91  91
316  92 RTN
317  76 LBL
318  11  A
319  08  8
320  42 STO
321  03  03
322  01  1
323  04  4
324  42 STO
325  01  01
326  01  1
327  94 +/-
328  64 PD*
329  01  01
330  69 OP
331  21  21
332  69 OP
333  21  21
334  97 DSZ
335  03  03
336  03  03
337  28  28
338  71 SBR
339  32 X:T
340  71 SBR
341  22 INV
342  71 SBR
343  32 X:T
344  71 SBR
345  23 LNX
346  71 SBR
347  32 X:T
348  04  4
349  42 STO
350  03  03
351  01  1
352  04  4
353  42 STO
354  01  01
355  01  1
356  94 +/-
357  64 PD*
358  01  01
359  69 OP
360  21  21
361  69 OP
362  21  21
363  97 DSZ
364  03  03
365  03  03
366  57  57
367  61 GTO
368  24 CE
369  76 LBL
370  12  B
371  71 SBR
372  32 X:T
373  71 SBR
374  22 INV
375  71 SBR
376  32 X:T
377  71 SBR
378  23 LNX
379  71 SBR
380  32 X:T
381  61 GTO
382  24 CE
383  00  0
384  00  0
385  00  0
386  00  0
```

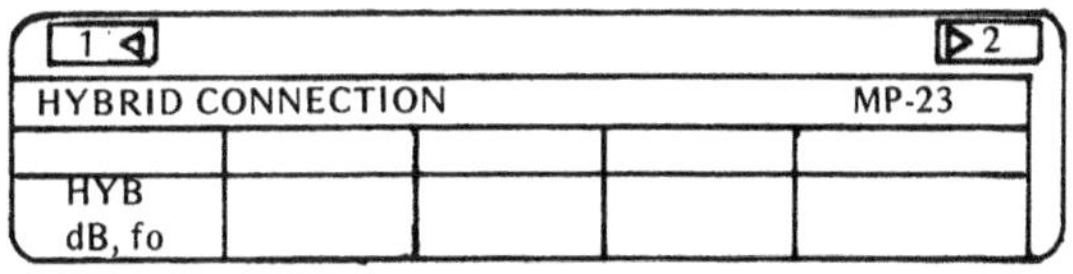


```
000  76 LBL
001  85  +
002  86 STF
003  00  00
004  76 LBL
005  75  -
006  42 STO
007  01  01
008  73 RC*
009  01  01
010  87 IFF
011  00  00
012  00  00
013  15  15
014  94 +/-
015  32 X⇅T
016  42 STO
017  02  02
018  69 OP
019  21  21
020  73 RC*
021  01  01
022  37 P/R
023  42 STO
024  04  04
025  73 RC*
026  02  02
027  32 X⇅T
028  42 STO
029  03  03
030  69 OP
031  22  22
032  73 RC*
033  02  02
034  37 P/R
035  44 SUM
036  04  04
037  32 X⇅T
038  44 SUM
039  03  03
040  43 RCL
041  03  03
042  32 X⇅T
043  43 RCL
044  04  04
045  22 INV
046  37 P/R
047  32 X⇅T
048  22 INV
049  86 STF
050  00  00
051  92 RTN
052  76 LBL
053  65  ×
054  42 STO
055  01  01
056  73 RC*
057  01  01
058  42 STO
059  03  03
060  69 OP
061  21  21
062  73 RC*
063  01  01
064  42 STO
065  04  04
066  32 X⇅T
067  42 STO
068  01  01
069  73 RC*
070  01  01
071  49 PRD
072  03  03
073  69 OP
074  21  21
075  73 RC*
076  01  01
077  44 SUM
078  04  04
079  43 RCL
080  04  04
081  32 X⇅T
082  43 RCL
083  03  03
084  92 RTN
085  76 LBL
086  11  A
087  42 STO
088  05  05
089  32 X⇅T
090  94 +/-
091  55  ÷
092  02  2
093  00  0
094  95  =
095  22 INV
096  28 LOG
097  42 STO
098  06  06
099  09  9
100  00  0
101  65  ×
102  43 RCL
103  00  00
104  55  ÷
105  43 RCL
106  05  05
107  95  =
108  42 STO
109  05  05
110  01  1
111  75  -
112  43 RCL
113  06  06
114  33 X²
115  95  =
116  55  ÷
117  53  (
118  01  1
119  75  -
120  53  (
121  43 RCL
122  06  06
123  33 X²
124  65  ×
125  43 RCL
126  05  05
127  39 COS
128  33 X²
129  95  =
130  34 √X
131  42 STO
132  07  07
133  01  1
134  75  -
135  43 RCL
136  06  06
137  33 X²
138  95  =
139  34 √X
140  65  ×
141  43 RCL
142  05  05
143  39 COS
144  95  =
145  32 X⇅T
146  43 RCL
147  05  05
148  38 SIN
149  22 INV
```

```
150  37 P/R
151  42 STO
152  08  08
153  01  1
154  75  -
155  43 RCL
156  07  07
157  33 X²
158  95  =
159  34 Γx
160  42 STO
161  09  09
162  43 RCL
163  07  07
164  33 X²
165  49 PRD
166  22  22
167  49 PRD
168  20  20
169  43 RCL
170  09  09
171  33 X²
172  49 PRD
173  14  14
174  49 PRD
175  28  28
176  43 RCL
177  07  07
178  65  ×
179  43 RCL
180  09  09
181  95  =
182  49 PRD
183  16  16
184  49 PRD
185  24  24
186  49 PRD
187  18  18
188  49 PRD
189  26  26
190  02  2
191  02  2
192  32 X↕T
193  01  1
194  04  4
195  71 SBR
196  75  -
197  42 STO
198  14  14
199  01  1
200  06  6
201  32 X↕T
202  42 STO
203  15  15
204  02  2
205  04  4
206  71 SBR
207  85  +
208  42 STO
209  16  16
210  01  1
211  08  8
212  32 X↕T
213  42 STO
214  17  17
215  02  2
216  06  6
217  71 SBR
218  85  +
219  42 STO
220  18  18
221  02  2
222  00  0
223  32 X↕T
224  42 STO
225  19  19
226  02  2
227  08  8
228  71 SBR
229  75  -
230  42 STO
231  20  20
232  32 X↕T
233  42 STO
234  21  21
235  01  1
236  42 STO
237  10  10
238  43 RCL
239  08  08
240  94 +/-
241  65  ×
242  02  2
243  95  =
244  42 STO
245  11  11
246  01  1
247  00  0
248  32 X↕T
249  01  1
250  04  4
251  71 SBR
252  65  ×
253  42 STO
254  14  14
255  01  1
256  00  0
257  32 X↕T
258  42 STO
259  15  15
260  02  2
261  00  0
262  71 SBR
263  65  ×
264  42 STO
265  20  20
266  32 X↕T
267  42 STO
268  21  21
269  09  9
270  00  0
271  44 SUM
272  11  11
273  01  1
274  00  0
275  32 X↕T
276  01  1
277  06  6
278  71 SBR
279  65  ×
280  42 STO
281  16  16
282  01  1
283  00  0
284  32 X↕T
285  42 STO
286  17  17
287  01  1
288  08  8
289  71 SBR
290  65  ×
291  42 STO
292  18  18
293  32 X↕T
294  42 STO
295  19  19
296  92 RTN
297  00  0
298  00  0
299  00  0
```

1 ◁				▷ 2
SERIES·PARALLEL & PARALLEL·SERIES				MP-24
SER·PAR	PAR·SER			

```
000  76 LBL
001  85  +
002  86 STF
003  00  00
004  76 LBL
005  75  -
006  42 STO
007  01  01
008  73 RC*
009  01  01
010  87 IFF
011  00  00
012  00  00
013  15  15
014  94 +/-
015  32 X:T
016  42 STO
017  02  02
018  69 OP
019  21  21
020  73 RC*
021  01  01
022  37 P/R
023  42 STO
024  04  04
025  73 RC*
026  02  02
027  32 X:T
028  42 STO
029  03  03
030  69 OP
031  22  22
032  73 RC*
033  02  02
034  37 P/R
035  44 SUM
036  04  04
037  32 X:T
038  44 SUM
039  03  03
040  43 RCL
041  03  03
042  32 X:T
043  43 RCL
044  04  04
045  22 INV
046  37 P/R
047  32 X:T
048  22 INV
049  86 STF
050  00  00
051  92 RTN
052  76 LBL
053  65  ×
054  42 STO
055  01  01
056  73 RC*
057  01  01
058  42 STO
059  03  03
060  69 OP
061  21  21
062  73 RC*
063  01  01
064  42 STO
065  04  04
066  32 X:T
067  42 STO
068  01  01
069  73 RC*
070  01  01
071  49 PRD
072  03  03
073  69 OP
074  21  21
075  73 RC*
076  01  01
077  44 SUM
078  04  04
079  43 RCL
080  04  04
081  32 X:T
082  43 RCL
083  03  03
084  92 RTN
085  76 LBL
086  32 X:T
087  01  1
088  42 STO
089  05  05
090  00  0
091  42 STO
092  06  06
093  05  5
094  32 X:T
095  01  1
096  04  4
097  71 SBR
098  85  +
099  42 STO
100  07  07
101  05  5
102  32 X:T
103  42 STO
104  08  08
105  02  2
106  00  0
107  71 SBR
108  85  +
109  42 STO
110  09  09
111  01  1
112  06  6
113  32 X:T
114  42 STO
115  10  10
116  01  1
117  08  8
118  71 SBR
119  65  ×
120  42 STO
121  11  11
122  07  7
123  32 X:T
124  42 STO
125  12  12
126  09  9
127  71 SBR
128  65  ×
129  42 STO
130  30  30
131  03  3
132  00  0
133  32 X:T
134  42 STO
135  31  31
136  01  1
137  01  1
138  71 SBR
139  75  -
140  35 1/X
141  65  ×
142  02  2
143  94 +/-
144  95  =
145  42 STO
146  11  11
147  01  1
148  01  1
149  32 X:T
```

```
150  94  +/-
151  42  STO
152  12   12
153  09   9
154  71  SBR
155  65   ×
156  42  STO
157  32   32
158  03   3
159  02   2
160  32  X:T
161  42  STO
162  33   33
163  05   5
164  71  SBR
165  85   +
166  94  +/-
167  42  STO
168  14   14
169  01   1
170  01   1
171  32  X:T
172  42  STO
173  15   15
174  01   1
175  06   6
176  71  SBR
177  65   ×
178  42  STO
179  16   16
180  01   1
181  01   1
182  32  X:T
183  42  STO
184  17   17
185  01   1
186  08   8
187  71  SBR
188  65   ×
189  42  STO
190  18   18
191  01   1
192  01   1
193  32  X:T
194  42  STO
195  19   19
196  07   7
197  71  SBR
198  65   ×
199  42  STO
200  32   32
201  03   3
202  02   2
203  32  X:T
204  42  STO
205  33   33
206  05   5
207  71  SBR
208  85   +
209  94  +/-
210  42  STO
211  20   20
212  32  X:T
213  42  STO
214  21   21
215  92  RTN
216  76  LBL
217  22  INV
218  02   2
219  02   2
220  42  STO
221  01   01
222  01   1
223  04   4
224  42  STO
225  02   02
226  08   8
227  42  STO
228  03   03
229  73  RC*
230  01   01
231  63  EX*
232  02   02
233  72  ST*
234  01   01
235  69  OP
236  21   21
237  69  OP
238  22   22
239  97  DSZ
240  03   03
241  02   02
242  29   29
243  92  RTN
244  76  LBL
245  23  LNX
246  04   4
247  42  STO
248  07   07
249  01   1
250  04   4
251  42  STO
252  05   05
253  02   2
254  02   2
255  42  STO
256  06   06
257  43  RCL
258  05   05
259  32  X:T
260  43  RCL
261  06   06
262  71  SBR
263  85   +
264  72  ST*
265  05   05
266  69  OP
267  25   25
268  32  X:T
269  72  ST*
270  05   05
271  69  OP
272  25   25
273  69  OP
274  26   26
275  69  OP
276  26   26
277  97  DSZ
278  07   07
279  02   02
280  57   57
281  92  RTN
282  76  LBL
283  24  CE
284  04   4
285  42  STO
286  03   03
287  01   1
288  04   4
289  42  STO
290  01   01
291  73  RC*
292  01   01
293  32  X:T
294  69  OP
295  21   21
296  73  RC*
297  01   01
298  37  P/R
299  22  INV
```

```
37 P/R
72 ST*
01 01
69 OP
31 31
32 X:T
72 ST*
01 01
69 OP
21 21
69 OP
21 21
97 DSZ
03 03
02 02
91 91
92 RTN
76 LBL
33 X²
01 1
04 4
32 X:T
02 2
00 0
71 SBR
65 ×
42 STO
05 05
01 1
06 6
32 X:T
42 STO
06 06
01 1
08 8
71 SBR
65 ×
42 STO
07 07
05 5
32 X:T
42 STO
08 08
07 7
71 SBR
75 -
35 1/X
42 STO
05 05
32 X:T
94 +/-
42 STO
06 06
43 RCL
14 14
48 EXC
20 20
42 STO
14 14
43 RCL
15 15
48 EXC
21 21
42 STO
15 15
01 1
94 +/-
49 PRD
16 16
49 PRD
18 18
01 1
04 4
32 X:T
05 5
71 SBR
65 ×
42 STO
14 14
01 1
06 6
32 X:T
42 STO
15 15
05 5
71 SBR
65 ×
42 STO
16 16
01 1
08 8
32 X:T
42 STO
17 17
05 5
71 SBR
65 ×
42 STO
18 18
02 2
00 0
32 X:T
42 STO
19 19
05 5
71 SBR
65 ×
42 STO
20 20
32 X:T
42 STO
21 21
92 RTN
76 LBL
11 A
86 STF
01 01
76 LBL
12 B
01 1
94 +/-
49 PRD
14 14
49 PRD
16 16
49 PRD
22 22
49 PRD
24 24
71 SBR
32 X:T
87 IFF
01 01
04 04
37 37
71 SBR
33 X²
71 SBR
22 INV
71 SBR
32 X:T
87 IFF
01 01
04 04
47 47
71 SBR
33 X²
71 SBR
23 LNX
87 IFF
```

```
450  01  01
451  04  04
452  55  55
453  71 SBR
454  33 X²
455  71 SBR
456  32 X:T
457  01  1
458  94 +/-
459  49 PRD
460  14  14
461  49 PRD
462  16  16
463  22 INV
464  86 STF
465  01  01
466  61 GTO
467  24 CE
468  00  0
469  00  0
470  00  0
471  00  0
472  00  0
```

1 ◁				▷ 2
PARAMETER CONVERSIONS				MP-25
Y ⇒ S	Z ⇒ S	H ⇒ S	G ⇒ S	G ⇌ H
S ⇒ Y	S ⇒ Z	S ⇒ H	S ⇒ G	Z ⇌ Y

```
000  76 LBL
001  85  +
002  86 STF
003  00  00
004  76 LBL
005  75  -
006  42 STO
007  01  01
008  73 RC*
009  01  01
010  87 IFF
011  00  00
012  00  00
013  15  15
014  94 +/-
015  32 X:T
016  42 STO
017  02  02
018  69 OP
019  21  21
020  73 RC*
021  01  01
022  37 P/R
023  42 STO
024  04  04
025  73 RC*
026  02  02
027  32 X:T
028  42 STO
029  03  03
030  69 OP
031  22  22
032  73 RC*
033  02  02
034  37 P/R
035  44 SUM
036  04  04
037  32 X:T
038  44 SUM
039  03  03
040  43 RCL
041  03  03
042  32 X:T
043  43 RCL
044  04  04
045  22 INV
046  37 P/R
047  32 X:T
048  22 INV
049  86 STF
050  00  00
051  92 RTN
052  76 LBL
053  65  ×
054  42 STO
055  01  01
056  73 RC*
057  01  01
058  42 STO
059  03  03
060  69 OP
061  21  21
062  73 RC*
063  01  01
064  42 STO
065  04  04
066  32 X:T
067  42 STO
068  01  01
069  73 RC*
070  01  01
071  49 PRD
072  03  03
073  69 OP
074  21  21
075  73 RC*
076  01  01
077  44 SUM
078  04  04
079  43 RCL
080  04  04
081  32 X:T
082  43 RCL
083  03  03
084  92 RTN
085  76 LBL
086  32 X:T
087  01  1
088  42 STO
089  05  05
090  00  0
091  42 STO
092  06  06
093  05  5
094  32 X:T
095  01  1
096  04  4
097  71 SBR
098  85  +
099  42 STO
100  07  07
101  05  5
102  32 X:T
103  42 STO
104  08  08
105  02  2
106  00  0
107  71 SBR
108  85  +
109  42 STO
110  09  09
111  01  1
112  06  6
113  32 X:T
114  42 STO
115  10  10
116  01  1
117  08  8
118  71 SBR
119  65  ×
120  42 STO
121  11  11
122  07  7
123  32 X:T
124  42 STO
125  12  12
126  09  9
127  71 SBR
128  65  ×
129  42 STO
130  30  30
131  03  3
132  00  0
133  32 X:T
134  42 STO
135  31  31
136  01  1
137  01  1
138  71 SBR
139  75  -
140  35 1/X
141  65  ×
142  02  2
143  94 +/-
144  95  =
145  42 STO
146  11  11
147  01  1
148  01  1
149  32 X:T
```

```
150  94  +/-
151  42  STO
152  12   12
153  09   9
154  71  SBR
155  65   ×
156  42  STO
157  32   32
158  03   3
159  02   2
160  32  X:T
161  42  STO
162  33   33
163  05   5
164  71  SBR
165  85   +
166  94  +/-
167  42  STO
168  14   14
169  01   1
170  01   1
171  32  X:T
172  42  STO
173  15   15
174  01   1
175  06   6
176  71  SBR
177  65   ×
178  42  STO
179  16   16
180  01   1
181  01   1
182  32  X:T
183  42  STO
184  17   17
185  01   1
186  08   8
187  71  SBR
188  65   ×
189  42  STO
190  18   18
191  01   1
192  01   1
193  32  X:T
194  42  STO
195  19   19
196  07   7
197  71  SBR
198  65   ×
199  42  STO
200  32   32
201  03   3
202  02   2
203  32  X:T
204  42  STO
205  33   33
206  05   5
207  71  SBR
208  85   +
209  94  +/-
210  42  STO
211  20   20
212  32  X:T
213  42  STO
214  21   21
215  92  RTN
216  76  LBL
217  24  CE
218  04   4
219  42  STO
220  03   03
221  01   1
222  04   4
223  42  STO
224  01   01
225  73  RC*
226  01   01
227  32  X:T
228  69  OP
229  21   21
230  73  RC*
231  01   01
232  37  P/R
233  22  INV
234  37  P/R
235  72  ST*
236  01   01
237  69  OP
238  31   31
239  32  X:T
240  72  ST*
241  01   01
242  69  OP
243  21   21
244  69  OP
245  21   21
246  97  DSZ
247  03   03
248  02   02
249  25   25
250  92  RTN
251  76  LBL
252  33  X²
253  01   1
254  04   4
255  32  X:T
256  02   2
257  00   0
258  71  SBR
259  65   ×
260  42  STO
261  05   05
262  01   1
263  06   6
264  32  X:T
265  42  STO
266  06   06
267  01   1
268  08   8
269  71  SBR
270  65   ×
271  42  STO
272  07   07
273  05   5
274  32  X:T
275  42  STO
276  08   08
277  07   7
278  71  SBR
279  75   -
280  35  1/X
281  42  STO
282  05   05
283  32  X:T
284  94  +/-
285  42  STO
286  06   06
287  43  RCL
288  14   14
289  48  EXC
290  20   20
291  42  STO
292  14   14
293  43  RCL
294  15   15
295  48  EXC
296  21   21
297  42  STO
298  15   15
299  01   1
```

```
300   94  +/-
301   49  PRD
302   16   16
303   49  PRD
304   18   18
305   01   1
306   04   4
307   32  X:T
308   05   5
309   71  SBR
310   65   x
311   42  STO
312   14   14
313   01   1
314   06   6
315   32  X:T
316   42  STO
317   15   15
318   05   5
319   71  SBR
320   65   x
321   42  STO
322   16   16
323   01   1
324   08   8
325   32  X:T
326   42  STO
327   17   17
328   05   5
329   71  SBR
330   65   x
331   42  STO
332   18   18
333   02   2
334   00   0
335   32  X:T
336   42  STO
337   19   19
338   05   5
339   71  SBR
340   65   x
341   42  STO
342   20   20
343   32  X:T
344   42  STO
345   21   21
346   92  RTN
347   76  LBL
348   16  A'
349   76  LBL
350   11   A
351   71  SBR
352   32  X:T
353   61  GTO
354   24  CE
355   76  LBL
356   12   B
357   01   1
358   94  +/-
359   49  PRD
360   14   14
361   49  PRD
362   16   16
363   49  PRD
364   18   18
365   49  PRD
366   20   20
367   71  SBR
368   32  X:T
369   61  GTO
370   24  CE
371   76  LBL
372   17  B'
373   71  SBR
374   32  X:T
375   01   1
376   94  +/-
377   49  PRD
378   14   14
379   49  PRD
380   16   16
381   49  PRD
382   18   18
383   49  PRD
384   20   20
385   61  GTO
386   24  CE
387   76  LBL
388   13   C
389   01   1
390   94  +/-
391   49  PRD
392   14   14
393   49  PRD
394   16   16
395   71  SBR
396   32  X:T
397   61  GTO
398   24  CE
399   76  LBL
400   18  C'
401   71  SBR
402   32  X:T
403   01   1
404   94  +/-
405   49  PRD
406   14   14
407   49  PRD
408   16   16
409   61  GTO
410   24  CE
411   76  LBL
412   14   D
413   01   1
414   94  +/-
415   49  PRD
416   14   14
417   49  PRD
418   16   16
419   71  SBR
420   32  X:T
421   71  SBR
422   33  X²
423   61  GTO
424   24  CE
425   76  LBL
426   19  D'
427   71  SBR
428   33  X²
429   71  SBR
430   32  X:T
431   01   1
432   94  +/-
433   49  PRD
434   14   14
435   49  PRD
436   16   16
437   61  GTO
438   24  CE
439   76  LBL
440   15   E
441   76  LBL
442   10  E'
443   71  SBR
444   33  X²
445   61  GTO
446   24  CE
447   00   0
448   00   0
449   00   0
```

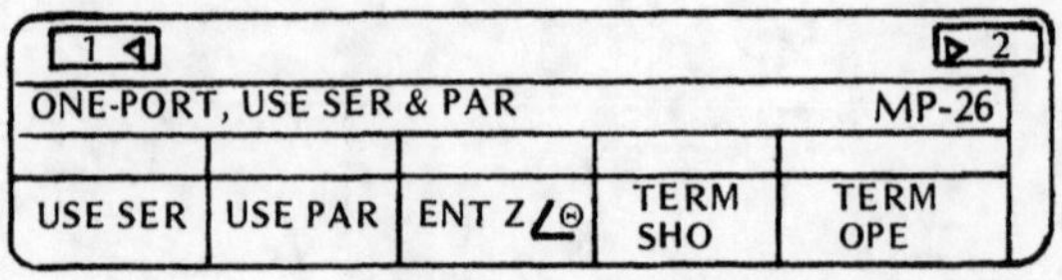


```
000  76 LBL
001  85  +
002  86 STF
003  00  00
004  76 LBL
005  75  -
006  42 STO
007  01  01
008  73 RC*
009  01  01
010  87 IFF
011  00  00
012  00  00
013  15  15
014  94 +/-
015  32 X:T
016  42 STO
017  02  02
018  69 OP
019  21  21
020  73 RC*
021  01  01
022  37 P/R
023  42 STO
024  04  04
025  73 RC*
026  02  02
027  32 X:T
028  42 STO
029  03  03
030  69 OP
031  22  22
032  73 RC*
033  02  02
034  37 P/R
035  44 SUM
036  04  04
037  32 X:T
038  44 SUM
039  03  03
040  43 RCL
041  03  03
042  32 X:T
043  43 RCL
044  04  04
045  22 INV
046  37 P/R
047  32 X:T
048  22 INV
049  86 STF
050  00  00
051  92 RTN
052  76 LBL
053  65  ×
054  42 STO
055  01  01
056  73 RC*
057  01  01
058  42 STO
059  03  03
060  69 OP
061  21  21
062  73 RC*
063  01  01
064  42 STO
065  04  04
066  32 X:T
067  42 STO
068  01  01
069  73 RC*
070  01  01
071  49 PRD
072  03  03
073  69 OP
074  21  21
075  73 RC*
076  01  01
077  44 SUM
078  04  04
079  43 RCL
080  04  04
081  32 X:T
082  43 RCL
083  03  03
084  92 RTN
085  76 LBL
086  13  C
087  42 STO
088  06  06
089  32 X:T
090  55  ÷
091  05  5
092  00  0
093  95  =
094  42 STO
095  05  05
096  01  1
097  42 STO
098  07  07
099  00  0
100  42 STO
101  08  08
102  05  5
103  32 X:T
104  07  7
105  71 SBR
106  75  -
107  35 1/X
108  65  ×
109  02  2
110  95  =
111  42 STO
112  09  09
113  09  9
114  32 X:T
115  94 +/-
116  42 STO
117  10  10
118  07  7
119  71 SBR
120  85  +
121  35 1/X
122  42 STO
123  14  14
124  32 X:T
125  94 +/-
126  42 STO
127  15  15
128  92 RTN
129  76 LBL
130  14  D
131  01  1
132  94 +/-
133  42 STO
134  09  09
135  00  0
136  42 STO
137  10  10
138  61 GTO
139  24 CE
140  76 LBL
141  15  E
142  01  1
143  42 STO
144  09  09
145  00  0
146  42 STO
147  10  10
148  76 LBL
149  24 CE
```

```
150  01  1
151  42  STO
152  05  05
153  00  0
154  42  STO
155  06  06
156  02  2
157  00  0
158  32  X:T
159  09  9
160  71  SBR
161  65  ×
162  42  STO
163  07  07
164  05  5
165  32  X:T
166  42  STO
167  08  08
168  07  7
169  71  SBR
170  75  -
171  35  1/X
172  42  STO
173  07  07
174  01  1
175  06  6
176  32  X:T
177  94  +/-
178  42  STO
179  08  08
180  07  7
181  71  SBR
182  65  ×
183  42  STO
184  07  07
185  01  1
186  08  8
187  32  X:T
188  42  STO
189  08  08
190  07  7
191  71  SBR
192  65  ×
193  42  STO
194  07  07
195  09  9
196  32  X:T
197  42  STO
198  08  08
199  07  7
200  71  SBR
201  65  ×
202  42  STO
203  07  07
204  01  1
205  04  4
206  32  X:T
207  42  STO
208  08  08
209  07  7
210  71  SBR
211  85  +
212  42  STO
213  14  14
214  32  X:T
215  42  STO
216  15  15
217  92  RTN
218  76  LBL
219  11  A
220  86  STF
221  01  01
222  76  LBL
223  12  B
224  01  1
225  42  STO
226  05  05
227  00  0
228  42  STO
229  06  06
230  05  5
231  32  X:T
232  01  1
233  04  4
234  71  SBR
235  75  -
236  35  1/X
237  42  STO
238  07  07
239  05  5
240  32  X:T
241  94  +/-
242  42  STO
243  08  08
244  01  1
245  04  4
246  71  SBR
247  85  +
248  42  STO
249  09  09
250  07  7
251  32  X:T
252  42  STO
253  10  10
254  09  9
255  71  SBR
256  65  ×
257  42  STO
258  07  07
259  32  X:T
260  42  STO
261  08  08
262  87  IFF
263  01  01
264  02  02
265  76  76
266  43  RCL
267  07  07
268  35  1/X
269  42  STO
270  07  07
271  43  RCL
272  08  08
273  94  +/-
274  42  STO
275  08  08
276  02  2
277  42  STO
278  05  05
279  05  5
280  32  X:T
281  07  7
282  71  SBR
283  85  +
284  35  1/X
285  42  STO
286  09  09
287  32  X:T
288  94  +/-
289  42  STO
290  10  10
291  87  IFF
292  01  01
293  02  02
294  99  99
295  01  1
296  94  +/-
297  49  PRD
298  07  07
299  22  INV
```

```
300  86  STF
301  01   01
302  05   5
303  32  X:T
304  09   9
305  71  SBR
306  65   x
307  42  STO
308  24   24
309  42  STO
310  26   26
311  32  X:T
312  42  STO
313  25   25
314  42  STO
315  27   27
316  07   7
317  32  X:T
318  09   9
319  71  SBR
320  65   x
321  42  STO
322  22   22
323  42  STO
324  28   28
325  32  X:T
326  42  STO
327  23   23
328  42  STO
329  29   29
330  92  RTN
331  00   0
332  00   0
333  00   0
334  00   0
335  00   0
```

1 ◀				▶ 2
SERIES R-L-C				MP-31
SE SLC	SE SRLC			USE PA
SE R	SE L	SE C	SE SRL	SE SRC

```
76 LBL
11 A
42 STO
05 05
00 0
42 STO
06 06
01 1
52 EE
05 5
00 0
42 STO
07 07
22 INV
52 EE
61 GTO
23 LNX
76 LBL
12 B
42 STO
06 06
00 0
42 STO
05 05
01 1
52 EE
05 5
00 0
42 STO
07 07
22 INV
52 EE
61 GTO
23 LNX
76 LBL
13 C
42 STO
07 07
00 0
42 STO
05 05
42 STO
06 06
61 GTO
23 LNX
76 LBL
14 D
42 STO
06 06
32 X⇄T
42 STO
05 05
01 1
52 EE
05 5
00 0
42 STO
07 07
22 INV
52 EE
61 GTO
23 LNX
76 LBL
15 E
42 STO
07 07
32 X⇄T
42 STO
05 05
00 0
42 STO
06 06
61 GTO
23 LNX
76 LBL
16 A'
42 STO
07 07
32 X⇄T
42 STO
06 06
00 0
42 STO
05 05
61 GTO
23 LNX
76 LBL
17 B'
42 STO
06 06
32 X⇄T
42 STO
05 05
91 R/S
42 STO
07 07
76 LBL
23 LNX
43 RCL
05 05
55 ÷
01 1
00 0
00 0
95 =
42 STO
08 08
02 2
65 ×
43 RCL
00 00
65 ×
89 π
95 =
42 STO
12 12
65 ×
43 RCL
06 06
75 -
53 (
01 1
00 0
00 0
00 0
55 ÷
53 (
43 RCL
12 12
65 ×
43 RCL
07 07
95 =
55 ÷
01 1
00 0
00 0
95 =
42 STO
09 09
22 INV
87 IFF
01 01
01 01
65 65
43 RCL
08 08
32 X⇄T
43 RCL
09 09
```

Step	Code	Key
150	22	INV
151	37	P/R
152	94	+/-
153	32	X:T
154	35	1/X
155	55	÷
156	04	4
157	95	=
158	32	X:T
159	37	P/R
160	42	STO
161	09	09
162	32	X:T
163	42	STO
164	08	08
165	43	RCL
166	08	08
167	85	+
168	01	1
169	95	=
170	42	STO
171	10	10
172	43	RCL
173	09	09
174	42	STO
175	11	11
176	43	RCL
177	08	08
178	32	X:T
179	43	RCL
180	09	09
181	22	INV
182	37	P/R
183	42	STO
184	09	09
185	32	X:T
186	42	STO
187	08	08
188	43	RCL
189	10	10
190	32	X:T
191	43	RCL
192	11	11
193	22	INV
194	37	P/R
195	42	STO
196	11	11
197	32	X:T
198	42	STO
199	10	10
200	87	IFF
201	01	01
202	02	02
203	50	50
204	43	RCL
205	08	08
206	55	÷
207	43	RCL
208	10	10
209	95	=
210	42	STO
211	22	22
212	42	STO
213	28	28
214	43	RCL
215	09	09
216	75	-
217	43	RCL
218	11	11
219	95	=
220	42	STO
221	23	23
222	42	STO
223	29	29
224	43	RCL
225	22	22
226	94	+/-
227	32	X:T
228	43	RCL
229	23	23
230	37	P/R
231	32	X:T
232	85	+
233	01	1
234	95	=
235	32	X:T
236	22	INV
237	37	P/R
238	42	STO
239	25	25
240	42	STO
241	27	27
242	32	X:T
243	42	STO
244	24	24
245	42	STO
246	26	26
247	09	9
248	09	9
249	92	RTN
250	22	INV
251	86	STF
252	01	01
253	43	RCL
254	08	08
255	94	+/-
256	55	÷
257	43	RCL
258	10	10
259	95	=
260	32	X:T
261	43	RCL
262	09	09
263	75	-
264	43	RCL
265	11	11
266	95	=
267	37	P/R
268	22	INV
269	37	P/R
270	42	STO
271	23	23
272	42	STO
273	29	29
274	32	X:T
275	42	STO
276	22	22
277	42	STO
278	28	28
279	43	RCL
280	22	22
281	32	X:T
282	43	RCL
283	23	23
284	37	P/R
285	32	X:T
286	85	+
287	01	1
288	95	=
289	32	X:T
290	22	INV
291	37	P/R
292	42	STO
293	25	25
294	42	STO
295	27	27
296	32	X:T
297	42	STO
298	24	24
299	42	STO

```
300  26  26
301  09   9
302  09   9
303  92 RTN
304  76 LBL
305  10 E'
306  86 STF
307  01  01
308  92 RTN
309  00   0
310  00   0
311  00   0
312  00   0
```

1 ◁				▷ 2
PARALLEL R-L-C				MP-32
PA PLC	PA PRLC			USE SE
PA R	PA L	PA C	PA PRL	PA PRC

```
76 LBL
11  A
42 STO
05  05
00  0
42 STO
07  07
01  1
52 EE
05  5
00  0
42 STO
06  06
61 GTO
23 LNX
76 LBL
12  B
42 STO
06  06
00  0
42 STO
07  07
01  1
52 EE
05  5
00  0
42 STO
05  05
61 GTO
23 LNX
76 LBL
13  C
42 STO
07  07
01  1
52 EE
05  5
00  0
42 STO
05  05
42 STO
06  06
61 GTO
23 LNX
76 LBL
14  D
42 STO
06  06
32 X:T
42 STO
05  05
00  0
42 STO
07  07
61 GTO
23 LNX
76 LBL
15  E
42 STO
07  07
32 X:T
42 STO
05  05
01  1
52 EE
05  5
00  0
42 STO
06  06
61 GTO
23 LNX
76 LBL
16 A'
42 STO
07  07
32 X:T
42 STO
06  06
01  1
52 EE
05  5
00  0
42 STO
05  05
61 GTO
23 LNX
76 LBL
17 B'
42 STO
06  06
32 X:T
42 STO
05  05
91 R/S
42 STO
07  07
76 LBL
23 LNX
22 INV
52 EE
02  2
05  5
55  ÷
43 RCL
05  05
95  =
42 STO
08  08
02  2
65  ×
43 RCL
00  00
65  ×
89  π
95  =
42 STO
12  12
65  ×
43 RCL
07  07
55  ÷
04  4
00  0
95  =
42 STO
09  09
02  2
05  5
55  ÷
53  (
43 RCL
12  12
65  ×
43 RCL
06  06
95  =
22 INV
44 SUM
09  09
22 INV
87 IFF
01  01
24 CE
43 RCL
08  08
32 X:T
43 RCL
09  09
22 INV
37 P/R
```

Step	Code	Key
150	94	+/-
151	32	X:T
152	35	1/X
153	55	÷
154	04	4
155	95	=
156	32	X:T
157	37	P/R
158	42	STO
159	09	09
160	32	X:T
161	42	STO
162	08	08
163	76	LBL
164	24	CE
165	43	RCL
166	08	08
167	85	+
168	01	1
169	95	=
170	42	STO
171	10	10
172	43	RCL
173	09	09
174	42	STO
175	11	11
176	43	RCL
177	08	08
178	32	X:T
179	43	RCL
180	09	09
181	22	INV
182	37	P/R
183	42	STO
184	09	09
185	32	X:T
186	42	STO
187	08	08
188	43	RCL
189	10	10
190	32	X:T
191	43	RCL
192	11	11
193	22	INV
194	37	P/R
195	42	STO
196	11	11
197	32	X:T
198	42	STO
199	10	10
200	43	RCL
201	08	08
202	94	+/-
203	55	÷
204	43	RCL
205	10	10
206	95	=
207	42	STO
208	08	08
209	43	RCL
210	09	09
211	75	-
212	43	RCL
213	11	11
214	95	=
215	42	STO
216	09	09
217	87	IFF
218	01	01
219	22	INV
220	43	RCL
221	08	08
222	32	X:T
223	43	RCL
224	09	09
225	37	P/R
226	22	INV
227	37	P/R
228	42	STO
229	23	23
230	42	STO
231	29	29
232	32	X:T
233	42	STO
234	22	22
235	42	STO
236	28	28
237	43	RCL
238	08	08
239	32	X:T
240	43	RCL
241	09	09
242	37	P/R
243	32	X:T
244	85	+
245	01	1
246	95	=
247	32	X:T
248	22	INV
249	37	P/R
250	42	STO
251	25	25
252	42	STO
253	27	27
254	32	X:T
255	42	STO
256	24	24
257	42	STO
258	26	26
259	09	9
260	09	9
261	92	RTN
262	76	LBL
263	22	INV
264	22	INV
265	86	STF
266	01	01
267	43	RCL
268	08	08
269	94	+/-
270	32	X:T
271	43	RCL
272	09	09
273	37	P/R
274	22	INV
275	37	P/R
276	42	STO
277	23	23
278	42	STO
279	29	29
280	32	X:T
281	42	STO
282	22	22
283	42	STO
284	28	28
285	43	RCL
286	22	22
287	94	+/-
288	32	X:T
289	43	RCL
290	23	23
291	37	P/R
292	32	X:T
293	85	+
294	01	1
295	95	=
296	32	X:T
297	22	INV
298	37	P/R
299	42	STO

```
300  25  25
301  42  STO
302  27  27
303  32  X:T
304  42  STO
305  24  24
306  42  STO
307  26  26
308  09  9
309  09  9
310  92  RTN
311  76  LBL
312  10  E'
313  86  STF
314  01  01
315  92  RTN
316  00  0
317  00  0
318  00  0
319  00  0
```

1 ◁				▷ 2
TRANSMISSION LINES				MP-33
				USE SE
TRL	PA OTRL	PA STRL		

Step	Code	Key
000	76	LBL
001	11	A
002	65	×
003	43	RCL
004	00	00
005	65	×
006	03	3
007	00	0
008	93	.
009	04	4
010	08	8
011	95	=
012	94	+/-
013	42	STO
014	05	05
015	32	X:T
016	42	STO
017	06	06
018	53	(
019	43	RCL
020	06	06
021	75	-
022	05	5
023	00	0
024	54	)
025	55	÷
026	53	(
027	43	RCL
028	06	06
029	85	+
030	05	5
031	00	0
032	54	)
033	95	=
034	42	STO
035	06	06
036	02	2
037	49	PRD
038	05	05
039	43	RCL
040	05	05
041	39	COS
042	42	STO
043	07	07
044	43	RCL
045	05	05
046	38	SIN
047	42	STO
048	08	08
049	01	1
050	75	-
051	53	(
052	43	RCL
053	06	06
054	33	X²
055	65	×
056	43	RCL
057	07	07
058	95	=
059	42	STO
060	12	12
061	43	RCL
062	06	06
063	33	X²
064	65	×
065	43	RCL
066	08	08
067	95	=
068	94	+/-
069	32	X:T
070	43	RCL
071	12	12
072	32	X:T
073	22	INV
074	37	P/R
075	94	+/-
076	42	STO
077	13	13
078	32	X:T
079	35	1/X
080	42	STO
081	12	12
082	01	1
083	75	-
084	43	RCL
085	07	07
086	95	=
087	42	STO
088	07	07
089	43	RCL
090	06	06
091	49	PRD
092	07	07
093	94	+/-
094	49	PRD
095	08	08
096	43	RCL
097	07	07
098	32	X:T
099	43	RCL
100	08	08
101	22	INV
102	37	P/R
103	85	+
104	43	RCL
105	13	13
106	95	=
107	42	STO
108	23	23
109	42	STO
110	29	29
111	32	X:T
112	65	×
113	43	RCL
114	12	12
115	95	=
116	42	STO
117	22	22
118	42	STO
119	28	28
120	02	2
121	22	INV
122	49	PRD
123	05	05
124	43	RCL
125	05	05
126	39	COS
127	42	STO
128	07	07
129	43	RCL
130	05	05
131	38	SIN
132	42	STO
133	08	08
134	01	1
135	75	-
136	43	RCL
137	06	06
138	33	X²
139	95	=
140	49	PRD
141	07	07
142	49	PRD
143	08	08
144	43	RCL
145	07	07
146	32	X:T
147	43	RCL
148	08	08
149	22	INV

```
37 P/R
85  +
43 RCL
13  13
95  =
42 STO
25  25
42 STO
27  27
32 X⇌T
65  ×
43 RCL
12  12
95  =
42 STO
24  24
42 STO
26  26
09  9
09  9
92 RTN
76 LBL
10 E'
86 STF
01  01
92 RTN
76 LBL
12  B
65  ×
03  3
00  0
93  .
04  4
08  8
65  ×
43 RCL
00  00
95  =
30 TAN
42 STO
05  05
32 X⇌T
55  ÷
02  2
05  5
95  =
22 INV
49 PRD
05  05
09  9
00  0
42 STO
06  06
87 IFF
01  01
24 CE
61 GTO
23 LNX
76 LBL
13  C
65  ×
03  3
00  0
93  .
04  4
08  8
65  ×
43 RCL
00  00
95  =
30 TAN
35 1/X
42 STO
05  05
32 X⇌T
55  ÷
02  2
05  5
95  =
22 INV
49 PRD
05  05
09  9
00  0
94 +/-
42 STO
06  06
87 IFF
01  01
24 CE
76 LBL
23 LNX
43 RCL
05  05
32 X⇌T
43 RCL
06  06
37 P/R
32 X⇌T
85  +
01  1
95  =
32 X⇌T
22 INV
37 P/R
42 STO
08  08
32 X⇌T
42 STO
07  07
43 RCL
05  05
94 +/-
55  ÷
43 RCL
07  07
95  =
32 X⇌T
43 RCL
06  06
75  -
43 RCL
08  08
95  =
37 P/R
22 INV
37 P/R
42 STO
23  23
42 STO
29  29
32 X⇌T
42 STO
22  22
42 STO
28  28
32 X⇌T
37 P/R
32 X⇌T
85  +
01  1
95  =
32 X⇌T
22 INV
37 P/R
42 STO
25  25
42 STO
27  27
32 X⇌T
```

```
42  STO
24   24
42  STO
26   26
09   9
09   9
92  RTN
76  LBL
24  CE
22  INV
86  STF
01   01
43  RCL
05   05
65   ×
04   4
95   =
35  1/X
42  STO
05   05
01   1
94  +/-
49  PRD
06   06
43  RCL
05   05
32  X⇌T
43  RCL
06   06
37  P/R
32  X⇌T
85   +
01   1
95   =
32  X⇌T
22  INV
37  P/R
42  STO
08   08
32  X⇌T
42  STO
07   07
43  RCL
05   05
55   ÷
43  RCL
07   07
95   =
42  STO
22   22
42  STO
28   28
43  RCL
06   06
75   -
43  RCL
08   08
95   =
42  STO
23   23
42  STO
29   29
43  RCL
22   22
94  +/-
32  X⇌T
43  RCL
23   23
37  P/R
32  X⇌T
85   +
01   1
95   =
32  X⇌T
22  INV
37  P/R
42  STO
25   25
42  STO
27   27
32  X⇌T
42  STO
24   24
42  STO
26   26
09   9
09   9
92  RTN
00   0
00   0
00   0
00   0
```

1 ◁				▷ 2
SPECIAL TWO-PORTS				MP-41
EQU	CDVS a, Θ	VDVS μ, Θ	CDCS β, Θ	VDCS gm, Θ
TRF Z_1, Z_2	TRF N_1, N_2	SHO	OPE	IMG

```
000  76  LBL
001  11   A
002  42  STO
003  01   01
004  32  X:T
005  55   ÷
006  43  RCL
007  01   01
008  95   =
009  34  √X
010  42  STO
011  01   01
012  61  GTO
013  22  INV
014  76  LBL
015  12   B
016  42  STO
017  01   01
018  32  X:T
019  55   ÷
020  43  RCL
021  01   01
022  95   =
023  42  STO
024  01   01
025  76  LBL
026  22  INV
027  43  RCL
028  01   01
029  33  X²
030  75   -
031  01   1
032  95   =
033  55   ÷
034  53   (
035  43  RCL
036  01   01
037  33  X²
038  85   +
039  01   1
040  95   =
041  32  X:T
042  00   0
043  37  P/R
044  22  INV
045  37  P/R
046  42  STO
047  23   23
048  32  X:T
049  42  STO
050  22   22
051  94  +/-
052  32  X:T
053  37  P/R
054  22  INV
055  37  P/R
056  42  STO
057  29   29
058  32  X:T
059  42  STO
060  28   28
061  02   2
062  65   ×
063  43  RCL
064  01   01
065  95   =
066  55   ÷
067  53   (
068  43  RCL
069  01   01
070  33  X²
071  85   +
072  01   1
073  95   =
074  32  X:T
075  00   0
076  37  P/R
077  22  INV
078  37  P/R
079  42  STO
080  25   25
081  42  STO
082  27   27
083  32  X:T
084  42  STO
085  24   24
086  42  STO
087  26   26
088  09   9
089  09   9
090  92  RTN
091  76  LBL
092  13   C
093  01   1
094  42  STO
095  22   22
096  42  STO
097  28   28
098  01   1
099  08   8
100  00   0
101  42  STO
102  23   23
103  42  STO
104  29   29
105  00   0
106  42  STO
107  24   24
108  42  STO
109  25   25
110  42  STO
111  26   26
112  42  STO
113  27   27
114  09   9
115  09   9
116  92  RTN
117  76  LBL
118  14   D
119  01   1
120  42  STO
121  22   22
122  42  STO
123  28   28
124  00   0
125  42  STO
126  23   23
127  42  STO
128  24   24
129  42  STO
130  25   25
131  42  STO
132  26   26
133  42  STO
134  27   27
135  42  STO
136  29   29
137  09   9
138  09   9
139  92  RTN
140  76  LBL
141  15   E
142  43  RCL
143  14   14
144  42  STO
145  28   28
146  43  RCL
147  15   15
148  42  STO
149  29   29
```

```
43 RCL
16  16
42 STO
26  26
43 RCL
17  17
42 STO
27  27
43 RCL
18  18
42 STO
24  24
43 RCL
19  19
42 STO
25  25
43 RCL
20  20
42 STO
22  22
43 RCL
21  21
42 STO
23  23
09  9
09  9
92 RTN
76 LBL
16 A'
43 RCL
14  14
42 STO
22  22
43 RCL
15  15
42 STO
23  23
43 RCL
16  16
42 STO
24  24
43 RCL
17  17
42 STO
25  25
43 RCL
18  18
42 STO
26  26
43 RCL
19  19
42 STO
27  27
43 RCL
20  20
42 STO
28  28
43 RCL
21  21
42 STO
29  29
92 RTN
76 LBL
17 B'
42 STO
27  27
32 X:T
65  ×
02  2
55  ÷
05  5
00  0
95  =
42 STO
26  26
01  1
42 STO
22  22
42 STO
28  28
01  1
08  8
00  0
42 STO
23  23
42 STO
29  29
00  0
42 STO
24  24
42 STO
25  25
09  9
09  9
92 RTN
76 LBL
18 C'
42 STO
27  27
32 X:T
65  ×
02  2
95  =
42 STO
26  26
01  1
42 STO
22  22
42 STO
28  28
00  0
42 STO
23  23
42 STO
24  24
42 STO
25  25
01  1
08  8
00  0
42 STO
29  29
09  9
09  9
92 RTN
76 LBL
19 D'
32 X:T
65  ×
02  2
94 +/-
95  =
32 X:T
37 P/R
22 INV
37 P/R
42 STO
27  27
32 X:T
42 STO
26  26
01  1
42 STO
22  22
42 STO
28  28
01  1
08  8
00  0
42 STO
```

```
300  23  23
301  00   0
302  42 STO
303  24  24
304  42 STO
305  25  25
306  42 STO
307  29  29
308  09   9
309  09   9
310  92 RTN
311  76 LBL
312  10 E'
313  32 X⇌T
314  65   ×
315  02   2
316  65   ×
317  05   5
318  00   0
319  94 +/-
320  95   =
321  32 X⇌T
322  37 P/R
323  22 INV
324  37 P/R
325  42 STO
326  27  27
327  32 X⇌T
328  42 STO
329  26  26
330  01   1
331  42 STO
332  22  22
333  42 STO
334  28  28
335  00   0
336  42 STO
337  23  23
338  42 STO
339  24  24
340  42 STO
341  25  25
342  42 STO
343  29  29
344  09   9
345  09   9
346  92 RTN
347  00   0
348  00   0
349  00   0
```

1 ◀				▶ 2
COUPLED LINES				MP-42.1
CPL	DIR	ISO		

Step	Code	Key
000	76	LBL
001	85	+
002	86	STF
003	00	00
004	76	LBL
005	75	-
006	42	STO
007	01	01
008	73	RC*
009	01	01
010	87	IFF
011	00	00
012	00	00
013	15	15
014	94	+/-
015	32	X↕T
016	42	STO
017	02	02
018	69	OP
019	21	21
020	73	RC*
021	01	01
022	37	P/R
023	42	STO
024	04	04
025	73	RC*
026	02	02
027	32	X↕T
028	42	STO
029	03	03
030	69	OP
031	22	22
032	73	RC*
033	02	02
034	37	P/R
035	44	SUM
036	04	04
037	32	X↕T
038	44	SUM
039	03	03
040	43	RCL
041	03	03
042	32	X↕T
043	43	RCL
044	04	04
045	22	INV
046	37	P/R
047	32	X↕T
048	22	INV
049	86	STF
050	00	00
051	92	RTN
052	76	LBL
053	65	×
054	42	STO
055	01	01
056	73	RC*
057	01	01
058	42	STO
059	03	03
060	69	OP
061	21	21
062	73	RC*
063	01	01
064	42	STO
065	04	04
066	32	X↕T
067	42	STO
068	01	01
069	73	RC*
070	01	01
071	49	PRD
072	03	03
073	69	OP
074	21	21
075	73	RC*
076	01	01
077	44	SUM
078	04	04
079	43	RCL
080	04	04
081	32	X↕T
082	43	RCL
083	03	03
084	92	RTN
085	76	LBL
086	12	B
087	86	STF
088	01	01
089	61	GTO
090	11	A
091	76	LBL
092	13	C
093	86	STF
094	02	02
095	76	LBL
096	11	A
097	42	STO
098	05	05
099	32	X↕T
100	42	STO
101	06	06
102	43	RCL
103	00	00
104	91	R/S
105	42	STO
106	07	07
107	43	RCL
108	06	06
109	75	-
110	43	RCL
111	05	05
112	95	=
113	55	÷
114	53	(
115	43	RCL
116	06	06
117	85	+
118	43	RCL
119	05	05
120	95	=
121	42	STO
122	08	08
123	43	RCL
124	05	05
125	65	×
126	43	RCL
127	06	06
128	95	=
129	34	√X
130	42	STO
131	05	05
132	09	9
133	00	0
134	65	×
135	43	RCL
136	00	00
137	55	÷
138	43	RCL
139	07	07
140	95	=
141	42	STO
142	06	06
143	38	SIN
144	32	X↕T
145	01	1
146	75	-
147	43	RCL
148	08	08
149	33	X^2

```
150  95  =
151  34  ΓX
152  65  ×
153  43  RCL
154  06   06
155  39  COS
156  95  =
157  32  X⇅T
158  22  INV
159  37  P/R
160  42  STO
161  07   07
162  43  RCL
163  06   06
164  39  COS
165  65  ×
166  43  RCL
167  08   08
168  95  =
169  33  X²
170  94  +/-
171  85  +
172  01  1
173  95  =
174  35  1/X
175  65  ×
176  53  (
177  01  1
178  75  -
179  43  RCL
180  08   08
181  33  X²
182  95  =
183  34  ΓX
184  42  STO
185  12   12
186  33  X²
187  94  +/-
188  85  +
189  01  1
190  95  =
191  34  ΓX
192  42  STO
193  10   10
194  43  RCL
195  07   07
196  94  +/-
197  42  STO
198  13   13
199  85  +
200  09  9
201  00  0
202  95  =
203  42  STO
204  11   11
205  05  5
206  00  0
207  75  -
208  43  RCL
209  05   05
210  95  =
211  55  ÷
212  53  (
213  05  5
214  00  0
215  85  +
216  43  RCL
217  05   05
218  95  =
219  42  STO
220  05   05
221  00  0
222  42  STO
223  06   06
224  42  STO
225  08   08
226  01  1
227  42  STO
228  07   07
229  87  IFF
230  01   01
231  22  INV
232  87  IFF
233  02   02
234  23  LNX
235  03  3
236  03  3
237  91  R/S
238  42  STO
239  31   31
240  32  X⇅T
241  42  STO
242  30   30
243  04  4
244  04  4
245  91  R/S
246  42  STO
247  33   33
248  32  X⇅T
249  42  STO
250  32   32
251  61  GTO
252  24  CE
253  76  LBL
254  22  INV
255  22  INV
256  86  STF
257  01   01
258  02  2
259  02  2
260  91  R/S
261  42  STO
262  31   31
263  32  X⇅T
264  42  STO
265  30   30
266  04  4
267  04  4
268  91  R/S
269  42  STO
270  33   33
271  32  X⇅T
272  42  STO
273  32   32
274  43  RCL
275  10   10
276  48  EXC
277  12   12
278  42  STO
279  10   10
280  43  RCL
281  11   11
282  48  EXC
283  13   13
284  42  STO
285  11   11
286  76  LBL
287  24  CE
288  03  3
289  00  0
290  32  X⇅T
291  03  3
292  02  2
293  71  SBR
294  65  ×
295  42  STO
296  34   34
297  03  3
298  04  4
299  32  X⇅T
```

```
300  42  STO
301  35   35
302  43  RCL
303  10   10
304  33  X²
305  42  STO
306  36   36
307  43  RCL
308  11   11
309  65   ×
310  02   2
311  95   =
312  42  STO
313  37   37
314  03   3
315  06   6
316  71  SBR
317  65   ×
318  94  +/-
319  42  STO
320  36   36
321  03   3
322  06   6
323  32  X↕T
324  42  STO
325  37   37
326  01   1
327  42  STO
328  38   38
329  00   0
330  42  STO
331  39   39
332  03   3
333  08   8
334  71  SBR
335  85   +
336  35  1/X
337  42  STO
338  36   36
339  32  X↕T
340  94  +/-
341  42  STO
342  37   37
343  43  RCL
344  12   12
345  33  X²
346  42  STO
347  38   38
348  43  RCL
349  13   13
350  65   ×
351  02   2
352  95   =
353  42  STO
354  39   39
355  03   3
356  06   6
357  32  X↕T
358  01   1
359  00   0
360  71  SBR
361  65   ×
362  42  STO
363  40   40
364  04   4
365  00   0
366  32  X↕T
367  42  STO
368  41   41
369  03   3
370  08   8
371  71  SBR
372  65   ×
373  42  STO
374  40   40
375  04   4
376  00   0
377  32  X↕T
378  42  STO
379  41   41
380  03   3
381  00   0
382  71  SBR
383  65   ×
384  42  STO
385  40   40
386  04   4
387  00   0
388  32  X↕T
389  42  STO
390  41   41
391  03   3
392  02   2
393  71  SBR
394  65   ×
395  42  STO
396  40   40
397  04   4
398  00   0
399  32  X↕T
400  42  STO
401  41   41
402  01   1
403  00   0
404  71  SBR
405  85   +
406  42  STO
407  24   24
408  42  STO
409  26   26
410  32  X↕T
411  42  STO
412  25   25
413  42  STO
414  27   27
415  03   3
416  06   6
417  32  X↕T
418  03   3
419  08   8
420  71  SBR
421  65   ×
422  42  STO
423  40   40
424  04   4
425  00   0
426  32  X↕T
427  42  STO
428  41   41
429  03   3
430  00   0
431  71  SBR
432  65   ×
433  42  STO
434  22   22
435  32  X↕T
436  42  STO
437  23   23
438  03   3
439  06   6
440  32  X↕T
441  03   3
442  08   8
443  71  SBR
444  65   ×
445  42  STO
446  40   40
447  04   4
448  00   0
449  32  X↕T
```

```
450  42 STO
451  41  41
452  03  3
453  02  2
454  71 SBR
455  65  x
456  42 STO
457  28  28
458  32 X:T
459  42 STO
460  29  29
461  76 LBL
462  23 LNX
463  04  4
464  02  2
465  93  .
466  02  2
467  92 RTN
468  00  0
469  00  0
470  00  0
```

1 ◁				▷ 2
COUPLED LINES				MP-42.2
RUN				

Step	Code	Key
000	76	LBL
001	85	+
002	86	STF
003	00	00
004	76	LBL
005	75	-
006	42	STO
007	01	01
008	73	RC*
009	01	01
010	87	IFF
011	00	00
012	00	00
013	15	15
014	94	+/-
015	32	X:T
016	42	STO
017	02	02
018	69	OP
019	21	21
020	73	RC*
021	01	01
022	37	P/R
023	42	STO
024	04	04
025	73	RC*
026	02	02
027	32	X:T
028	42	STO
029	03	03
030	69	OP
031	22	22
032	73	RC*
033	02	02
034	37	P/R
035	44	SUM
036	04	04
037	32	X:T
038	44	SUM
039	03	03
040	43	RCL
041	03	03
042	32	X:T
043	43	RCL
044	04	04
045	22	INV
046	37	P/R
047	32	X:T
048	22	INV
049	86	STF
050	00	00
051	92	RTN
052	76	LBL
053	65	×
054	42	STO
055	01	01
056	73	RC*
057	01	01
058	42	STO
059	03	03
060	69	OP
061	21	21
062	73	RC*
063	01	01
064	42	STO
065	04	04
066	32	X:T
067	42	STO
068	01	01
069	73	RC*
070	01	01
071	49	PRD
072	03	03
073	69	OP
074	21	21
075	73	RC*
076	01	01
077	44	SUM
078	04	04
079	43	RCL
080	04	04
081	32	X:T
082	43	RCL
083	03	03
084	92	RTN
085	76	LBL
086	11	A
087	22	INV
088	87	IFF
089	02	02
090	22	INV
091	22	INV
092	86	STF
093	02	02
094	02	2
095	02	2
096	91	R/S
097	42	STO
098	31	31
099	32	X:T
100	42	STO
101	30	30
102	03	3
103	03	3
104	91	R/S
105	42	STO
106	33	33
107	32	X:T
108	42	STO
109	32	32
110	43	RCL
111	10	10
112	33	X^2
113	42	STO
114	34	34
115	02	2
116	65	×
117	43	RCL
118	11	11
119	95	=
120	42	STO
121	35	35
122	43	RCL
123	12	12
124	33	X^2
125	42	STO
126	36	36
127	02	2
128	65	×
129	43	RCL
130	13	13
131	95	=
132	42	STO
133	37	37
134	03	3
135	04	4
136	32	X:T
137	03	3
138	00	0
139	71	SBR
140	65	×
141	42	STO
142	38	38
143	03	3
144	06	6
145	32	X:T
146	42	STO
147	39	39
148	03	3
149	02	2

```
150  71  SBR
151  65  ×
152  42  STO
153  40  40
154  04  4
155  00  0
156  32  X↕T
157  42  STO
158  41  41
159  03  3
160  08  8
161  71  SBR
162  85  +
163  42  STO
164  22  22
165  03  3
166  04  4
167  32  X↕T
168  42  STO
169  23  23
170  03  3
171  02  2
172  71  SBR
173  65  ×
174  42  STO
175  38  38
176  03  3
177  06  6
178  32  X↕T
179  42  STO
180  39  39
181  03  3
182  00  0
183  71  SBR
184  65  ×
185  42  STO
186  40  40
187  04  4
188  00  0
189  32  X↕T
190  42  STO
191  41  41
192  03  3
193  08  8
194  71  SBR
195  85  +
196  42  STO
197  28  28
198  32  X↕T
199  42  STO
200  29  29
201  03  3
202  00  0
203  32  X↕T
204  03  3
205  02  2
206  71  SBR
207  85  +
208  42  STO
209  40  40
210  04  4
211  00  0
212  32  X↕T
213  42  STO
214  41  41
215  01  1
216  00  0
217  71  SBR
218  65  ×
219  42  STO
220  40  40
221  04  4
222  00  0
223  32  X↕T
224  42  STO
225  41  41
226  01  1
227  02  2
228  71  SBR
229  65  ×
230  42  STO
231  24  24
232  42  STO
233  26  26
234  32  X↕T
235  42  STO
236  25  25
237  42  STO
238  27  27
239  76  LBL
240  22  INV
241  05  5
242  32  X↕T
243  02  2
244  02  2
245  71  SBR
246  65  ×
247  94  +/-
248  42  STO
249  34  34
250  03  3
251  04  4
252  32  X↕T
253  42  STO
254  35  35
255  07  7
256  71  SBR
257  85  +
258  42  STO
259  34  34
260  05  5
261  32  X↕T
262  42  STO
263  35  35
264  02  2
265  08  8
266  71  SBR
267  65  ×
268  94  +/-
269  42  STO
270  36  36
271  03  3
272  06  6
273  32  X↕T
274  42  STO
275  37  37
276  07  7
277  71  SBR
278  85  +
279  42  STO
280  36  36
281  32  X↕T
282  42  STO
283  37  37
284  43  RCL
285  05  05
286  65  ×
287  43  RCL
288  24  24
289  33  X²
290  95  =
291  42  STO
292  09  09
293  65  ×
294  43  RCL
295  05  05
296  95  =
297  94  +/-
298  42  STO
299  11  11
```

```
300  43 RCL
301  25  25
302  65  ×
303  02  2
304  95  =
305  42 STO
306  10  10
307  42 STO
308  12  12
309  03  3
310  04  4
311  32 X:T
312  03  3
313  06  6
314  71 SBR
315  65  ×
316  42 STO
317  38  38
318  03  3
319  08  8
320  32 X:T
321  42 STO
322  39  39
323  01  1
324  01  1
325  71 SBR
326  85  +
327  35 1/X
328  42 STO
329  38  38
330  02  2
331  02  2
332  32 X:T
333  94 +/-
334  42 STO
335  39  39
336  01  1
337  94 +/-
338  49 PRD
339  05  05
340  05  5
341  71 SBR
342  85  +
343  42 STO
344  40  40
345  04  4
346  00  0
347  32 X:T
348  42 STO
349  41  41
350  03  3
351  06  6
352  71 SBR
353  65  ×
354  42 STO
355  42  42
356  04  4
357  02  2
358  32 X:T
359  42 STO
360  43  43
361  09  9
362  71 SBR
363  85  +
364  42 STO
365  42  42
366  04  4
367  02  2
368  32 X:T
369  42 STO
370  43  43
371  03  3
372  08  8
373  71 SBR
374  65  ×
375  32 X:T
376  37 P/R
377  22 INV
378  37 P/R
379  42 STO
380  23  23
381  32 X:T
382  42 STO
383  22  22
384  02  2
385  08  8
386  32 X:T
387  05  5
388  71 SBR
389  85  +
390  42 STO
391  40  40
392  04  4
393  00  0
394  32 X:T
395  42 STO
396  41  41
397  03  3
398  04  4
399  71 SBR
400  65  ×
401  42 STO
402  42  42
403  04  4
404  02  2
405  32 X:T
406  42 STO
407  43  43
408  09  9
409  71 SBR
410  85  +
411  42 STO
412  42  42
413  04  4
414  02  2
415  32 X:T
416  42 STO
417  43  43
418  03  3
419  08  8
420  71 SBR
421  65  ×
422  32 X:T
423  37 P/R
424  22 INV
425  37 P/R
426  42 STO
427  29  29
428  32 X:T
429  42 STO
430  28  28
431  02  2
432  06  6
433  32 X:T
434  03  3
435  08  8
436  71 SBR
437  65  ×
438  42 STO
439  42  42
440  32 X:T
441  42 STO
442  43  43
443  01  1
444  75  -
445  43 RCL
446  05  05
447  33 X²
448  95  =
449  49 PRD
```

```
450  42   42
451  43  RCL
452  42   42
453  32  X↕T
454  43  RCL
455  43   43
456  37  P/R
457  22  INV
458  37  P/R
459  42  STO
460  25   25
461  42  STO
462  27   27
463  32  X↕T
464  42  STO
465  24   24
466  42  STO
467  26   26
468  09   9
469  09   9
470  91  R/S
471  00   0
472  00   0
473  00   0
474  00   0
```

1 ◁			▷ 2
TWO to THREE-PORT CONVERSION			MP-51.1
SERIES FB	SHUNT FB	CONFIG	

Step	Code	Key
000	76	LBL
001	85	+
002	86	STF
003	00	00
004	76	LBL
005	75	-
006	42	STO
007	01	01
008	73	RC*
009	01	01
010	87	IFF
011	00	00
012	00	00
013	15	15
014	94	+/-
015	32	X:T
016	42	STO
017	02	02
018	69	OP
019	21	21
020	73	RC*
021	01	01
022	37	P/R
023	42	STO
024	04	04
025	73	RC*
026	02	02
027	32	X:T
028	42	STO
029	03	03
030	69	OP
031	22	22
032	73	RC*
033	02	02
034	37	P/R
035	44	SUM
036	04	04
037	32	X:T
038	44	SUM
039	03	03
040	43	RCL
041	03	03
042	32	X:T
043	43	RCL
044	04	04
045	22	INV
046	37	P/R
047	32	X:T
048	22	INV
049	86	STF
050	00	00
051	92	RTN
052	76	LBL
053	65	×
054	42	STO
055	01	01
056	73	RC*
057	01	01
058	42	STO
059	03	03
060	69	OP
061	21	21
062	73	RC*
063	01	01
064	42	STO
065	04	04
066	32	X:T
067	42	STO
068	01	01
069	73	RC*
070	01	01
071	49	PRD
072	03	03
073	69	OP
074	21	21
075	73	RC*
076	01	01
077	44	SUM
078	04	04
079	43	RCL
080	04	04
081	32	X:T
082	43	RCL
083	03	03
084	92	RTN
085	76	LBL
086	32	X:T
087	01	1
088	42	STO
089	05	05
090	00	0
091	42	STO
092	06	06
093	05	5
094	32	X:T
095	01	1
096	04	4
097	71	SBR
098	85	+
099	42	STO
100	07	07
101	05	5
102	32	X:T
103	42	STO
104	08	08
105	02	2
106	00	0
107	71	SBR
108	85	+
109	42	STO
110	09	09
111	01	1
112	06	6
113	32	X:T
114	42	STO
115	10	10
116	01	1
117	08	8
118	71	SBR
119	65	×
120	42	STO
121	11	11
122	07	7
123	32	X:T
124	42	STO
125	12	12
126	09	9
127	71	SBR
128	65	×
129	42	STO
130	30	30
131	03	3
132	00	0
133	32	X:T
134	42	STO
135	31	31
136	01	1
137	01	1
138	71	SBR
139	75	-
140	35	1/X
141	65	×
142	02	2
143	94	+/-
144	95	=
145	42	STO
146	11	11
147	01	1
148	01	1
149	32	X:T

```
150  94  +/-
151  42  STO
152  12   12
153  09  9
154  71  SBR
155  65   ×
156  42  STO
157  32   32
158  03  3
159  02  2
160  32  X:T
161  42  STO
162  33   33
163  05  5
164  71  SBR
165  85   +
166  94  +/-
167  42  STO
168  14   14
169  01  1
170  01  1
171  32  X:T
172  42  STO
173  15   15
174  01  1
175  06  6
176  71  SBR
177  65   ×
178  42  STO
179  16   16
180  01  1
181  01  1
182  32  X:T
183  42  STO
184  17   17
185  01  1
186  08  8
187  71  SBR
188  65   ×
189  42  STO
190  18   18
191  01  1
192  01  1
193  32  X:T
194  42  STO
195  19   19
196  07  7
197  71  SBR
198  65   ×
199  42  STO
200  32   32
201  03  3
202  02  2
203  32  X:T
204  42  STO
205  33   33
206  05  5
207  71  SBR
208  85   +
209  94  +/-
210  42  STO
211  20   20
212  32  X:T
213  42  STO
214  21   21
215  92  RTN
216  76  LBL
217  24  CE
218  01  1
219  42  STO
220  22   22
221  42  STO
222  24   24
223  42  STO
224  26   26
225  42  STO
226  28   28
227  00  0
228  42  STO
229  23   23
230  42  STO
231  25   25
232  42  STO
233  27   27
234  42  STO
235  29   29
236  92  RTN
237  76  LBL
238  23  LNX
239  01  1
240  04  4
241  42  STO
242  05   05
243  02  2
244  02  2
245  42  STO
246  06   06
247  04  4
248  42  STO
249  07   07
250  43  RCL
251  05   05
252  32  X:T
253  43  RCL
254  06   06
255  71  SBR
256  85   +
257  72  ST*
258  05   05
259  69  OP
260  25   25
261  32  X:T
262  72  ST*
263  05   05
264  69  OP
265  25   25
266  69  OP
267  26   26
268  69  OP
269  26   26
270  97  DSZ
271  07   07
272  02   02
273  50   50
274  92  RTN
275  76  LBL
276  13   C
277  76  LBL
278  11   A
279  01  1
280  94  +/-
281  49  PRD
282  14   14
283  49  PRD
284  16   16
285  49  PRD
286  18   18
287  49  PRD
288  20   20
289  71  SBR
290  32  X:T
291  71  SBR
292  24  CE
293  71  SBR
294  23  LNX
295  71  SBR
296  32  X:T
297  01  1
298  94  +/-
299  49  PRD
```

Step	Code	Key
300	14	14
301	49	PRD
302	16	16
303	49	PRD
304	18	18
305	49	PRD
306	20	20
307	61	GTO
308	14	D
309	76	LBL
310	12	B
311	86	STF
312	01	01
313	71	SBR
314	32	X:T
315	71	SBR
316	24	CE
317	01	1
318	94	+/-
319	49	PRD
320	24	24
321	49	PRD
322	26	26
323	71	SBR
324	23	LNX
325	71	SBR
326	32	X:T
327	76	LBL
328	14	D
329	43	RCL
330	20	20
331	42	STO
332	22	22
333	43	RCL
334	21	21
335	42	STO
336	23	23
337	43	RCL
338	18	18
339	42	STO
340	20	20
341	43	RCL
342	19	19
343	42	STO
344	21	21
345	01	1
346	42	STO
347	05	05
348	00	0
349	42	STO
350	06	06
351	87	IFF
352	01	01
353	15	E
354	05	5
355	32	X:T
356	01	1
357	04	4
358	71	SBR
359	75	-
360	42	STO
361	08	08
362	08	8
363	32	X:T
364	42	STO
365	09	09
366	02	2
367	00	0
368	71	SBR
369	75	-
370	42	STO
371	26	26
372	05	5
373	32	X:T
374	42	STO
375	27	27
376	01	1
377	06	6
378	71	SBR
379	75	-
380	42	STO
381	08	08
382	08	8
383	32	X:T
384	42	STO
385	09	09
386	02	2
387	02	2
388	71	SBR
389	75	-
390	42	STO
391	28	28
392	05	5
393	32	X:T
394	42	STO
395	29	29
396	02	2
397	06	6
398	71	SBR
399	75	-
400	42	STO
401	08	08
402	08	8
403	32	X:T
404	42	STO
405	09	09
406	02	2
407	08	8
408	71	SBR
409	75	-
410	42	STO
411	30	30
412	05	5
413	32	X:T
414	42	STO
415	31	31
416	01	1
417	04	4
418	71	SBR
419	75	-
420	42	STO
421	08	08
422	08	8
423	32	X:T
424	42	STO
425	09	09
426	01	1
427	06	6
428	71	SBR
429	75	-
430	42	STO
431	18	18
432	05	5
433	32	X:T
434	42	STO
435	19	19
436	02	2
437	00	0
438	71	SBR
439	75	-
440	42	STO
441	08	08
442	08	8
443	32	X:T
444	42	STO
445	09	09
446	02	2
447	02	2
448	71	SBR
449	75	-

```
450  42 STO
451  24  24
452  32 X:T
453  42 STO
454  25  25
455  09  9
456  09  9
457  92 RTN
458  76 LBL
459  15  E
460  22 INV
461  86 STF
462  01  01
463  05  5
464  01  1
465  93  .
466  02  2
467  92 RTN
468  00  0
469  00  0
470  00  0
471  00  0
472  00  0
473  00  0
474  00  0
475  00  0
476  00  0
477  00  0
478  00  0
479  00  0
```

1 ◁				▷ 2
TWO to THREE–PORT CONVERSION				MP-51.2
CONT SHUNT FB			DSP [S]	PRT [S]

Step	Code	Key
000	76	LBL
001	85	+
002	86	STF
003	00	00
004	76	LBL
005	75	-
006	42	STO
007	01	01
008	73	RC*
009	01	01
010	87	IFF
011	00	00
012	00	00
013	15	15
014	94	+/-
015	32	X⇄T
016	42	STO
017	02	02
018	69	OP
019	21	21
020	73	RC*
021	01	01
022	37	P/R
023	42	STO
024	04	04
025	73	RC*
026	02	02
027	32	X⇄T
028	42	STO
029	03	03
030	69	OP
031	22	22
032	73	RC*
033	02	02
034	37	P/R
035	44	SUM
036	04	04
037	32	X⇄T
038	44	SUM
039	03	03
040	43	RCL
041	03	03
042	32	X⇄T
043	43	RCL
044	04	04
045	22	INV
046	37	P/R
047	32	X⇄T
048	22	INV
049	86	STF
050	00	00
051	92	RTN
052	76	LBL
053	65	×
054	42	STO
055	01	01
056	73	RC*
057	01	01
058	42	STO
059	03	03
060	69	OP
061	21	21
062	73	RC*
063	01	01
064	42	STO
065	04	04
066	32	X⇄T
067	42	STO
068	01	01
069	73	RC*
070	01	01
071	49	PRD
072	03	03
073	69	OP
074	21	21
075	73	RC*
076	01	01
077	44	SUM
078	04	04
079	43	RCL
080	04	04
081	32	X⇄T
082	43	RCL
083	03	03
084	92	RTN
085	76	LBL
086	11	A
087	05	5
088	32	X⇄T
089	01	1
090	04	4
091	71	SBR
092	85	+
093	42	STO
094	08	08
095	08	8
096	32	X⇄T
097	42	STO
098	09	09
099	02	2
100	00	0
101	71	SBR
102	75	-
103	42	STO
104	26	26
105	01	1
106	06	6
107	32	X⇄T
108	42	STO
109	27	27
110	02	2
111	02	2
112	71	SBR
113	75	-
114	42	STO
115	08	08
116	08	8
117	32	X⇄T
118	42	STO
119	09	09
120	05	5
121	71	SBR
122	75	-
123	42	STO
124	28	28
125	02	2
126	06	6
127	32	X⇄T
128	42	STO
129	29	29
130	05	5
131	71	SBR
132	75	-
133	42	STO
134	08	08
135	08	8
136	32	X⇄T
137	42	STO
138	09	09
139	02	2
140	08	8
141	71	SBR
142	75	-
143	42	STO
144	30	30
145	05	5
146	32	X⇄T
147	42	STO
148	31	31
149	01	1

```
150  04   4
151  71  SBR
152  85   +
153  42  STO
154  08   08
155  08   8
156  32  X↕T
157  42  STO
158  09   09
159  01   1
160  06   6
161  71  SBR
162  75   -
163  42  STO
164  18   18
165  02   2
166  00   0
167  32  X↕T
168  42  STO
169  19   19
170  02   2
171  02   2
172  71  SBR
173  75   -
174  42  STO
175  08   08
176  08   8
177  32  X↕T
178  42  STO
179  09   09
180  05   5
181  71  SBR
182  75   -
183  42  STO
184  24   24
185  32  X↕T
186  42  STO
187  25   25
188  09   9
189  09   9
190  92  RTN
191  76  LBL
192  14   D
193  01   1
194  01   1
195  91  R/S
196  43  RCL
197  15   15
198  32  X↕T
199  43  RCL
200  14   14
201  91  R/S
202  01   1
203  02   2
204  91  R/S
205  43  RCL
206  17   17
207  32  X↕T
208  43  RCL
209  16   16
210  91  R/S
211  01   1
212  03   3
213  91  R/S
214  43  RCL
215  19   19
216  32  X↕T
217  43  RCL
218  18   18
219  91  R/S
220  02   2
221  01   1
222  91  R/S
223  43  RCL
224  21   21
225  32  X↕T
226  43  RCL
227  20   20
228  91  R/S
229  02   2
230  02   2
231  91  R/S
232  43  RCL
233  23   23
234  32  X↕T
235  43  RCL
236  22   22
237  91  R/S
238  02   2
239  03   3
240  91  R/S
241  43  RCL
242  25   25
243  32  X↕T
244  43  RCL
245  24   24
246  91  R/S
247  03   3
248  01   1
249  91  R/S
250  43  RCL
251  27   27
252  32  X↕T
253  43  RCL
254  26   26
255  91  R/S
256  03   3
257  02   2
258  91  R/S
259  43  RCL
260  29   29
261  32  X↕T
262  43  RCL
263  28   28
264  91  R/S
265  03   3
266  03   3
267  91  R/S
268  43  RCL
269  31   31
270  32  X↕T
271  43  RCL
272  30   30
273  91  R/S
274  09   9
275  09   9
276  92  RTN
277  76  LBL
278  15   E
279  04   4
280  02   2
281  00   0
282  03   3
283  03   3
284  03   3
285  02   2
286  69  OP
287  01   01
288  03   3
289  05   5
290  03   3
291  07   7
292  00   0
293  00   0
294  03   3
295  06   6
296  02   2
297  00   0
298  69  OP
299  02   02
```

Step	Code	Key
300	03	3
301	03	3
302	01	1
303	03	3
304	03	3
305	05	5
306	01	1
307	03	3
308	03	3
309	00	0
310	69	OP
311	03	03
312	01	1
313	07	7
314	03	3
315	07	7
316	01	1
317	07	7
318	03	3
319	05	5
320	03	3
321	06	6
322	69	OP
323	04	04
324	69	OP
325	05	05
326	03	3
327	06	6
328	00	0
329	02	2
330	00	0
331	02	2
332	42	STO
333	01	01
334	69	OP
335	04	04
336	43	RCL
337	14	14
338	69	OP
339	06	06
340	43	RCL
341	15	15
342	99	PRT
343	01	1
344	44	SUM
345	01	01
346	43	RCL
347	01	01
348	69	OP
349	04	04
350	43	RCL
351	16	16
352	69	OP
353	06	06
354	43	RCL
355	17	17
356	99	PRT
357	01	1
358	44	SUM
359	01	01
360	43	RCL
361	01	01
362	69	OP
363	04	04
364	43	RCL
365	18	18
366	69	OP
367	06	06
368	43	RCL
369	19	19
370	99	PRT
371	03	3
372	06	6
373	00	0
374	03	3
375	00	0
376	02	2
377	42	STO
378	01	01
379	69	OP
380	04	04
381	43	RCL
382	20	20
383	69	OP
384	06	06
385	43	RCL
386	21	21
387	99	PRT
388	01	1
389	44	SUM
390	01	01
391	43	RCL
392	01	01
393	69	OP
394	04	04
395	43	RCL
396	22	22
397	69	OP
398	06	06
399	43	RCL
400	23	23
401	99	PRT
402	01	1
403	44	SUM
404	01	01
405	43	RCL
406	01	01
407	69	OP
408	04	04
409	43	RCL
410	24	24
411	69	OP
412	06	06
413	43	RCL
414	25	25
415	99	PRT
416	03	3
417	06	6
418	00	0
419	04	4
420	00	0
421	02	2
422	42	STO
423	01	01
424	69	OP
425	04	04
426	43	RCL
427	26	26
428	69	OP
429	06	06
430	43	RCL
431	27	27
432	99	PRT
433	01	1
434	44	SUM
435	01	01
436	43	RCL
437	01	01
438	69	OP
439	04	04
440	43	RCL
441	28	28
442	69	OP
443	06	06
444	43	RCL
445	29	29
446	99	PRT
447	01	1
448	44	SUM
449	01	01

```
450   43  RCL
451   01   01
452   69  OP
453   04   04
454   43  RCL
455   30   30
456   69  OP
457   06   06
458   43  RCL
459   31   31
460   99  PRT
461   09   9
462   09   9
463   92  RTN
464   00   0
465   00   0
466   00   0
467   00   0
468   00   0
469   00   0
```

1 ◁			▷ 2
CONFIGURATION CONVERSION			MP-52
FROM:	CE	CB	CC
TO:	CE	CB	CC

```
000   76 LBL
001   34 ΓX
002   01  1
003   04  4
004   42 STO
005   05  05
006   03  3
007   02  2
008   42 STO
009   06  06
010   01  1
011   08  8
012   42 STO
013   07  07
014   73 RC*
015   05  05
016   72 ST*
017   06  06
018   69 OP
019   25  25
020   69 OP
021   26  26
022   97 DSZ
023   07  07
024   00  00
025   14  14
026   92 RTN
027   76 LBL
028   35 1/X
029   42 STO
030   06  06
031   73 RC*
032   06  06
033   72 ST*
034   05  05
035   69 OP
036   25  25
037   69 OP
038   26  26
039   73 RC*
040   06  06
041   72 ST*
042   05  05
043   69 OP
044   25  25
045   92 RTN
046   76 LBL
047   22 INV
048   71 SBR
049   34 ΓX
050   01  1
051   04  4
052   42 STO
053   05  05
054   04  4
055   08  8
056   71 SBR
057   35 1/X
058   04  4
059   06  6
060   71 SBR
061   35 1/X
062   04  4
063   04  4
064   71 SBR
065   35 1/X
066   04  4
067   02  2
068   71 SBR
069   35 1/X
070   04  4
071   00  0
072   71 SBR
073   35 1/X
074   03  3
075   08  8
076   71 SBR
077   35 1/X
078   03  3
079   06  6
080   71 SBR
081   35 1/X
082   03  3
083   04  4
084   71 SBR
085   35 1/X
086   03  3
087   02  2
088   71 SBR
089   35 1/X
090   61 GTO
091   33 X²
092   76 LBL
093   23 LNX
094   71 SBR
095   34 ΓX
096   01  1
097   04  4
098   42 STO
099   05  05
100   03  3
101   02  2
102   71 SBR
103   35 1/X
104   03  3
105   06  6
106   71 SBR
107   35 1/X
108   03  3
109   04  4
110   71 SBR
111   35 1/X
112   04  4
113   04  4
114   71 SBR
115   35 1/X
116   04  4
117   08  8
118   71 SBR
119   35 1/X
120   04  4
121   06  6
122   71 SBR
123   35 1/X
124   03  3
125   08  8
126   71 SBR
127   35 1/X
128   04  4
129   02  2
130   71 SBR
131   35 1/X
132   04  4
133   00  0
134   71 SBR
135   35 1/X
136   61 GTO
137   33 X²
138   76 LBL
139   24 CE
140   71 SBR
141   34 ΓX
142   01  1
143   04  4
144   42 STO
145   05  05
146   04  4
147   08  8
148   71 SBR
149   35 1/X
```

```
150  04  4
151  04  4
152  71  SBR
153  35  1/X
154  04  4
155  06  6
156  71  SBR
157  35  1/X
158  03  3
159  06  6
160  71  SBR
161  35  1/X
162  03  3
163  02  2
164  71  SBR
165  35  1/X
166  03  3
167  04  4
168  71  SBR
169  35  1/X
170  04  4
171  02  2
172  71  SBR
173  35  1/X
174  03  3
175  08  8
176  71  SBR
177  35  1/X
178  04  4
179  00  0
180  71  SBR
181  35  1/X
182  61  GTO
183  33  X²
184  76  LBL
185  32  X:T
186  71  SBR
187  34  √X
188  01  1
189  04  4
190  42  STO
191  05  05
192  04  4
193  00  0
194  71  SBR
195  35  1/X
196  04  4
197  02  2
198  71  SBR
199  35  1/X
200  03  3
201  08  8
202  71  SBR
203  35  1/X
204  04  4
205  06  6
206  71  SBR
207  35  1/X
208  04  4
209  08  8
210  71  SBR
211  35  1/X
212  04  4
213  04  4
214  71  SBR
215  35  1/X
216  03  3
217  04  4
218  71  SBR
219  35  1/X
220  03  3
221  06  6
222  71  SBR
223  35  1/X
224  03  3
225  02  2
226  71  SBR
227  35  1/X
228  61  GTO
229  33  X²
230  76  LBL
231  17  B'
232  86  STF
233  01  01
234  92  RTN
235  76  LBL
236  18  C'
237  86  STF
238  02  02
239  92  RTN
240  76  LBL
241  19  D'
242  86  STF
243  03  03
244  92  RTN
245  76  LBL
246  12  B
247  87  IFF
248  01  01
249  33  X²
250  87  IFF
251  02  02
252  22  INV
253  87  IFF
254  03  03
255  23  LNX
256  76  LBL
257  13  C
258  87  IFF
259  01  01
260  22  INV
261  87  IFF
262  02  02
263  33  X²
264  87  IFF
265  03  03
266  32  X:T
267  76  LBL
268  14  D
269  87  IFF
270  01  01
271  23  LNX
272  87  IFF
273  02  02
274  24  CE
275  87  IFF
276  03  03
277  33  X²
278  76  LBL
279  33  X²
280  22  INV
281  86  STF
282  01  01
283  22  INV
284  86  STF
285  02  02
286  22  INV
287  86  STF
288  03  03
289  09  9
290  09  9
291  92  RTN
292  00  0
293  00  0
294  00  0
295  00  0
296  00  0
297  00  0
```

1 ◁				▷ 2
THREE to TWO-PORT CONVERSION				MP-53
				PRT
3 ⇒ 2 PORT	LIST A			

```
000  76 LBL
001  85  +
002  86 STF
003  00  00
004  76 LBL
005  75  -
006  42 STO
007  01  01
008  73 RC*
009  01  01
010  87 IFF
011  00  00
012  00  00
013  15  15
014  94 +/-
015  32 X:T
016  42 STO
017  02  02
018  69 OP
019  21  21
020  73 RC*
021  01  01
022  37 P/R
023  42 STO
024  04  04
025  73 RC*
026  02  02
027  32 X:T
028  42 STO
029  03  03
030  69 OP
031  22  22
032  73 RC*
033  02  02
034  37 P/R
035  44 SUM
036  04  04
037  32 X:T
038  44 SUM
039  03  03
040  43 RCL
041  03  03
042  32 X:T
043  43 RCL
044  04  04
045  22 INV
046  37 P/R
047  32 X:T
048  22 INV
049  86 STF
050  00  00
051  92 RTN
052  76 LBL
053  65  ×
054  42 STO
055  01  01
056  73 RC*
057  01  01
058  42 STO
059  03  03
060  69 OP
061  21  21
062  73 RC*
063  01  01
064  42 STO
065  04  04
066  32 X:T
067  42 STO
068  01  01
069  73 RC*
070  01  01
071  49 PRD
072  03  03
073  69 OP
074  21  21
075  73 RC*
076  01  01
077  44 SUM
078  04  04
079  43 RCL
080  04  04
081  32 X:T
082  43 RCL
083  03  03
084  92 RTN
085  76 LBL
086  11  A
087  01  1
088  42 STO
089  05  05
090  00  0
091  42 STO
092  06  06
093  05  5
094  32 X:T
095  03  3
096  00  0
097  71 SBR
098  85  +
099  35 1/X
100  42 STO
101  05  05
102  05  5
103  32 X:T
104  94 +/-
105  42 STO
106  06  06
107  01  1
108  08  8
109  71 SBR
110  65  ×
111  42 STO
112  07  07
113  07  7
114  32 X:T
115  42 STO
116  08  08
117  02  2
118  06  6
119  71 SBR
120  65  ×
121  42 STO
122  07  07
123  01  1
124  04  4
125  32 X:T
126  42 STO
127  08  08
128  07  7
129  71 SBR
130  75  -
131  42 STO
132  14  14
133  05  5
134  32 X:T
135  42 STO
136  15  15
137  01  1
138  08  8
139  71 SBR
140  65  ×
141  42 STO
142  07  07
143  07  7
144  32 X:T
145  42 STO
146  08  08
147  02  2
148  08  8
149  71 SBR
```

Step	Code	Key
150	65	×
151	42	STO
152	07	07
153	01	1
154	06	6
155	32	X⇌T
156	42	STO
157	08	08
158	07	7
159	71	SBR
160	75	-
161	42	STO
162	16	16
163	32	X⇌T
164	42	STO
165	17	17
166	05	5
167	32	X⇌T
168	02	2
169	04	4
170	71	SBR
171	65	×
172	42	STO
173	07	07
174	02	2
175	06	6
176	32	X⇌T
177	42	STO
178	08	08
179	07	7
180	71	SBR
181	65	×
182	42	STO
183	07	07
184	02	2
185	00	0
186	32	X⇌T
187	42	STO
188	08	08
189	07	7
190	71	SBR
191	75	-
192	42	STO
193	18	18
194	32	X⇌T
195	42	STO
196	19	19
197	05	5
198	32	X⇌T
199	02	2
200	04	4
201	71	SBR
202	65	×
203	42	STO
204	07	07
205	02	2
206	08	8
207	32	X⇌T
208	42	STO
209	08	08
210	07	7
211	71	SBR
212	65	×
213	42	STO
214	07	07
215	02	2
216	02	2
217	32	X⇌T
218	42	STO
219	08	08
220	07	7
221	71	SBR
222	75	-
223	42	STO
224	20	20
225	32	X⇌T
226	42	STO
227	21	21
228	09	9
229	09	9
230	92	RTN
231	76	LBL
232	10	E'
233	86	STF
234	00	00
235	03	3
236	06	6
237	00	0
238	02	2
239	00	0
240	02	2
241	42	STO
242	05	05
243	85	+
244	01	1
245	95	=
246	42	STO
247	06	06
248	03	3
249	06	6
250	00	0
251	03	3
252	00	0
253	02	2
254	42	STO
255	07	07
256	85	+
257	01	1
258	95	=
259	42	STO
260	08	08
261	03	3
262	07	7
263	04	4
264	03	3
265	03	3
266	02	2
267	02	2
268	00	0
269	03	3
270	03	3
271	42	STO
272	09	09
273	03	3
274	02	2
275	03	3
276	05	5
277	03	3
278	07	7
279	00	0
280	00	0
281	01	1
282	03	3
283	42	STO
284	10	10
285	09	9
286	09	9
287	92	RTN
288	76	LBL
289	12	B
290	04	4
291	42	STO
292	03	03
293	00	0
294	42	STO
295	04	04
296	01	1
297	04	4
298	42	STO
299	01	01

```
300  01  1
301  05  5
302  42  STO
303  02  02
304  87  IFF
305  00  00
306  19  D'
307  61  GTO
308  35  1/X
309  76  LBL
310  19  D'
311  22  INV
312  58  FIX
313  69  OP
314  00  00
315  43  RCL
316  09  09
317  69  OP
318  02  02
319  43  RCL
320  10  10
321  85  +
322  43  RCL
323  04  04
324  95  =
325  69  OP
326  03  03
327  69  OP
328  05  05
329  43  RCL
330  05  05
331  69  OP
332  04  04
333  73  RC*
334  01  01
335  69  OP
336  06  06
337  73  RC*
338  02  02
339  99  PRT
340  43  RCL
341  06  06
342  69  OP
343  04  04
344  02  2
345  44  SUM
346  01  01
347  44  SUM
348  02  02
349  73  RC*
350  01  01
351  69  OP
352  06  06
353  73  RC*
354  02  02
355  99  PRT
356  43  RCL
357  07  07
358  69  OP
359  04  04
360  02  2
361  44  SUM
362  01  01
363  44  SUM
364  02  02
365  73  RC*
366  01  01
367  69  OP
368  06  06
369  73  RC*
370  02  02
371  99  PRT
372  43  RCL
373  08  08
374  69  OP
375  04  04
376  02  2
377  44  SUM
378  01  01
379  44  SUM
380  02  02
381  73  RC*
382  01  01
383  69  OP
384  06  06
385  73  RC*
386  02  02
387  99  PRT
388  98  ADV
389  22  INV
390  86  STF
391  00  00
392  92  RTN
393  76  LBL
394  35  1/X
395  73  RC*
396  02  02
397  32  X:T
398  73  RC*
399  01  01
400  91  R/S
401  02  2
402  44  SUM
403  01  01
404  44  SUM
405  02  02
406  97  DSZ
407  03  03
408  35  1/X
409  09  9
410  09  9
411  92  RTN
412  00  0
413  00  0
414  00  0
415  00  0
416  00  0
417  00  0
418  00  0
```

1 ◁				▷ 2
THREE-PORT MAPPING				MP-54.1
MAP:	CHART			
MAP:	S_{11}	S_{12}	S_{21}	S_{22}

```
000  76 LBL
001  85  +
002  86 STF
003  00  00
004  76 LBL
005  75  -
006  42 STO
007  01  01
008  73 RC*
009  01  01
010  87 IFF
011  00  00
012  00  00
013  15  15
014  94 +/-
015  32 X:T
016  42 STO
017  02  02
018  69 OP
019  21  21
020  73 RC*
021  01  01
022  37 P/R
023  42 STO
024  04  04
025  73 RC*
026  02  02
027  32 X:T
028  42 STO
029  03  03
030  69 OP
031  22  22
032  73 RC*
033  02  02
034  37 P/R
035  44 SUM
036  04  04
037  32 X:T
038  44 SUM
039  03  03
040  43 RCL
041  03  03
042  32 X:T
043  43 RCL
044  04  04
045  22 INV
046  37 P/R
047  32 X:T
048  22 INV
049  86 STF
050  00  00
051  92 RTN
052  76 LBL
053  65  X
054  42 STO
055  01  01
056  73 RC*
057  01  01
058  42 STO
059  03  03
060  69 OP
061  21  21
062  73 RC*
063  01  01
064  42 STO
065  04  04
066  32 X:T
067  42 STO
068  01  01
069  73 RC*
070  01  01
071  49 PRD
072  03  03
073  69 OP
074  21  21
075  73 RC*
076  01  01
077  44 SUM
078  04  04
079  43 RCL
080  04  04
081  32 X:T
082  43 RCL
083  03  03
084  92 RTN
085  76 LBL
086  17 B'
087  00  0
088  42 STO
089  08  08
090  42 STO
091  09  09
092  42 STO
093  11  11
094  42 STO
095  13  13
096  01  1
097  42 STO
098  10  10
099  42 STO
100  12  12
101  05  5
102  04  4
103  93  .
104  02  2
105  92 RTN
106  76 LBL
107  24 CE
108  01  1
109  42 STO
110  06  06
111  00  0
112  42 STO
113  07  07
114  01  1
115  94 +/-
116  49 PRD
117  31  31
118  06  6
119  32 X:T
120  03  3
121  00  0
122  71 SBR
123  85  +
124  42 STO
125  32  32
126  06  6
127  32 X:T
128  42 STO
129  33  33
130  03  3
131  00  0
132  71 SBR
133  75  -
134  35 1/X
135  42 STO
136  34  34
137  03  3
138  04  4
139  32 X:T
140  94 +/-
141  42 STO
142  35  35
143  03  3
144  02  2
145  71 SBR
146  65  X
147  42 STO
148  12  12
149  32 X:T
```

Step	Code	Key
150	42	STO
151	13	13
152	01	1
153	55	÷
154	53	(
155	01	1
156	75	-
157	43	RCL
158	30	30
159	33	X^2
160	95	=
161	42	STO
162	05	05
163	65	×
164	43	RCL
165	30	30
166	95	=
167	42	STO
168	36	36
169	43	RCL
170	31	31
171	42	STO
172	37	37
173	06	6
174	32	X⇌T
175	03	3
176	00	0
177	71	SBR
178	75	-
179	65	×
180	43	RCL
181	05	05
182	95	=
183	42	STO
184	32	32
185	06	6
186	32	X⇌T
187	42	STO
188	33	33
189	01	1
190	94	+/-
191	49	PRD
192	31	31
193	03	3
194	00	0
195	71	SBR
196	75	-
197	35	1/X
198	42	STO
199	34	34
200	03	3
201	04	4
202	32	X⇌T
203	94	+/-
204	42	STO
205	35	35
206	03	3
207	02	2
208	71	SBR
209	65	×
210	42	STO
211	32	32
212	32	X⇌T
213	42	STO
214	33	33
215	92	RTN
216	76	LBL
217	12	B
218	71	SBR
219	24	CE
220	01	1
221	08	8
222	32	X⇌T
223	02	2
224	06	6
225	71	SBR
226	65	×
227	42	STO
228	38	38
229	03	3
230	06	6
231	32	X⇌T
232	42	STO
233	39	39
234	03	3
235	08	8
236	71	SBR
237	65	×
238	42	STO
239	40	40
240	04	4
241	00	0
242	32	X⇌T
243	42	STO
244	41	41
245	01	1
246	04	4
247	76	LBL
248	34	√X
249	71	SBR
250	85	+
251	42	STO
252	08	08
253	03	3
254	08	8
255	32	X⇌T
256	42	STO
257	09	09
258	03	3
259	02	2
260	71	SBR
261	65	×
262	42	STO
263	10	10
264	32	X⇌T
265	42	STO
266	11	11
267	43	RCL
268	08	08
269	32	X⇌T
270	43	RCL
271	09	09
272	37	P/R
273	42	STO
274	09	09
275	32	X⇌T
276	42	STO
277	08	08
278	43	RCL
279	10	10
280	32	X⇌T
281	43	RCL
282	11	11
283	37	P/R
284	42	STO
285	11	11
286	32	X⇌T
287	42	STO
288	10	10
289	43	RCL
290	12	12
291	32	X⇌T
292	43	RCL
293	13	13
294	37	P/R
295	42	STO
296	13	13
297	32	X⇌T
298	42	STO
299	12	12

```
300  05   5
301  04   4
302  93   .
303  02   2
304  92  RTN
305  76  LBL
306  13   C
307  71  SBR
308  24  CE
309  01   1
310  08   8
311  32  X↕T
312  02   2
313  08   8
314  71  SBR
315  65   ×
316  42  STO
317  38   38
318  03   3
319  06   6
320  32  X↕T
321  42  STO
322  39   39
323  03   3
324  08   8
325  71  SBR
326  65   ×
327  42  STO
328  40   40
329  04   4
330  00   0
331  32  X↕T
332  42  STO
333  41   41
334  01   1
335  06   6
336  61  GTO
337  34  √X
338  76  LBL
339  14   D
340  71  SBR
341  24  CE
342  02   2
343  04   4
344  32  X↕T
345  02   2
346  06   6
347  71  SBR
348  65   ×
349  42  STO
350  38   38
351  03   3
352  08   8
353  32  X↕T
354  42  STO
355  39   39
356  03   3
357  06   6
358  71  SBR
359  65   ×
360  42  STO
361  40   40
362  04   4
363  00   0
364  32  X↕T
365  42  STO
366  41   41
367  02   2
368  00   0
369  61  GTO
370  34  √X
371  76  LBL
372  15   E
373  71  SBR
374  24  CE
375  02   2
376  04   4
377  32  X↕T
378  02   2
379  08   8
380  71  SBR
381  65   ×
382  42  STO
383  38   38
384  03   3
385  08   8
386  32  X↕T
387  42  STO
388  39   39
389  03   3
390  06   6
391  71  SBR
392  65   ×
393  42  STO
394  40   40
395  04   4
396  00   0
397  32  X↕T
398  42  STO
399  41   41
400  02   2
401  02   2
402  61  GTO
403  34  √X
404  00   0
405  00   0
406  00   0
407  00   0
408  00   0
409  00   0
```

1 ◁				▷ 2
THREE-PORT MAPPING				MP-54.2
				PRT ON
ENT: ± R	ENT: ± X			PRT OFF

```
000  76 LBL
001  10 E'
002  86 STF
003  01  01
004  92 RTN
005  76 LBL
006  15  E
007  22 INV
008  86 STF
009  01  01
010  92 RTN
011  76 LBL
012  11  A
013  55  ÷
014  05  5
015  00  0
016  95  =
017  42 STO
018  01  01
019  03  3
020  05  5
021  42 STO
022  05  05
023  43 RCL
024  01  01
025  55  ÷
026  53  (
027  43 RCL
028  01  01
029  85  +
030  43 RCL
031  12  12
032  95  =
033  42 STO
034  06  06
035  00  0
036  42 STO
037  07  07
038  43 RCL
039  08  08
040  85  +
041  53  (
042  43 RCL
043  10  10
044  65  ×
045  43 RCL
046  06  06
047  54  )
048  75  -
049  53  (
050  43 RCL
051  11  11
052  65  ×
053  43 RCL
054  07  07
055  95  =
056  42 STO
057  02  02
058  43 RCL
059  09  09
060  85  +
061  53  (
062  43 RCL
063  10  10
064  65  ×
065  43 RCL
066  07  07
067  54  )
068  85  +
069  53  (
070  43 RCL
071  11  11
072  65  ×
073  43 RCL
074  06  06
075  95  =
076  42 STO
077  03  03
078  43 RCL
079  02  02
080  32 X⇌T
081  43 RCL
082  03  03
083  22 INV
084  37 P/R
085  42 STO
086  03  03
087  32 X⇌T
088  42 STO
089  02  02
090  43 RCL
091  12  12
092  55  ÷
093  53  (
094  43 RCL
095  01  01
096  85  +
097  43 RCL
098  12  12
099  95  =
100  50 I×I
101  42 STO
102  04  04
103  43 RCL
104  10  10
105  32 X⇌T
106  43 RCL
107  11  11
108  22 INV
109  37 P/R
110  32 X⇌T
111  49 PRD
112  04  04
113  05  5
114  00  0
115  49 PRD
116  01  01
117  87 IFF
118  01  01
119  24 CE
120  43 RCL
121  03  03
122  32 X⇌T
123  43 RCL
124  02  02
125  91 R/S
126  43 RCL
127  01  01
128  32 X⇌T
129  43 RCL
130  04  04
131  92 RTN
132  76 LBL
133  12  B
134  55  ÷
135  05  5
136  00  0
137  95  =
138  42 STO
139  01  01
140  04  4
141  04  4
142  42 STO
143  05  05
144  01  1
145  42 STO
146  06  06
147  43 RCL
148  12  12
149  55  ÷
```

```
150  53  (
151  43  RCL
152  01  01
153  75  -
154  43  RCL
155  13  13
156  95  =
157  42  STO
158  07  07
159  43  RCL
160  08  08
161  85  +
162  53  (
163  43  RCL
164  10  10
165  65  ×
166  43  RCL
167  06  06
168  54  )
169  75  -
170  53  (
171  43  RCL
172  11  11
173  65  ×
174  43  RCL
175  07  07
176  95  =
177  42  STO
178  02  02
179  43  RCL
180  09  09
181  85  +
182  53  (
183  43  RCL
184  10  10
185  65  ×
186  43  RCL
187  07  07
188  54  )
189  85  +
190  53  (
191  43  RCL
192  11  11
193  65  ×
194  43  RCL
195  06  06
196  95  =
197  42  STO
198  03  03
199  43  RCL
200  02  02
201  32  X↕T
202  43  RCL
203  03  03
204  22  INV
205  37  P/R
206  42  STO
207  03  03
208  32  X↕T
209  42  STO
210  02  02
211  43  RCL
212  12  12
213  55  ÷
214  53  (
215  43  RCL
216  01  01
217  75  -
218  43  RCL
219  13  13
220  95  =
221  50  I×I
222  42  STO
223  04  04
224  43  RCL
225  10  10
226  32  X↕T
227  43  RCL
228  11  11
229  22  INV
230  37  P/R
231  32  X↕T
232  49  PRD
233  04  04
234  05  5
235  00  0
236  49  PRD
237  01  01
238  87  IFF
239  01  01
240  24  CE
241  43  RCL
242  03  03
243  32  X↕T
244  43  RCL
245  02  02
246  91  R/S
247  43  RCL
248  01  01
249  32  X↕T
250  43  RCL
251  04  04
252  92  RTN
253  76  LBL
254  24  CE
255  43  RCL
256  05  05
257  69  OP
258  04  04
259  43  RCL
260  01  01
261  69  OP
262  06  06
263  01  1
264  05  5
265  03  3
266  07  7
267  03  3
268  05  5
269  69  OP
270  04  04
271  43  RCL
272  02  02
273  69  OP
274  06  06
275  43  RCL
276  03  03
277  99  PRT
278  03  3
279  05  5
280  01  1
281  03  3
282  01  1
283  06  6
284  69  OP
285  04  04
286  43  RCL
287  04  04
288  69  OP
289  06  06
290  98  ADV
291  92  RTN
292  00  0
293  00  0
294  00  0
295  00  0
296  00  0
297  00  0
298  00  0
299  00  0
```

1 ◁			▷ 2
MAP-LOAD and MAP-SOURCE			MP-55.1
MAP-LOAD	MAP-SOURCE		

Step	Code	Key
000	76	LBL
001	85	+
002	86	STF
003	00	00
004	76	LBL
005	75	-
006	42	STO
007	01	01
008	73	RC*
009	01	01
010	87	IFF
011	00	00
012	00	00
013	15	15
014	94	+/-
015	32	X⇄T
016	42	STO
017	02	02
018	69	OP
019	21	21
020	73	RC*
021	01	01
022	37	P/R
023	42	STO
024	04	04
025	73	RC*
026	02	02
027	32	X⇄T
028	42	STO
029	03	03
030	69	OP
031	22	22
032	73	RC*
033	02	02
034	37	P/R
035	44	SUM
036	04	04
037	32	X⇄T
038	44	SUM
039	03	03
040	43	RCL
041	03	03
042	32	X⇄T
043	43	RCL
044	04	04
045	22	INV
046	37	P/R
047	32	X⇄T
048	22	INV
049	86	STF
050	00	00
051	92	RTN
052	76	LBL
053	65	×
054	42	STO
055	01	01
056	73	RC*
057	01	01
058	42	STO
059	03	03
060	69	OP
061	21	21
062	73	RC*
063	01	01
064	42	STO
065	04	04
066	32	X⇄T
067	42	STO
068	01	01
069	73	RC*
070	01	01
071	49	PRD
072	03	03
073	69	OP
074	21	21
075	73	RC*
076	01	01
077	44	SUM
078	04	04
079	43	RCL
080	04	04
081	32	X⇄T
082	43	RCL
083	03	03
084	92	RTN
085	76	LBL
086	24	CE
087	01	1
088	55	÷
089	53	(
090	01	1
091	75	-
092	43	RCL
093	20	20
094	33	X²
095	95	=
096	42	STO
097	05	05
098	43	RCL
099	20	20
100	65	×
101	43	RCL
102	05	05
103	95	=
104	42	STO
105	30	30
106	43	RCL
107	21	21
108	94	+/-
109	42	STO
110	31	31
111	01	1
112	42	STO
113	06	06
114	00	0
115	42	STO
116	07	07
117	01	1
118	94	+/-
119	49	PRD
120	21	21
121	06	6
122	32	X⇄T
123	02	2
124	00	0
125	71	SBR
126	75	-
127	42	STO
128	32	32
129	35	1/X
130	42	STO
131	34	34
132	06	6
133	32	X⇄T
134	42	STO
135	33	33
136	94	+/-
137	42	STO
138	35	35
139	02	2
140	00	0
141	71	SBR
142	85	+
143	42	STO
144	36	36
145	03	3
146	04	4
147	32	X⇄T
148	42	STO
149	37	37

```
150  03  3
151  06  6
152  71  SBR
153  65  ×
154  32  X↕T
155  37  P/R
156  42  STO
157  13  13
158  32  X↕T
159  42  STO
160  12  12
161  01  1
162  94  +/-
163  49  PRD
164  21  21
165  06  6
166  32  X↕T
167  02  2
168  00  0
169  71  SBR
170  75  -
171  35  1/X
172  42  STO
173  34  34
174  03  3
175  04  4
176  32  X↕T
177  94  +/-
178  42  STO
179  35  35
180  03  3
181  02  2
182  71  SBR
183  65  ×
184  42  STO
185  32  32
186  32  X↕T
187  42  STO
188  33  33
189  43  RCL
190  05  05
191  49  PRD
192  32  32
193  01  1
194  06  6
195  32  X↕T
196  01  1
197  08  8
198  71  SBR
199  65  ×
200  42  STO
201  34  34
202  03  3
203  04  4
204  32  X↕T
205  42  STO
206  35  35
207  03  3
208  02  2
209  71  SBR
210  65  ×
211  32  X↕T
212  37  P/R
213  42  STO
214  11  11
215  32  X↕T
216  42  STO
217  10  10
218  03  3
219  04  4
220  32  X↕T
221  03  3
222  00  0
223  71  SBR
224  65  ×
225  42  STO
226  36  36
227  03  3
228  06  6
229  32  X↕T
230  42  STO
231  37  37
232  01  1
233  04  4
234  71  SBR
235  85  +
236  32  X↕T
237  37  P/R
238  42  STO
239  09  09
240  32  X↕T
241  42  STO
242  08  08
243  92  RTN
244  76  LBL
245  11  A
246  71  SBR
247  24  CE
248  05  5
249  05  5
250  93  .
251  02  2
252  92  RTN
253  76  LBL
254  12  B
255  43  RCL
256  14  14
257  48  EXC
258  20  20
259  42  STO
260  14  14
261  43  RCL
262  15  15
263  48  EXC
264  21  21
265  42  STO
266  15  15
267  71  SBR
268  24  CE
269  43  RCL
270  14  14
271  48  EXC
272  20  20
273  42  STO
274  14  14
275  43  RCL
276  15  15
277  48  EXC
278  21  21
279  42  STO
280  15  15
281  05  5
282  05  5
283  93  .
284  02  2
285  92  RTN
286  00  0
287  00  0
288  00  0
289  00  0
290  00  0
291  00  0
292  00  0
293  00  0
294  00  0
```

1 ◀				▶ 2
MAP-LOAD AND MAP-SOURCE				MP-55.2
				PRT ON
ENT: ± R	ENT: ± X			PRT OFF

```
000  76 LBL
001  10 E'
002  86 STF
003  01  01
004  92 RTN
005  76 LBL
006  15  E
007  22 INV
008  86 STF
009  01  01
010  92 RTN
011  76 LBL
012  11  A
013  55  ÷
014  05  5
015  00  0
016  95  =
017  42 STO
018  01  01
019  03  3
020  05  5
021  42 STO
022  05  05
023  43 RCL
024  01  01
025  55  ÷
026  53  (
027  43 RCL
028  01  01
029  85  +
030  43 RCL
031  12  12
032  95  =
033  42 STO
034  06  06
035  00  0
036  42 STO
037  07  07
038  43 RCL
039  08  08
040  85  +
041  53  (
042  43 RCL
043  10  10
044  65  ×
045  43 RCL
046  06  06
047  54  )
048  75  -
049  53  (
050  43 RCL
051  11  11
052  65  ×
053  43 RCL
054  07  07
055  95  =
056  42 STO
057  02  02
058  43 RCL
059  09  09
060  85  +
061  53  (
062  43 RCL
063  10  10
064  65  ×
065  43 RCL
066  07  07
067  54  )
068  85  +
069  53  (
070  43 RCL
071  11  11
072  65  ×
073  43 RCL
074  06  06
075  95  =
076  42 STO
077  03  03
078  43 RCL
079  02  02
080  32 X⇄T
081  43 RCL
082  03  03
083  22 INV
084  37 P/R
085  42 STO
086  03  03
087  32 X⇄T
088  42 STO
089  02  02
090  43 RCL
091  12  12
092  55  ÷
093  53  (
094  43 RCL
095  01  01
096  85  +
097  43 RCL
098  12  12
099  95  =
100  50 I×I
101  42 STO
102  04  04
103  43 RCL
104  10  10
105  32 X⇄T
106  43 RCL
107  11  11
108  22 INV
109  37 P/R
110  32 X⇄T
111  49 PRD
112  04  04
113  05  5
114  00  0
115  49 PRD
116  01  01
117  87 IFF
118  01  01
119  24 CE
120  43 RCL
121  03  03
122  32 X⇄T
123  43 RCL
124  02  02
125  91 R/S
126  43 RCL
127  01  01
128  32 X⇄T
129  43 RCL
130  04  04
131  92 RTN
132  76 LBL
133  12  B
134  55  ÷
135  05  5
136  00  0
137  95  =
138  42 STO
139  01  01
140  04  4
141  04  4
142  42 STO
143  05  05
144  01  1
145  42 STO
146  06  06
147  43 RCL
148  12  12
149  55  ÷
```

```
150  53  (
151  43  RCL
152  01  01
153  75  -
154  43  RCL
155  13  13
156  95  =
157  42  STO
158  07  07
159  43  RCL
160  08  08
161  85  +
162  53  (
163  43  RCL
164  10  10
165  65  ×
166  43  RCL
167  06  06
168  54  )
169  75  -
170  53  (
171  43  RCL
172  11  11
173  65  ×
174  43  RCL
175  07  07
176  95  =
177  42  STO
178  02  02
179  43  RCL
180  09  09
181  85  +
182  53  (
183  43  RCL
184  10  10
185  65  ×
186  43  RCL
187  07  07
188  54  )
189  85  +
190  53  (
191  43  RCL
192  11  11
193  65  ×
194  43  RCL
195  06  06
196  95  =
197  42  STO
198  03  03
199  43  RCL
200  02  02
201  32  X:T
202  43  RCL
203  03  03
204  22  INV
205  37  P/R
206  42  STO
207  03  03
208  32  X:T
209  42  STO
210  02  02
211  43  RCL
212  12  12
213  55  ÷
214  53  (
215  43  RCL
216  01  01
217  75  -
218  43  RCL
219  13  13
220  95  =
221  50  I×I
222  42  STO
223  04  04
224  43  RCL
225  10  10
226  32  X:T
227  43  RCL
228  11  11
229  22  INV
230  37  P/R
231  32  X:T
232  49  PRD
233  04  04
234  05  5
235  00  0
236  49  PRD
237  01  01
238  87  IFF
239  01  01
240  24  CE
241  43  RCL
242  03  03
243  32  X:T
244  43  RCL
245  02  02
246  91  R/S
247  43  RCL
248  01  01
249  32  X:T
250  43  RCL
251  04  04
252  92  RTN
253  76  LBL
254  24  CE
255  43  RCL
256  05  05
257  69  OP
258  04  04
259  43  RCL
260  01  01
261  69  OP
262  06  06
263  01  1
264  05  5
265  03  3
266  07  7
267  03  3
268  05  5
269  69  OP
270  04  04
271  43  RCL
272  02  02
273  69  OP
274  06  06
275  43  RCL
276  03  03
277  99  PRT
278  03  3
279  05  5
280  01  1
281  03  3
282  01  1
283  06  6
284  69  OP
285  04  04
286  43  RCL
287  04  04
288  69  OP
289  06  06
290  98  ADV
291  92  RTN
292  00  0
293  00  0
294  00  0
295  00  0
296  00  0
297  00  0
298  00  0
```

1 ◀				▶ 2
LADDER ANALYSIS				MP-61
TRL	Z ⇒ Γ	Γ ⇒ VSWR/dB	PRT OUTPUT	USE PA
SE R	SE SLC	SE PLC	SE SST	SE OST

```
000  76  LBL
001  11   A
002  42  STO
003  05   05
004  00   0
005  42  STO
006  06   06
007  61  GTO
008  00   00
009  96   96
010  76  LBL
011  13   C
012  86  STF
013  02   02
014  76  LBL
015  12   B
016  42  STO
017  07   07
018  02   2
019  65   ×
020  43  RCL
021  00   00
022  65   ×
023  89   π
024  95   =
025  48  EXC
026  07   07
027  65   ×
028  43  RCL
029  07   07
030  95   =
031  35  1/X
032  65   ×
033  01   1
034  00   0
035  00   0
036  00   0
037  95   =
038  42  STO
039  08   08
040  32  X⇌T
041  65   ×
042  43  RCL
043  07   07
044  95   =
045  87  IFF
046  02   02
047  00   00
048  56   56
049  75   -
050  43  RCL
051  08   08
052  95   =
053  61  GTO
054  00   00
055  90   90
056  35  1/X
057  75   -
058  43  RCL
059  08   08
060  35  1/X
061  95   =
062  35  1/X
063  22  INV
064  86  STF
065  02   02
066  61  GTO
067  00   00
068  90   90
069  76  LBL
070  14   D
071  71  SBR
072  01   01
073  63   63
074  65   ×
075  32  X⇌T
076  95   =
077  61  GTO
078  00   00
079  90   90
080  76  LBL
081  15   E
082  71  SBR
083  01   01
084  63   63
085  94  +/-
086  35  1/X
087  65   ×
088  32  X⇌T
089  95   =
090  42  STO
091  05   05
092  09   9
093  00   0
094  42  STO
095  06   06
096  22  INV
097  87  IFF
098  01   01
099  01   01
100  21   21
101  43  RCL
102  03   03
103  35  1/X
104  42  STO
105  03   03
106  43  RCL
107  04   04
108  94  +/-
109  42  STO
110  04   04
111  43  RCL
112  05   05
113  35  1/X
114  42  STO
115  05   05
116  43  RCL
117  06   06
118  94  +/-
119  42  STO
120  06   06
121  43  RCL
122  03   03
123  32  X⇌T
124  43  RCL
125  04   04
126  37  P/R
127  48  EXC
128  05   05
129  32  X⇌T
130  48  EXC
131  06   06
132  37  P/R
133  44  SUM
134  05   05
135  32  X⇌T
136  44  SUM
137  06   06
138  43  RCL
139  06   06
140  32  X⇌T
141  43  RCL
142  05   05
143  22  INV
144  37  P/R
145  22  INV
146  87  IFF
147  01   01
148  01   01
149  54   54
```

```
150  94  +/-
151  32  X↕T
152  35  1/X
153  32  X↕T
154  42  STO
155  04   04
156  32  X↕T
157  42  STO
158  03   03
159  22  INV
160  86  STF
161  01   01
162  92  RTN
163  65   ×
164  03   3
165  00   0
166  93   .
167  04   4
168  08   8
169  65   ×
170  43  RCL
171  00   00
172  95   =
173  30  TAN
174  92  RTN
175  76  LBL
176  16  A'
177  71  SBR
178  01   01
179  63   63
180  42  STO
181  05   05
182  32  X↕T
183  42  STO
184  06   06
185  43  RCL
186  03   03
187  32  X↕T
188  43  RCL
189  04   04
190  37  P/R
191  42  STO
192  08   08
193  32  X↕T
194  42  STO
195  07   07
196  32  X↕T
197  43  RCL
198  06   06
199  65   ×
200  43  RCL
201  05   05
202  85   +
203  43  RCL
204  08   08
205  95   =
206  22  INV
207  37  P/R
208  42  STO
209  04   04
210  32  X↕T
211  65   ×
212  43  RCL
213  06   06
214  95   =
215  42  STO
216  03   03
217  43  RCL
218  06   06
219  75   -
220  43  RCL
221  08   08
222  65   ×
223  43  RCL
224  05   05
225  95   =
226  32  X↕T
227  43  RCL
228  07   07
229  65   ×
230  43  RCL
231  05   05
232  95   =
233  22  INV
234  37  P/R
235  22  INV
236  44  SUM
237  04   04
238  32  X↕T
239  22  INV
240  49  PRD
241  03   03
242  43  RCL
243  04   04
244  32  X↕T
245  43  RCL
246  03   03
247  92  RTN
248  76  LBL
249  17  B'
250  32  X↕T
251  37  P/R
252  32  X↕T
253  75   -
254  05   5
255  00   0
256  95   =
257  32  X↕T
258  22  INV
259  37  P/R
260  94  +/-
261  32  X↕T
262  35  1/X
263  65   ×
264  01   1
265  00   0
266  00   0
267  95   =
268  32  X↕T
269  37  P/R
270  32  X↕T
271  85   +
272  01   1
273  95   =
274  32  X↕T
275  22  INV
276  37  P/R
277  94  +/-
278  32  X↕T
279  35  1/X
280  92  RTN
281  76  LBL
282  18  C'
283  42  STO
284  05   05
285  33  X²
286  94  +/-
287  85   +
288  01   1
289  95   =
290  28  LOG
291  65   ×
292  01   1
293  00   0
294  95   =
295  32  X↕T
296  43  RCL
297  05   05
298  35  1/X
299  75   -
```

```
300  01  1
301  95  =
302  35  1/X
303  65  ×
304  02  2
305  85  +
306  01  1
307  95  =
308  86  STF
309  03  03
310  92  RTN
311  76  LBL
312  19  D'
313  22  INV
314  58  FIX
315  42  STO
316  05  05
317  02  2
318  01  1
319  03  3
320  05  5
321  01  1
322  07  7
323  03  3
324  04  4
325  69  OP
326  04  04
327  43  RCL
328  00  00
329  69  OP
330  06  06
331  87  IFF
332  03  03
333  03  03
334  61  61
335  03  3
336  00  0
337  01  1
338  03  3
339  02  2
340  02  2
341  69  OP
342  04  04
343  43  RCL
344  05  05
345  69  OP
346  06  06
347  01  1
348  03  3
349  03  3
350  01  1
351  02  2
352  02  2
353  69  OP
354  04  04
355  32  X:T
356  69  OP
357  06  06
358  61  GTO
359  03  03
360  84  84
361  04  4
362  02  2
363  03  3
364  06  6
365  04  4
366  03  3
367  03  3
368  05  5
369  69  OP
370  04  04
371  43  RCL
372  05  05
373  69  OP
374  06  06
375  01  1
376  06  6
377  01  1
378  04  4
379  69  OP
380  04  04
381  32  X:T
382  69  OP
383  06  06
384  32  X:T
385  43  RCL
386  05  05
387  98  ADV
388  22  INV
389  86  STF
390  03  03
391  92  RTN
392  76  LBL
393  10  E'
394  86  STF
395  01  01
396  92  RTN
397  22  INV
398  86  STF
399  04  04
400  91  R/S
401  76  LBL
402  95  =
403  22  INV
404  87  IFF
405  04  04
406  85  +
407  43  RCL
408  02  02
409  44  SUM
410  00  00
411  43  RCL
412  00  00
413  32  X:T
414  43  RCL
415  01  01
416  22  INV
417  77  GE
418  03  03
419  97  97
420  76  LBL
421  85  +
422  86  STF
423  04  04
424  76  LBL
425  25  CLR
```

APPENDIX B: HP-67/97 Program Listings

◁	ENTER S_{11}	ENTER S_{12}	ENTER S_{21}	ENTER S_{22}		MP-01
	RECALL S_{11}	RECALL S_{12}	RECALL S_{21}	RECALL S_{22}	USE SB	

```
01.0

*LBLA      21 11
F2?     16 23 02
GTO0       22 00
RCL5       36 05
RCL1       36 01
GSB9       23 09
RTN           24
*LBLB      21 12
F2?     16 23 02
GTO1       22 01
RCL6       36 06
RCL2       36 02
GSB9       23 09
RTN           24
*LBLC      21 13
F2?     16 23 02
GTO2       22 02
RCL7       36 07
RCL3       36 03
GSB9       23 09
RTN           24
*LBLD      21 14
F2?     16 23 02
GTO3       22 03
RCL8       36 08
RCL4       36 04
GSB9       23 09
RTN           24
*LBL0      21 00
P⇄S        16-51
GSBA       23 11
P⇄S        16-51
RTN           24
*LBL1      21 01
P⇄S        16-51
GSBB       23 12
P⇄S        16-51
RTN           24
*LBL2      21 02
P⇄S        16-51
GSBC       23 13
P⇄S        16-51
RTN           24
*LBL3      21 03
P⇄S        16-51
GSBD       23 14
P⇄S        16-51
RTN           24
*LBLa   21 16 11
F2?     16 23 02
GTO4       22 04
STO5       35 05
R↓           -31
STO1       35 01
RTN           24
*LBLb   21 16 12
F2?     16 23 02
GTO5       22 05
STO6       35 06
R↓           -31
STO2       35 02
RTN           24
*LBLc   21 16 13
F2?     16 23 02
GTO6       22 06
STO7       35 07
R↓           -31
STO3       35 03
RTN           24
*LBLd   21 16 14
F2?     16 23 02
GTO7       22 07
STO8       35 08
R↓           -31
STO4       35 04
RTN           24
*LBL4      21 04
P⇄S        16-51
GSBa    23 16 11
P⇄S        16-51
RTN           24
*LBL5      21 05
P⇄S        16-51
GSBb    23 16 12
P⇄S        16-51
RTN           24
*LBL6      21 06
P⇄S        16-51
GSBc    23 16 13
P⇄S        16-51
RTN           24
*LBL7      21 07
P⇄S        16-51
GSBd    23 16 14
P⇄S        16-51
RTN           24
*LBLE      21 15
SF2     16 21 02
RTN           24
*LBL9      21 09
→R            44
→P            34
PSE        16 51
X⇄Y          -41
PSE        16 51
X⇄Y          -41
RTN           24
R/S           51
```

◁	RECALL Z	RECALL Ymmho	RECALL H	RECALL G		MP-02
	ENTER Z	ENTER Ymmho	ENTER H	ENTER G		

```
        02.0
001  *LBLA    21 11
002   GSB0    23 00
003   STOI    35 46
004  *LBL1    21 01
005      1       01
006      1       01
007    R/S       51
008   STO5    35 05
009     R↓      -31
010   RCLI    36 46
011      ÷      -24
012   STO1    35 01
013      1       01
014      2       02
015    R/S       51
016   STO6    35 06
017     R↓      -31
018   RCLI    36 46
019      ÷      -24
020   STO2    35 02
021      2       02
022      1       01
023    R/S       51
024   STO7    35 07
025     R↓      -31
026   RCLI    36 46
027      ÷      -24
028   STO3    35 03
029      2       02
030      2       02
031    R/S       51
032   STO8    35 08
033     R↓      -31
034   RCLI    36 46
035      ÷      -24
036   STO4    35 04
037    RTN       24
038  *LBLa  21 16 11
039   GSB0    23 00
040   STOI    35 46
041  *LBL2    21 02
042   RCL5    36 05
043   RCL1    36 01
044   RCLI    36 46
045      x      -35
046     →R       44
047     →P       34
048      1       01
049      1       01
050    R/S       51
051   RCL6    36 06
052   RCL2    36 02
053   RCLI    36 46
054      x      -35
055     →R       44
056     →P       34
057      1       01
058      2       02
059    R/S       51
060   RCL7    36 07
061   RCL3    36 03
062   RCLI    36 46
063      x      -35
064     →R       44
065     →P       34
066      2       02
067      1       01
068    R/S       51
069   RCL8    36 08
070   RCL4    36 04
071   RCLI    36 46
072      x      -35
073     →R       44
074     →P       34
075      2       02
076      2       02
077    RTN       24
078  *LBLB    21 12
079   GSB0    23 00
080    EEX      -23
081    CHS      -22
082      3       03
083      x      -35
084    1/X       52
085   STOI    35 46
086   GTO1    22 01
087  *LBLb  21 16 12
088   GSB0    23 00
089    EEX      -23
090    CHS      -22
091      3       03
092      x      -35
093    1/X       52
094   STOI    35 46
095   GTO2    22 02
096  *LBLC    21 13
097   GSB0    23 00
098   STOI    35 46
099  *LBL3    21 03
100      1       01
101      1       01
102    R/S       51
103   STO5    35 05
104     R↓      -31
105   RCLI    36 46
106      ÷      -24
107   STO1    35 01
108      1       01
109      2       02
110    R/S       51
111   STO6    35 06
112     R↓      -31
113   STO2    35 02
114      2       02
115      1       01
116    R/S       51
117   STO7    35 07
118     R↓      -31
119   STO3    35 03
120      2       02
121      2       02
122    R/S       51
123   STO8    35 08
124     R↓      -31
125   RCLI    36 46
126      x      -35
127   STO4    35 04
128    RTN       24
129  *LBLc  21 16 13
130   GSB0    23 00
131   STOI    35 46
132  *LBL4    21 04
133   RCL5    36 05
134   RCL1    36 01
135   RCLI    36 46
136      x      -35
137     →R       44
138     →P       34
139      1       01
140      1       01
141    R/S       51
142   RCL6    36 06
143   RCL2    36 02
144     →R       44
145     →P       34
146      1       01
147      2       02
148    R/S       51
149   RCL7    36 07
```

```
 RCL3      36 03
   →R         44
   →P         34
    2         02
    1         01
  R/S         51
 RCL8      36 08
 RCL4      36 04
 RCLI      36 46
    ÷        -24
   →R         44
   →P         34
    2         02
    2         02
  R/S         51
  RTN         24
*LBLD      21 14
 GSB0      23 00
  1/X         52
 STOI      35 46
 GTO3      22 03
*LBLd   21 16 14
 GSB0      23 00
  1/X         52
 STOI      35 46
 GTO4      22 04
*LBL0      21 00
    5         05
    0         00
  RTN         24
  R/S         51
```

◁	$Z \Rightarrow \Gamma$	Ymmho $\Rightarrow \Gamma$	RFL LOSS $\rightleftharpoons \Gamma$	MW $\Rightarrow$ dBm	Rp, Xp $\Rightarrow$ Rs, Xs	MP-03
	$\Gamma \Rightarrow Z$	$\Gamma \Rightarrow$ Ymmho	RTN LOSS $\rightleftharpoons \Gamma$	VSWR $\rightleftharpoons \Gamma$	Rs, Xs $\Rightarrow$ Rp, Xp	

```
03.0

001  *LBLA      21 11
002   X⇄Y         -41
003    →R          44
004     1          01
005     +         -55
006    →P          34
007   1/X          52
008     2          02
009     ×         -35
010   X⇄Y         -41
011   CHS         -22
012   X⇄Y         -41
013   CHS         -22
014    →R          44
015     1          01
016     +         -55
017    →P          34
018   1/X          52
019     5          05
020     0          00
021     ×         -35
022   CHS         -22
023   X⇄Y         -41
024   CHS         -22
025   X⇄Y         -41
026    →R          44
027    →P          34
028   RTN          24
029  *LBLB      21 12
030  GSBA      23 11
031   1/X          52
032   EEX         -23
033     3          03
034     ×         -35
035   X⇄Y         -41
036   CHS         -22
037   X⇄Y         -41
038   RTN          24
039  *LBLa   21 16 11
040   X⇄Y         -41
041     5          05
042     0          00
043     ÷         -24
044  *LBL0      21 00
045    →R          44
046     1          01
047     -         -45
048    →P          34
049   1/X          52
050     2          02
051     ×         -35
052   X⇄Y         -41
053   CHS         -22
054   X⇄Y         -41
055    →R          44
056     1          01
057     +         -55
058    →P          34
059   1/X          52
060   X⇄Y         -41
061   CHS         -22
062   X⇄Y         -41
063    →R          44
064    →P          34
065   RTN          24
066  *LBLb   21 16 12
067   X⇄Y         -41
068     2          02
069     0          00
070     ÷         -24
071   1/X          52
072   X⇄Y         -41
073   CHS         -22
074   X⇄Y         -41
075  GTO0       22 00
076  *LBLE      21 15
077   X⇄Y         -41
078    →P          34
079   1/X          52
080    →R          44
081   1/X          52
082   X⇄Y         -41
083   1/X          52
084   X⇄Y         -41
085   RTN          24
086  *LBLC      21 13
087   F3?    16 23 03
088  GTO1       22 01
089  RCLI       36 46
090  *LBL1      21 01
091  X<0?       16-45
092  GTO2       22 02
093  STOI       35 46
094   LOG       16 32
095     2          02
096     0          00
097     ×         -35
098   CF3    16 22 03
099   RTN          24
100  *LBL2      21 02
101     2          02
102     0          00
103     ÷         -24
104   10ˣ       16 33
105  STOI       35 46
106   CF3    16 22 03
107   RTN          24
108  *LBLc   21 16 13
109   F3?    16 23 03
110  GTO3       22 03
111  RCLI       36 46
112  *LBL3      21 03
113  X<0?       16-45
114  GTO4       22 04
115  STOI       35 46
116    X²          53
117   CHS         -22
118     1          01
119     +         -55
120   LOG       16 32
121     1          01
122     0          00
123     ×         -35
124   CF3    16 22 03
125   RTN          24
126  *LBL4      21 04
127     1          01
128     0          00
129     ÷         -24
130   10ˣ       16 33
131   CHS         -22
132     1          01
133     +         -55
134    √X          54
135  STOI       35 46
136   CF3    16 22 03
137   RTN          24
138  *LBLD      21 14
139   F3?    16 23 03
140  GTO5       22 05
141  RCLI       36 46
142  *LBL5      21 05
143     1          01
144  X≤Y?       16-35
145  GTO6       22 06
146    R↓         -31
147  STOI       35 46
148     1          01
149     +         -55
```

```
1          01
RCLI    36 46
-         -45
÷         -24
CF3  16 22 03
RTN        24
*LBL6   21 06
R↓        -31
STOI    35 46
1          01
-         -45
RCLI    36 46
1          01
+         -55
÷         -24
STOI    35 46
CF3  16 22 03
RTN        24
*LBLd 21 16 14
LOG     16 32
1          01
0          00
x         -35
RTN        24
*LBLe 21 16 15
1/X        52
X⇄Y       -41
1/X        52
→P         34
1/X        52
→R         44
RTN        24
R/S        51
```

◁	MSG					MP-11
	ANALYZE	Ga max	Γml	Γms	K	

11.6

```
001  *LBLA      21 11
002    CF0   16 22 00
003   RCL7      36 07
004   RCL3      36 03
005   RCL6      36 06
006   RCL2      36 02
007   GSB9      23 09
008   STOI      35 46
009    CHS        -22
010   STO0      35 00
011     R↓        -31
012   STO9      35 09
013   RCL8      36 08
014   RCL4      36 04
015   RCL5      36 05
016   RCL1      36 01
017   GSB9      23 09
018   RCL9      36 09
019   RCL0      36 00
020   GSB8      23 08
021    CHS        -22
022   STO0      35 00
023     R↓        -31
024   STO9      35 09
025      1         01
026   RCL0      36 00
027     X²         53
028      +        -55
029   RCL1      36 01
030     X²         53
031      -        -45
032   RCL4      36 04
033     X²         53
034      -        -45
035      2         02
036      ÷        -24
037   RCLI      36 46
038    ABS      16 31
039      ÷        -24
040   STOI      35 46
041      1         01
042   X>Y?      16-34
043   GTO1      22 01
044   RCL8      36 08
045    CHS        -22
046   RCL4      36 04
047   RCL9      36 09
048   RCL0      36 00
049   GSB9      23 09
050   RCL5      36 05
051   RCL1      36 01
052   GSB8      23 08
053   STOB      35 12
054     R↓        -31
055    CHS        -22
056   STOC      35 13
057   RCL5      36 05
058    CHS        -22
059   RCL1      36 01
060   RCL9      36 09
061   RCL0      36 00
062   GSB9      23 09
063   RCL8      36 08
064   RCL4      36 04
065   GSB8      23 08
066   STOD      35 14
067     R↓        -31
068    CHS        -22
069   STOE      35 15
070      1         01
071   RCL1      36 01
072     X²         53
073      +        -55
074   RCL4      36 04
075     X²         53
076      -        -45
077   RCL0      36 00
078     X²         53
079   STO0      35 00
080      -        -45
081   STO9      35 09
082      1         01
083   RCL4      36 04
084     X²         53
085      +        -55
086   RCL1      36 01
087     X²         53
088      -        -45
089   RCL0      36 00
090      -        -45
091   STO0      35 00
092   RCL9      36 09
093     X²         53
094   RCLB      36 12
095     X²         53
096      4         04
097      x        -35
098      -        -45
099     √X         54
100   RCL9      36 09
101   GSB2      23 02
102   RCLB      36 12
103     X²         53
104      ÷        -24
105   RCLB      36 12
106      x        -35
107   STOB      35 12
108   RCL0      36 00
109     X²         53
110   RCLD      36 14
111     X²         53
112      4         04
113      x        -35
114      -        -45
115     √X         54
116   RCL0      36 00
117   GSB2      23 02
118   RCLD      36 14
119     X²         53
120      ÷        -24
121   RCLD      36 14
122      x        -35
123   STOD      35 14
124   RCLI      36 46
125     X²         53
126      1         01
127      -        -45
128     √X         54
129   RCL9      36 09
130   X<0?      16-45
131   GTO4      22 04
132     R↓        -31
133   RCLI      36 46
134    X⇄Y        -41
135      -        -45
136   GTO1      22 01
137  *LBL4      21 04
138     R↓        -31
139   RCLI      36 46
140      +        -55
141  *LBL1      21 01
142   RCL3      36 03
143      x        -35
144   RCL2      36 02
145      ÷        -24
146    ABS      16 31
147    LOG      16 32
148      1         01
149      0         00
```

```
x          -35
STO0       35 00
RCLI       36 46
1          01
X>Y?       16-34
SF0        16 21 00
X⇄Y        -41
RTN        24
*LBLB      21 12
0          00
F0?        16 23 00
1/X        52
RCL0       36 00
RTN        24
*LBLC      21 13
0          00
F0?        16 23 00
1/X        52
RCLD       36 14
RCLE       36 15
X⇄Y        -41
RTN        24
*LBLD      21 14
0          00
F0?        16 23 00
1/X        52
RCLB       36 12
RCLC       36 13
X⇄Y        -41
RTN        24
*LBLE      21 15
RCLI       36 46
RTN        24
*LBLa      21 16 11
RCL0       36 00
RTN        24
*LBL8      21 08
→R         44
R↓         -31
R↓         -31
→R         44
X⇄Y        -41
R↓         -31
+          -55
R↓         -31
+          -55
R↑         16-31
→P         34
RTN        24
*LBL9      21 09
X⇄Y        -41
R↓         -31
x          -35
R↓         -31
+          -55
R↑         16-31
RTN        24
*LBL2      21 02
X<0?       16-45
GTO3       22 03
X⇄Y        -41
-          -45
2          02
÷          -24
RTN        24
*LBL3      21 03
+          -55
2          02
÷          -24
RTN        24
R/S        51
```

◁				0 = outside 1 = inside	MP-12
	ANALYZE	INPUT PLANE	OUTPUT PLANE	10 = uncond.	

12.0

```
001  *LBLA    21 11
002   RCL7    36 07
003   RCL3    36 03
004   RCL6    36 06
005   RCL2    36 02
006   GSB9    23 09
007   STO0    35 00
008     R↓      -31
009   STO9    35 09
010   RCL8    36 08
011   RCL4    36 04
012   RCL5    36 05
013   RCL1    36 01
014   GSB9    23 09
015   RCL9    36 09
016   RCL0    36 00
017    CHS      -22
018   GSB8    23 08
019   STOB    35 12
020    X≷Y      -41
021   STOC    35 13
022    X≷Y      -41
023   RCL8    36 08
024    CHS      -22
025   RCL4    36 04
026   GSB9    23 09
027    CHS      -22
028   RCL5    36 05
029   RCL1    36 01
030   GSB8    23 08
031   STOD    35 14
032     R↓      -31
033    CHS      -22
034   STOE    35 15
035   RCLC    36 13
036   RCLB    36 12
037   RCL5    36 05
038    CHS      -22
039   RCL1    36 01
040   GSB9    23 09
041    CHS      -22
042   RCL8    36 08
043   RCL4    36 04
044   GSB8    23 08
045   STO9    35 09
046     R↓      -31
047    CHS      -22
048   STOI    35 46
049   RCL1    36 01

050     X²       53
051   RCLB    36 12
052     X²       53
053      -      -45
054   STOC    35 13
055   RCL4    36 04
056     X²       53
057   RCLB    36 12
058     X²       53
059      -      -45
060   RCL0    36 00
061   RCL9    36 09
062    P≷S    16-51
063   STO1    35 01
064     R↓      -31
065   STO2    35 02
066     R↓      -31
067   STO0    35 00
068   RCLI    36 46
069   STO3    35 03
070    P≷S    16-51
071      1       01
072   RCLB    36 12
073     X²       53
074      +      -55
075   RCL1    36 01
076     X²       53
077      -      -45
078   RCL4    36 04
079     X²       53
080      -      -45
081      2       02
082      ÷      -24
083   RCL0    36 00
084      ÷      -24
085   STOI    35 46
086    RTN       24
087  *LBL8    21 08
088     →R       44
089     R↓      -31
090     R↓      -31
091     →R       44
092    X≷Y      -41
093     R↓      -31
094      +      -55
095     R↓      -31
096      +      -55
097     R↑    16-31
098     →P       34
099    RTN       24

100  *LBL9    21 09
101    X≷Y      -41
102     R↓      -31
103      x      -35
104     R↓      -31
105      +      -55
106     R↑    16-31
107    RTN       24
108  *LBLB    21 12
109    P≷S    16-51
110   RCLE    36 15
111   STO8    35 08
112   RCLD    36 14
113   RCLC    36 13
114      ÷      -24
115   STO7    35 07
116   RCL2    36 02
117   RCLC    36 13
118      ÷      -24
119   STO9    35 09
120   GTOa  22 16 11
121  *LBLC    21 13
122    SF1  16 21 01
123    P≷S    16-51
124   RCL3    36 03
125   STO8    35 08
126   RCL1    36 01
127   RCL0    36 00
128      ÷      -24
129   STO7    35 07
130   RCL2    36 02
131   RCL0    36 00
132      ÷      -24
133   STO9    35 09
134  *LBLa  21 16 11
135    CF2  16 22 02
136    CF0  16 22 00
137    P≷S    16-51
138      1       01
139   RCL1    36 01
140    P≷S    16-51
141   X>Y?    16-34
142    SF2  16 21 02
143   X>Y?    16-34
144   GTO0    22 00
145      1       01
146    P≷S    16-51
147   RCL4    36 04
148    P≷S    16-51
149   X>Y?    16-34
```

```
150     SF0   16 21 00
151    X>Y?      16-34
152    GTO0      22 00
153    RCL2      36 02
154    RCLD      36 14
155       -        -45
156    RCLC      36 13
157       ÷        -24
158     ABS      16 31
159       1         01
160    X>Y?      16-34
161    GTO0      22 00
162    RCL2      36 02
163    RCL1      36 01
164       -        -45
165    RCL0      36 00
166       ÷        -24
167     ABS      16 31
168       1         01
169    X>Y?      16-34
170    GTO0      22 00
171       1         01
172       0         00
173    STO5      35 05
174    GTO5      22 05
175   *LBL0      21 00
176    RCL7      36 07
177     ABS      16 31
178    RCL9      36 09
179     ABS      16 31
180       -        -45
181    X>0?      16-44
182    GTO1      22 01
183       1         01
184    GTO2      22 02
185   *LBL1      21 01
186       0         00
187   *LBL2      21 02
188    STO5      35 05
189     F1?   16 23 01
190    GTO6      22 06
191     F0?   16 23 00
192    GTO7      22 07
193    GTO5      22 05
194   *LBL6      21 06
195     F2?   16 23 02
196    GTO7      22 07
197    GTO5      22 05
198   *LBL7      21 07
199    RCL5      36 05
200    X≠0?      16-42
201    GTO3      22 03
202       1         01
203    STO5      35 05
204    GTO5      22 05
205   *LBL3      21 03
206       0         00
207    STO5      35 05
208   *LBL5      21 05
209    RCL5      36 05
210    RCL9      36 09
211     ABS      16 31
212    RCL8      36 08
213    RCL7      36 07
214      →R         44
215      →P         34
216     P⇄S      16-51
217     CF1   16 22 01
218     CF0   16 22 00
219     RTN         24
220     R/S         51
```

	Ga max	MSG				MP-13
◁	INPUT GAIN CIRCLES	OUTPUT GAIN CIRCLES	S'_{11}	S'_{22}	K	

13.0

```
*LBLB     21 12
SF0    16 21 00
*LBLA     21 11
F0?    16 23 00
GSB1      23 01
GSB0      23 00
F0?    16 23 00
GSB1      23 01
CF0    16 22 00
2            02
1            01
R/S          51
*LBL5     21 05
1            01
0            00
÷           -24
10^X      16 33
RCL3      36 03
X²           53
÷           -24
P⇄S       16-51
STO0      35 00
1            01
RCLB      36 12
2            02
x           -35
RCLI      36 46
x           -35
RCL0      36 00
x           -35
-           -45
RCLB      36 12
X²           53
RCL0      36 00
X²           53
x           -35
+           -55
√X           54
RCL0      36 00
RCLE      36 15
x           -35
1            01
+           -55
STO9      35 09
÷           -24
RCL0      36 00
RCL9      36 09
÷           -24
P⇄S       16-51
RCL0      36 00
x           -35
RCL9      36 09
X⇄Y         -41
R/S          51
GTO5      22 05
*LBLC     21 13
STO0      35 00
X⇄Y         -41
STO9      35 09
RCL8      36 08
RCL4      36 04
GSB9      23 09
CHS         -22
→R           44
1            01
+           -55
→P           34
1/X          52
X⇄Y         -41
CHS         -22
X⇄Y         -41
RCL0      36 00
RCL9      36 09
GSB9      23 09
RCL7      36 07
RCL3      36 03
GSB9      23 09
RCL6      36 06
RCL2      36 02
GSB9      23 09
RCL5      36 05
RCL1      36 01
GSB8      23 08
RTN          24
*LBLD     21 14
GSB1      23 01
GSBC      23 13
GTO1      22 01
*LBLE     21 15
GSB0      23 00
RCLI      36 46
RTN          24
*LBLa  21 16 11
GSB0      23 00
1            01
RCL1      36 01
X²           53
+           -55
RCL4      36 04
X²           53
-           -45
RCLD      36 14
X²           53
-           -45
X>0?      16-44
SF2    16 21 02
0            00
RCLI      36 46
RCLI      36 46
X²           53
1            01
-           -45
√X           54
F2?    16 23 02
-           -45
+           -55
GTO6      22 06
*LBLb  21 16 12
1            01
*LBL6     21 06
RCL3      36 03
x           -35
RCL2      36 02
÷           -24
ABS       16 31
LOG       16 32
1            01
0            00
x           -35
RTN          24
*LBL0     21 00
RCL7      36 07
RCL3      36 03
RCL6      36 06
RCL2      36 02
GSB9      23 09
CHS         -22
STOB      35 12
R↓          -31
STOC      35 13
RCL8      36 08
RCL4      36 04
RCL5      36 05
RCL1      36 01
GSB9      23 09
RCLC      36 13
RCLB      36 12
GSB8      23 08
STOD      35 14
```

```
 R↓        -31
 STOE    35 15
 1          01
 RCLD    36 14
 X²         53
 +         -55
 RCL1    36 01
 X²         53
 -         -45
 RCL4    36 04
 X²         53
 -         -45
 2          02
 ÷         -24
 RCLB    36 12
 ABS     16 31
 STOB    35 12
 ÷         -24
 STOI    35 46
 RCL8    36 08
 CHS       -22
 RCL4    36 04
 RCLE    36 15
 RCLD    36 14
 GSB9    23 09
 CHS       -22
 RCL5    36 05
 RCL1    36 01
 GSB8    23 08
 STO0    35 00
 R↓        -31
 CHS       -22
 STO9    35 09
 RCL1    36 01
 X²         53
 RCLD    36 14
 X²         53
 -         -45
 STOE    35 15
 RTN        24
*LBL1    21 01
 RCL1    36 01
 RCL4    36 04
 STO1    35 01
 R↓        -31
 STO4    35 04
 R↓        -31
 RCL5    36 05
 RCL8    36 08
 STO5    35 05
 R↓        -31
 STO8    35 08
 R↓        -31
 RTN        24
*LBL8    21 08
 →R         44
 R↓        -31
 R↓        -31
 →R         44
 X⇄Y       -41
 R↓        -31
 +         -55
 R↓        -31
 +         -55
 R↑      16-31
 →P         34
 RTN        24
*LBL9    21 09
 X⇄Y       -41
 R↓        -31
 X         -35
 R↓        -31
 +         -55
 R↑      16-31
 RTN        24
```

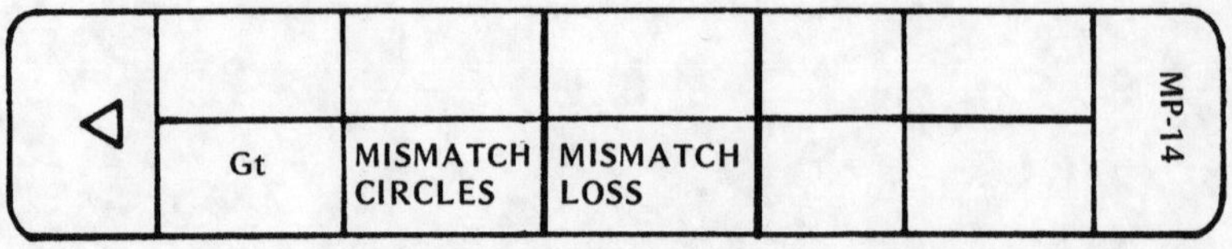

14.0

```
*LBLA    21 11
1           01
1           01
R/S         51
STO0     35 00
X⇄Y        -41
STO9     35 09
RCL5     36 05
RCL1     36 01
GSB9     23 09
CHS        -22
GSB7     23 07
STOD     35 14
R↓         -31
STOE     35 15
2           02
2           02
R/S         51
STOB     35 12
X⇄Y        -41
STOC     35 13
RCL8     36 08
RCL4     36 04
GSB9     23 09
CHS        -22
GSB7     23 07
RCLE     36 15
RCLD     36 14
GSB9     23 09
STOD     35 14
R↓         -31
STOE     35 15
RCL0     36 00
RCL9     36 09
RCLB     36 12
RCLC     36 13
GSB9     23 09
RCL6     36 06
RCL2     36 02
GSB9     23 09
RCL7     36 07
RCL3     36 03
GSB9     23 09
CHS        -22
RCLE     36 15
RCLD     36 14
GSB8     23 08
X²          53
1/X         52
RCL3     36 03
X²          53
x          -35
1           01
RCL9     36 09
X²          53
-          -45
x          -35
1           01
RCLC     36 13
X²          53
-          -45
x          -35
LOG      16 32
1           01
0           00
x          -35
RTN         24
*LBLB    21 12
CHS        -22
STO9     35 09
X⇄Y        -41
STO0     35 00
2           02
1           01
R/S         51
*LBL2    21 02
CHS        -22
1           01
0           00
÷          -24
10^x     16 33
STOB     35 12
1           01
RCL0     36 00
X²          53
1           01
RCLB     36 12
-          -45
x          -35
-          -45
1/X         52
STOC     35 13
RCLB     36 12
x          -35
RCL0     36 00
x          -35
STOD     35 14
1           01
RCL0     36 00
X²          53
-          -45
1           01
RCLB     36 12
-          -45
√X          54
x          -35
RCLC     36 13
x          -35
ABS      16 31
RCL9     36 09
RCLD     36 14
→R          44
→P          34
R/S         51
GTO2     22 02
*LBLC    21 13
1           01
1           01
R/S         51
STO9     35 09
X⇄Y        -41
STO0     35 00
2           02
2           02
R/S         51
STOC     35 13
X⇄Y        -41
STOB     35 12
RCL9     36 09
RCL0     36 00
GSB9     23 09
CHS        -22
GSB7     23 07
X²          53
1/X         52
1           01
RCLB     36 12
X²          53
-          -45
x          -35
1           01
RCL0     36 00
X²          53
-          -45
x          -35
LOG      16 32
1           01
0           00
x          -35
```

```
   RTN        24
 *LBL7     21 07
    →R        44
     1        01
     +       -55
    →P        34
   RTN        24
 *LBL8     21 08
    →R        44
    R↓       -31
    R↓       -31
    →R        44
   X⇄Y       -41
    R↓       -31
     +       -55
    R↓       -31
     +       -55
    R↑     16-31
    →P        34
   RTN        24
 *LBL9     21 09
   X⇄Y       -41
    R↓       -31
     ×       -35
    R↓       -31
     +       -55
    R↑     16-31
   RTN        24
   R/S        51
```

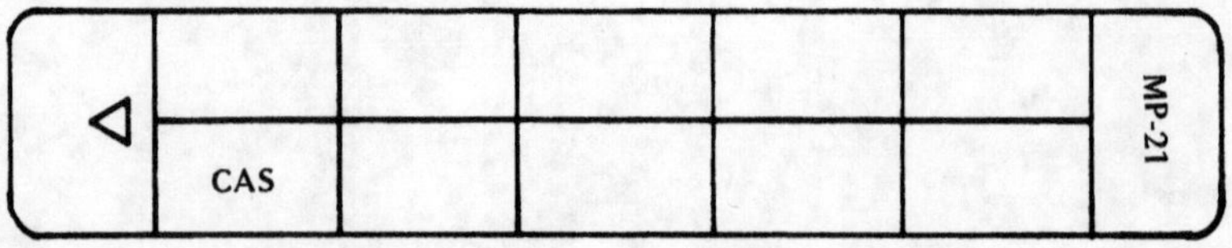


```
21.0

*LBLA    21 11
RCL8     36 08
RCL4     36 04
P⇄S      16-51
RCL5     36 05
RCL1     36 01
GSB9     23 09
CHS        -22
GSB7     23 07
GSB6     23 06
STOB     35 12
X⇄Y        -41
STOC     35 13
X⇄Y        -41
RCL5     36 05
RCL1     36 01
GSB9     23 09
P⇄S      16-51
RCL7     36 07
RCL3     36 03
GSB9     23 09
RCL6     36 06
RCL2     36 02
GSB9     23 09
RCL5     36 05
RCL1     36 01
GSB8     23 08
STO1     35 01
R↓         -31
STO5     35 05
RCLC     36 13
RCLB     36 12
RCL8     36 08
RCL4     36 04
GSB9     23 09
P⇄S      16-51
RCL7     36 07
RCL3     36 03
GSB9     23 09
RCL6     36 06
RCL2     36 02
GSB9     23 09
RCL8     36 08
RCL4     36 04
GSB8     23 08
P⇄S      16-51
STO4     35 04
R↓         -31
STO8     35 08
RCLC     36 13
RCLB     36 12
RCL6     36 06
RCL2     36 02
GSB9     23 09
P⇄S      16-51
RCL6     36 06
RCL2     36 02
GSB9     23 09
P⇄S      16-51
STO2     35 02
R↓         -31
STO6     35 06
RCL7     36 07
RCL3     36 03
RCLC     36 13
RCLB     36 12
GSB9     23 09
P⇄S      16-51
RCL7     36 07
RCL3     36 03
GSB9     23 09
P⇄S      16-51
STO3     35 03
R↓         -31
STO7     35 07
RTN         24
*LBL6    21 06
1/X         52
X⇄Y        -41
CHS        -22
X⇄Y        -41
RTN         24
*LBL7    21 07
→R          44
1           01
+          -55
→P          34
RTN         24
*LBL8    21 08
→R          44
R↓         -31
R↓         -31
→R          44
X⇄Y        -41
R↓         -31
+          -55
R↓         -31
+          -55
R↑       16-31
→P          34
RTN         24
*LBL9    21 09
X⇄Y        -41
R↓         -31
×          -35
R↓         -31
+          -55
R↑       16-31
RTN         24
R/S         51
```

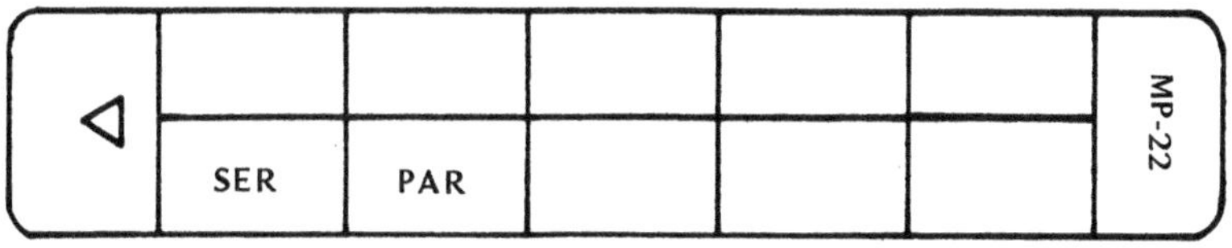

22.0

```
*LBLA    21 11
GSB4     23 04
GSB1     23 01
P⇄S      16-51
GSB4     23 04
GSB1     23 01
P⇄S      16-51
GSB2     23 02
GSB1     23 01
GSB4     23 04
RTN         24
*LBLB    21 12
GSB1     23 01
P⇄S      16-51
GSB1     23 01
P⇄S      16-51
GSB2     23 02
GSB1     23 01
RTN         24
*LBL4    21 04
1           01
8           08
0           00
ST-5  35-45 05
ST-6  35-45 06
ST-7  35-45 07
ST-8  35-45 08
RTN         24
*LBL2    21 02
P⇄S      16-51
RCL5     36 05
RCL1     36 01
P⇄S      16-51
RCL5     36 05
RCL1     36 01
GSB8     23 08
STO1     35 01
R↓         -31
STO5     35 05
P⇄S      16-51
RCL6     36 06
RCL2     36 02
P⇄S      16-51
RCL6     36 06
RCL2     36 02
GSB8     23 08
STO2     35 02
R↓         -31
STO6     35 06
P⇄S      16-51
RCL7     36 07
RCL3     36 03
P⇄S      16-51
RCL7     36 07
RCL3     36 03
GSB8     23 08
STO3     35 03
R↓         -31
STO7     35 07
P⇄S      16-51
RCL8     36 08
RCL4     36 04
P⇄S      16-51
RCL8     36 08
RCL4     36 04
GSB8     23 08
STO4     35 04
R↓         -31
STO8     35 08
RTN         24
*LBL1    21 01
RCL7     36 07
RCL3     36 03
RCL6     36 06
RCL2     36 02
GSB9     23 09
CHS        -22
STO0     35 00
R↓         -31
STO9     35 09
RCL5     36 05
RCL1     36 01
GSB7     23 07
STOB     35 12
R↓         -31
STOC     35 13
RCL8     36 08
RCL4     36 04
GSB7     23 07
RCLC     36 13
RCLB     36 12
GSB9     23 09
RCL9     36 09
RCL0     36 00
GSB8     23 08
1/X         52
2           02
×          -35
STO0     35 00
X⇄Y        -41
CHS        -22
1           01
8           08
0           00
STOI     35 46
-          -45
STO9     35 09
X⇄Y        -41
RCL6     36 06
RCL2     36 02
GSB9     23 09
STO2     35 02
R↓         -31
STO6     35 06
RCL9     36 09
RCL0     36 00
RCL7     36 07
RCL3     36 03
GSB9     23 09
STO3     35 03
R↓         -31
STO7     35 07
RCL8     36 08
RCL4     36 04
GSB7     23 07
RCL9     36 09
RCL0     36 00
GSB9     23 09
GSB7     23 07
STOD     35 14
R↓         -31
STOE     35 15
RCL5     36 05
RCL1     36 01
GSB7     23 07
RCL9     36 09
RCL0     36 00
GSB9     23 09
GSB7     23 07
STO4     35 04
R↓         -31
STO8     35 08
RCLD     36 14
STO1     35 01
RCLE     36 15
STO5     35 05
RCLI     36 46
ST-5  35-45 05
ST-8  35-45 08
```

```
150    RTN      24
151  *LBL7   21 07
152     →R      44
153      1      01
154      +     -55
155     →P      34
156    RTN      24
157  *LBL8   21 08
158     →R      44
159     R↓     -31
160     R↓     -31
161     →R      44
162    X⇄Y     -41
163     R↓     -31
164      +     -55
165     R↓     -31
166      +     -55
167     R↑   16-31
168     →P      34
169    RTN      24
170  *LBL9   21 09
171    X⇄Y     -41
172     R↓     -31
173      ÷     -35
174     R↓     -31
175      +     -55
176     R↑   16-31
177    RTN      24
178    R/S      51
```

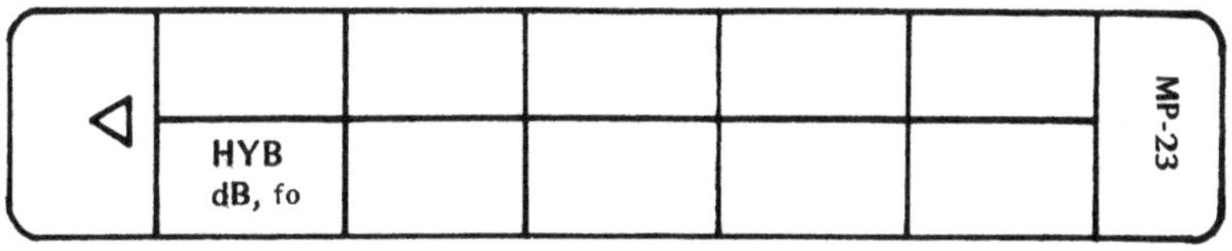

23.0

Step	Key	Code
001	*LBLA	21 11
002	9	09
003	0	00
004	X⇄Y	-41
005	÷	-24
006	EEX	-23
007	CHS	-22
008	8	08
009	+	-55
010	RCLA	36 11
011	×	-35
012	STOB	35 12
013	R↓	-31
014	CHS	-22
015	2	02
016	0	00
017	÷	-24
018	10ˣ	16 33
019	STOC	35 13
020	X²	53
021	CHS	-22
022	1	01
023	+	-55
024	STOD	35 14
025	√X	54
026	RCLB	36 12
027	COS	42
028	×	-35
029	RCLB	36 12
030	SIN	41
031	X⇄Y	-41
032	→P	34
033	R↓	-31
034	STOE	35 15
035	RCLD	36 14
036	1	01
037	RCLB	36 12
038	COS	42
039	RCLC	36 13
040	×	-35
041	X²	53
042	-	-45
043	÷	-24
044	√X	54
045	STOB	35 12
046	X²	53
047	CHS	-22
048	1	01
049	+	-55
050	√X	54
051	STOC	35 13
052	RCLE	36 15
053	CHS	-22
054	2	02
055	×	-35
056	STOD	35 14
057	9	09
058	0	00
059	+	-55
060	STOE	35 15
061	RCL5	36 05
062	RCL1	36 01
063	RCLC	36 13
064	X²	53
065	×	-35
066	CHS	-22
067	P⇄S	16-51
068	RCL5	36 05
069	RCLB	36 12
070	X²	53
071	ST×1	35-35 01
072	R↓	-31
073	RCL1	36 01
074	P⇄S	16-51
075	GSB8	23 08
076	STO1	35 01
077	R↓	-31
078	RCLD	36 14
079	+	-55
080	STO5	35 05
081	P⇄S	16-51
082	RCL8	36 08
083	RCL4	36 04
084	RCLC	36 13
085	X²	53
086	×	-35
087	CHS	-22
088	P⇄S	16-51
089	RCL8	36 08
090	RCLB	36 12
091	X²	53
092	ST×4	35-35 04
093	R↓	-31
094	RCL4	36 04
095	GSB8	23 08
096	STO4	35 04
097	R↓	-31
098	RCLD	36 14
099	+	-55
100	STO8	35 08
101	P⇄S	16-51
102	RCL6	36 06
103	RCL2	36 02
104	P⇄S	16-51
105	RCL6	36 06
106	RCL2	36 02
107	GSB8	23 08
108	RCLC	36 13
109	×	-35
110	RCLB	36 12
111	×	-35
112	STO2	35 02
113	R↓	-31
114	RCLE	36 15
115	+	-55
116	STO6	35 06
117	P⇄S	16-51
118	RCL7	36 07
119	RCL3	36 03
120	P⇄S	16-51
121	RCL7	36 07
122	RCL3	36 03
123	GSB8	23 08
124	RCLC	36 13
125	×	-35
126	RCLB	36 12
127	×	-35
128	STO3	35 03
129	R↓	-31
130	RCLE	36 15
131	+	-55
132	STO7	35 07
133	RTN	24
134	*LBL8	21 08
135	→R	44
136	R↓	-31
137	R↓	-31
138	→R	44
139	X⇄Y	-41
140	R↓	-31
141	+	-55
142	R↓	-31
143	+	-55
144	R↑	16-31
145	→P	34
146	RTN	24
147	R/S	51

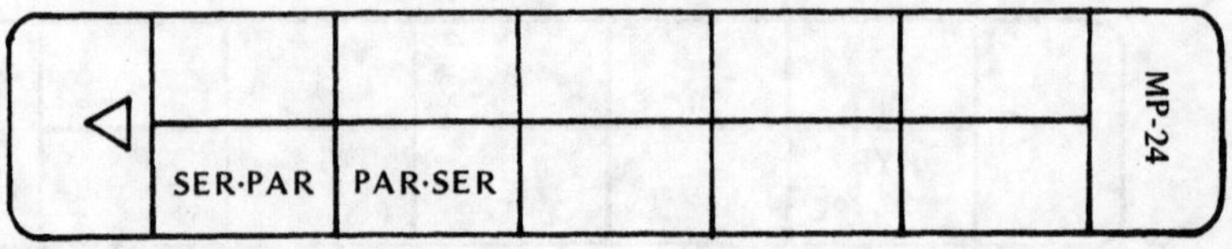

24.0

```
001  *LBLB       21 12
002    SF0    16 21 00
003  *LBLA       21 11
004    P⇄S       16-51
005   GSB3       23 03
006   GSB1       23 01
007    F0?    16 23 00
008   GSB4       23 04
009    P⇄S       16-51
010   GSB3       23 03
011   GSB1       23 01
012    F0?    16 23 00
013   GSB4       23 04
014   GSB2       23 02
015    F0?    16 23 00
016   GSB4       23 04
017   GSB1       23 01
018   GSB3       23 03
019    CF0    16 22 00
020    RTN          24
021  *LBL1       21 01
022   RCL7       36 07
023   RCL3       36 03
024   RCL6       36 06
025   RCL2       36 02
026   GSB9       23 09
027    CHS         -22
028   STO0       35 00
029     R↓         -31
030   STO9       35 09
031   RCL5       36 05
032   RCL1       36 01
033   GSB7       23 07
034   STOB       35 12
035     R↓         -31
036   STOC       35 13
037   RCL8       36 08
038   RCL4       36 04
039   GSB7       23 07
040   RCLC       36 13
041   RCLB       36 12
042   GSB9       23 09
043   RCL9       36 09
044   RCL0       36 00
045   GSB8       23 08
046    1/X          52
047      2          02
048      ×         -35
049   STO0       35 00
050    X⇄Y         -41
051    CHS         -22
052      1          01
053      8          08
054      0          00
055   STOI       35 46
056      -         -45
057   STO9       35 09
058    X⇄Y         -41
059   RCL6       36 06
060   RCL2       36 02
061   GSB9       23 09
062   STO2       35 02
063     R↓         -31
064   STO6       35 06
065   RCL9       36 09
066   RCL0       36 00
067   RCL7       36 07
068   RCL3       36 03
069   GSB9       23 09
070   STO3       35 03
071     R↓         -31
072   STO7       35 07
073   RCL8       36 08
074   RCL4       36 04
075   GSB7       23 07
076   RCL9       36 09
077   RCL0       36 00
078   GSB9       23 09
079   GSB7       23 07
080   STOD       35 14
081     R↓         -31
082   STOE       35 15
083   RCL5       36 05
084   RCL1       36 01
085   GSB7       23 07
086   RCL9       36 09
087   RCL0       36 00
088   GSB9       23 09
089   GSB7       23 07
090   STO4       35 04
091     R↓         -31
092   STO8       35 08
093   RCLD       36 14
094   STO1       35 01
095   RCLE       36 15
096   STO5       35 05
097   RCLI       36 46
098   ST-5    35-45 05
099   ST-8    35-45 08
100    RTN          24
101  *LBL2       21 02
102    P⇄S       16-51
103   RCL5       36 05
104   RCL1       36 01
105    P⇄S       16-51
106   RCL5       36 05
107   RCL1       36 01
108   GSB8       23 08
109   STO1       35 01
110     R↓         -31
111   STO5       35 05
112    P⇄S       16-51
113   RCL6       36 06
114   RCL2       36 02
115    P⇄S       16-51
116   RCL6       36 06
117   RCL2       36 02
118   GSB8       23 08
119   STO2       35 02
120     R↓         -31
121   STO6       35 06
122    P⇄S       16-51
123   RCL7       36 07
124   RCL3       36 03
125    P⇄S       16-51
126   RCL7       36 07
127   RCL3       36 03
128   GSB8       23 08
129   STO3       35 03
130     R↓         -31
131   STO7       35 07
132    P⇄S       16-51
133   RCL8       36 08
134   RCL4       36 04
135    P⇄S       16-51
136   RCL8       36 08
137   RCL4       36 04
138   GSB8       23 08
139   STO4       35 04
140     R↓         -31
141   STO8       35 08
142    RTN          24
143  *LBL3       21 03
144      1          01
145      8          08
146      0          00
147   ST-5    35-45 05
148   ST-6    35-45 06
149    RTN          24
```

```
150  *LBL4      21 04
151   RCL1      36 01
152   RCL4      36 04
153   STO1      35 01
154     R↓        -31
155   STO4      35 04
156   RCL5      36 05
157   RCL8      36 08
158   STO5      35 05
159     R↓        -31
160   STO8      35 08
161   RCLI      36 46
162   ST-6   35-45 06
163   ST-7   35-45 07
164   RCL7      36 07
165   RCL3      36 03
166   RCL6      36 06
167   RCL2      36 02
168   GSB9      23 09
169    CHS        -22
170   STO0      35 00
171     R↓        -31
172   STO9      35 09
173   RCL8      36 08
174   RCL4      36 04
175   RCL5      36 05
176   RCL1      36 01
177   GSB9      23 09
178   RCL9      36 09
179   RCL0      36 00
180   GSB8      23 08
181   ST÷1   35-24 01
182   ST÷2   35-24 02
183   ST÷3   35-24 03
184   ST÷4   35-24 04
185    X⇄Y        -41
186   ST-5   35-45 05
187   ST-6   35-45 06
188   ST-7   35-45 07
189   ST-8   35-45 08
190    RTN         24
191  *LBL7      21 07
192     →R         44
193      1         01
194      +        -55
195     →P         34
196    RTN         24
197  *LBL8      21 08
198     →R         44
199     R↓        -31
200     R↓        -31
201     →R         44
202    X⇄Y        -41
203     R↓        -31
204      ÷        -55
205     R↓        -31
206      +        -55
207     R↑     16-31
208     →P         34
209    RTN         24
210  *LBL9      21 09
211    X⇄Y        -41
212     R↓        -31
213      ×        -35
214     R↓        -31
215      +        -55
216     R↑     16-31
217    RTN         24
218    R/S         51
```

◁	Y ⇒ S	Z ⇒ S	H ⇒ S	G ⇒ S	G ⇌ H	MP-25
	S ⇒ Y	S ⇒ Z	S ⇒ H	S ⇒ G	Z ⇌ Y	

25.6

```
*LBLA      21 11
GSB1       23 01
RTN           24
*LBLa   21 16 11
GSB1       23 01
RTN           24
*LBLB      21 12
GSB2       23 02
GSB1       23 01
RTN           24
*LBLb   21 16 12
GSB1       23 01
GSB2       23 02
RTN           24
*LBLC      21 13
GSB3       23 03
GSB1       23 01
RTN           24
*LBLc   21 16 13
GSB1       23 01
GSB3       23 03
RTN           24
*LBLD      21 14
GSB3       23 03
GSB1       23 01
GSB4       23 04
RTN           24
*LBLd   21 16 14
GSB4       23 04
GSB1       23 01
GSB3       23 03
RTN           24
*LBLE      21 15
GSB4       23 04
RTN           24
*LBLe   21 16 15
GSB4       23 04
RTN           24
*LBL1      21 01
RCL7       36 07
RCL3       36 03
RCL6       36 06
RCL2       36 02
GSB9       23 09
CHS          -22
STO0       35 00
R↓           -31
STO9       35 09
RCL5       36 05
RCL1       36 01
GSB7       23 07
STOB       35 12
R↓           -31
STOC       35 13
RCL8       36 08
RCL4       36 04
GSB7       23 07
RCLC       36 13
RCLB       36 12
GSB9       23 09
RCL9       36 09
RCL0       36 00
GSB8       23 08
1/X           52
2             02
x            -35
STO0       35 00
X⇌Y          -41
CHS          -22
1             01
8             08
0             00
STOI       35 46
-            -45
STO9       35 09
X⇌Y          -41
RCL6       36 06
RCL2       36 02
GSB9       23 09
STO2       35 02
R↓           -31
STO6       35 06
RCL9       36 09
RCL0       36 00
RCL7       36 07
RCL3       36 03
GSB9       23 09
STO3       35 03
R↓           -31
STO7       35 07
RCL8       36 08
RCL4       36 04
GSB7       23 07
RCL9       36 09
RCL0       36 00
GSB9       23 09
GSB7       23 07
STOD       35 14
R↓           -31
STOE       35 15
RCL5       36 05
RCL1       36 01
GSB7       23 07
RCL9       36 09
RCL0       36 00
GSB9       23 09
GSB7       23 07
STO4       35 04
R↓           -31
STO8       35 08
RCLD       36 14
STO1       35 01
RCLE       36 15
STO5       35 05
RCLI       36 46
ST-5    35-45 05
ST-8    35-45 08
RTN           24
*LBL2      21 02
1             01
8             08
0             00
ST-5    35-45 05
ST-6    35-45 06
ST-7    35-45 07
ST-8    35-45 08
RTN           24
*LBL3      21 03
1             01
8             08
0             00
ST-5    35-45 05
ST-6    35-45 06
RTN           24
*LBL4      21 04
RCL1       36 01
RCL4       36 04
STO1       35 01
R↓           -31
STO4       35 04
RCL5       36 05
RCL8       36 08
STO5       35 05
R↓           -31
STO8       35 08
RCLI       36 46
ST-6    35-45 06
ST-7    35-45 07
RCL7       36 07
```

```
RCL3     36 03
RCL6     36 06
RCL2     36 02
GSB9     23 09
CHS        -22
STO0     35 00
R↓         -31
STO9     35 09
RCL8     36 08
RCL4     36 04
RCL5     36 05
RCL1     36 01
GSB9     23 09
RCL9     36 09
RCL0     36 00
GSB8     23 08
ST÷1  35-24 01
ST÷2  35-24 02
ST÷3  35-24 03
ST÷4  35-24 04
X⇄Y        -41
ST-5  35-45 05
ST-6  35-45 06
ST-7  35-45 07
ST-8  35-45 08
RTN         24
*LBL7    21 07
→R          44
1           01
+          -55
→P          34
RTN         24
*LBL8    21 08
→R          44
R↓         -31
R↓         -31
→R          44
X⇄Y        -41
R↓         -31
+          -55
R↓         -31
+          -55
R↑       16-31
→P          34
RTN         24
*LBL9    21 09
X⇄Y        -41
R↓         -31
x          -35
R↓         -31
+          -55
R↑       16-31
RTN         24
R/S         51
```

◁				[SA]	[SA]	MP-26
	USE SER	USE PAR	ENTER Z∠Θ	TERM SHO	TERM OPE	

26.0

```
*LBLA   21 11
RCL5    36 05
STOC    35 13
RCL1    36 01
STOB    35 12
P⇄S     16-51
GSB7    23 07
STOD    35 14
R↓        -31
STOE    35 15
RCLC    36 13
RCLB    36 12
CHS       -22
GSB7    23 07
1/X        52
X⇄Y       -41
CHS       -22
X⇄Y       -41
RCLE    36 15
RCLD    36 14
GSB9    23 09
2          02
÷         -24
STOB    35 12
X⇄Y       -41
STOC    35 13
X⇄Y       -41
GSB7    23 07
1/X        52
STO2    35 02
STO3    35 03
X⇄Y       -41
CHS       -22
STO6    35 06
STO7    35 07
X⇄Y       -41
RCLC    36 13
RCLB    36 12
GSB9    23 09
STO1    35 01
STO4    35 04
R↓        -31
STO5    35 05
STO8    35 08
GTO5    22 05
*LBLB   21 12
RCL5    36 05
STOC    35 13
RCL1    36 01
STOB    35 12
P⇄S     16-51
CHS       -22
GSB7    23 07
STOD    35 14
R↓        -31
STOE    35 15
RCLC    36 13
RCLB    36 12
GSB7    23 07
1/X        52
X⇄Y       -41
CHS       -22
X⇄Y       -41
RCLE    36 15
RCLD    36 14
GSB9    23 09
2          02
÷         -24
STOB    35 12
X⇄Y       -41
STOC    35 13
X⇄Y       -41
GSB7    23 07
1/X        52
STO2    35 02
STO3    35 03
X⇄Y       -41
CHS       -22
STO6    35 06
STO7    35 07
X⇄Y       -41
RCLC    36 13
RCLB    36 12
CHS       -22
GSB9    23 09
STO1    35 01
STO4    35 04
R↓        -31
STO5    35 05
STO8    35 08
*LBL5   21 05
P⇄S     16-51
2          02
6          06
RTN        24
*LBLD   21 14
RCL8    36 08
RCL4    36 04
CHS       -22
→R         44
1          01
CHS       -22
+         -55
→P         34
1/X        52
X⇄Y       -41
CHS       -22
X⇄Y       -41
RCL7    36 07
RCL3    36 03
GSB9    23 09
RCL6    36 06
RCL2    36 02
GSB9    23 09
RCL5    36 05
RCL1    36 01
GSB8    23 08
STO1    35 01
R↓        -31
STO5    35 05
2          02
6          06
RTN        24
*LBLE   21 15
RCL8    36 08
RCL4    36 04
CHS       -22
GSB7    23 07
1/X        52
X⇄Y       -41
CHS       -22
X⇄Y       -41
RCL7    36 07
RCL3    36 03
GSB9    23 09
RCL6    36 06
RCL2    36 02
GSB9    23 09
RCL5    36 05
RCL1    36 01
GSB8    23 08
STO1    35 01
R↓        -31
STO5    35 05
2          02
6          06
RTN        24
*LBLC   21 13
X⇄Y       -41
```

```
GSB1     23 01
÷          -24
→R          44
1           01
-          -45
→P          34
1/X         52
2           02
X          -35
X⇄Y        -41
CHS        -22
X⇄Y        -41
GSB7     23 07
1/X         52
STO1     35 01
X⇄Y        -41
CHS        -22
STO5     35 05
2           02
6           06
RTN         24
*LBL7    21 07
→R          44
1           01
+          -55
→P          34
RTN         24
*LBL8    21 08
→R          44
R↓         -31
R↓         -31
→R          44
X⇄Y        -41
R↓         -31
+          -55
R↓         -31
+          -55
R↑       16-31
→P          34
RTN         24
*LBL9    21 09
X⇄Y        -41
R↓         -31
X          -35
R↓         -31
+          -55
R↑       16-31
RTN         24
*LBL1    21 01
5           05
0           00
RTN         24
R/S         51
```

◁	SE SLC	SE SRLC			USE PA	MP-31
	SE R	SE L	SE C	SE SRL	SE SRC	

31.0

```
*LBLA     21 11
STOB      35 12
EEX         -23
CHS         -22
5            05
0            00
STOC      35 13
EEX         -23
5            05
0            00
STOD      35 14
GTO0      22 00
*LBLB     21 12
STOC      35 13
EEX         -23
CHS         -22
5            05
0            00
STOB      35 12
EEX         -23
5            05
0            00
STOD      35 14
GTO0      22 00
*LBLC     21 13
STOD      35 14
EEX         -23
CHS         -22
5            05
0            00
STOB      35 12
STOC      35 13
GTO0      22 00
*LBLD     21 14
STOC      35 13
R↓          -31
STOB      35 12
EEX         -23
5            05
0            00
STOD      35 14
GTO0      22 00
*LBLE     21 15
STOD      35 14
R↓          -31
STOB      35 12
EEX         -23
CHS         -22
5            05
0            00
STOC      35 13
GTO0      22 00
*LBLa  21 16 11
STOD      35 14
R↓          -31
STOC      35 13
EEX         -23
CHS         -22
5            05
0            00
STOB      35 12
GTO0      22 00
*LBLb  21 16 12
STOD      35 14
R↓          -31
STOC      35 13
R↓          -31
STOB      35 12
*LBL0     21 00
RCLA      36 11
2            02
×           -35
Pi        16-24
×           -35
STOE      35 15
RCLC      36 13
×           -35
STOC      35 13
EEX         -23
3            03
RCLE      36 15
÷           -24
RCLD      36 14
÷           -24
STOD      35 14
F0?    16 23 00
GSB9      23 09
F2?    16 23 02
GSB8      23 08
P⇄S       16-51
RCLC      36 13
RCLD      36 14
-           -45
RCLB      36 12
→P           34
5            05
0            00
÷           -24
STOB      35 12
R↓          -31
STOC      35 13
F1?    16 23 01
GSB7      23 07
RCLB      36 12
2            02
÷           -24
STOB      35 12
RCLC      36 13
RCLB      36 12
→R           44
1            01
+           -55
→P           34
1/X          52
X⇄Y         -41
CHS         -22
X⇄Y         -41
STO2      35 02
STO3      35 03
R↓          -31
STO6      35 06
STO7      35 07
F1?    16 23 01
GSB6      23 06
CF1    16 22 01
RCL7      36 07
RCLC      36 13
+           -55
RCL3      36 03
RCLB      36 12
×           -35
→R           44
→P           34
STO1      35 01
STO4      35 04
R↓          -31
STO5      35 05
STO8      35 08
P⇄S       16-51
RTN          24
*LBLc  21 16 13
SF0    16 21 00
÷           -24
RCLA      36 11
×           -35
STO0      35 00
RTN          24
*LBLd  21 16 14
SF2    16 21 02
```

```
150      ÷          -24
151    RCLA       36 11
152      X          -35
153    STO9       35 09
154     RTN          24
155   *LBLe    21 16 15
156     SF1    16 21 01
157     RTN          24
158   *LBL9       21 09
159    RCLC       36 13
160    RCL0       36 06
161      ÷          -24
162    RCLB       36 12
163      +          -55
164    STOB       35 12
165     CF0    16 22 00
166     RTN          24
167   *LBL6       21 08
168    RCLD       36 14
169    RCL9       36 09
170      ÷          -24
171    RCLB       36 12
172      +          -55
173    STOB       35 12
174     RTN          24
175   *LBL7       21 07
176    RCLC       36 13
177     CHS         -22
178    STOC       35 13
179    RCLB       36 12
180     1/X          52
181    STOB       35 12
182     RTN          24
183   *LBL6       21 06
184    RCLB       36 12
185     CHS         -22
186    STOB       35 12
187     RTN          24
188     R/S          51
```

◁	PA PLC	PA PRLC			USE SE	MP-32
	PA R	PA L	PA C	PA PRL	PA PRC	

32.0

```
*LBLA     21 11
STOB      35 12
EEX         -23
5            05
0            00
STOC      35 13
EEX         -23
CHS         -22
5            05
0            00
STOD      35 14
GTO0      22 00
*LBLB     21 12
STOC      35 13
EEX         -23
5            05
0            00
STOB      35 12
EEX         -23
CHS         -22
5            05
0            00
STOD      35 14
GTO0      22 00
*LBLC     21 13
STOD      35 14
EEX         -23
5            05
0            00
STOB      35 12
STOC      35 13
GTO0      22 00
*LBLD     21 14
STOC      35 13
R↓          -31
STOB      35 12
EEX         -23
CHS         -22
5            05
0            00
STOD      35 14
GTO0      22 00
*LBLE     21 15
STOD      35 14
R↓          -31
STOB      35 12
EEX         -23
5            05
0            00
STOC      35 13
GTO0      22 00
*LBLa  21 16 11
STOD      35 14
R↓          -31
STOC      35 13
EEX         -23
5            05
0            00
STOB      35 12
GTO0      22 00
*LBLb  21 16 12
STOD      35 14
R↓          -31
STOC      35 13
R↓          -31
STOB      35 12
*LBL0     21 00
RCLA      36 11
2            02
x           -35
Pi        16-24
x           -35
STOE      35 15
RCLC      36 13
x           -35
STOC      35 13
EEX         -23
3            03
RCLE      36 15
÷           -24
RCLD      36 14
÷           -24
STOD      35 14
F0?    16 23 00
GSB9      23 09
F2?    16 23 02
GSB8      23 08
P⇄S       16-51
RCLD      36 14
1/X          52
RCLC      36 13
1/X          52
-           -45
RCLB      36 12
1/X          52
→P           34
5            05
0            00
x           -35
STOB      35 12
X⇄Y         -41
STOC      35 13
F1?    16 23 01
GSB7      23 07
RCLB      36 12
2            02
÷           -24
STOB      35 12
RCLC      36 13
RCLB      36 12
→R           44
1            01
+           -55
→P           34
1/X          52
X⇄Y         -41
CHS         -22
X⇄Y         -41
STO2      35 02
STO3      35 03
R↓          -31
STO6      35 06
STO7      35 07
F1?    16 23 01
GSB6      23 06
GSB6      23 06
CF1    16 22 01
RCL7      36 07
RCLC      36 13
+           -55
RCL3      36 03
RCLB      36 12
x           -35
→R           44
→P           34
STO1      35 01
STO4      35 04
R↓          -31
STO5      35 05
STO8      35 08
P⇄S       16-51
RTN          24
*LBLc  21 16 13
SF0    16 21 00
÷           -24
RCLA      36 11
x           -35
STO0      35 00
RTN          24
```

```
150  *LBLd  21 16 14
151    SF2  16 21 02
152      ÷       -24
153   RCLA     36 11
154      x       -35
155   STO9     35 09
156    RTN        24
157  *LBLe  21 16 15
158    SF1  16 21 01
159    RTN        24
160  *LBL9     21 09
161   RCL0     36 00
162    1/X        52
163   RCL0     36 00
164      +       -55
165   RCLC     36 13
166      x       -35
167    1/X        52
168   RCLB     36 12
169    1/X        52
170      +       -55
171    1/X        52
172   STOB     35 12
173   RCL0     36 00
174     X²        53
175    1/X        52
176      1        01
177      +       -55
178   RCLC     36 13
179      x       -35
180   STOC     35 13
181    CF0  16 22 00
182    RTN        24
183  *LBL8     21 08
184   RCL9     36 09
185    1/X        52
186   RCL9     36 09
187      +       -55
188   RCLD     36 14
189      x       -35
190    1/X        52
191   RCLB     36 12
192    1/X        52
193      +       -55
194    1/X        52
195   STOB     35 12
196   RCL9     36 09
197     X²        53
198    1/X        52
199      1        01
200      +       -55
201   RCLD     36 14
202      x       -35
203   STOD     35 14
204    RTN        24
205  *LBL7     21 07
206   RCLC     36 13
207    CHS       -22
208   STOC     35 13
209   RCLB     36 12
210    1/X        52
211   STOB     35 12
212    RTN        24
213  *LBL6     21 06
214   RCLB     36 12
215    CHS       -22
216   STOB     35 12
217    RTN        24
218    R/S        51
```

◁					USE SE	MP-33
	TRL	OST	SST			

33.0

```
*LBLA      21 11
GSBa    23 16 11
eˣ            33
RCLD       36 14
X²            53
CHS          -22
x            -35
GSB8       23 08
GSB9       23 09
STOB       35 12
R↓           -31
STOE       35 15
RCLC       36 13
2             02
x            -35
RCLI       36 46
2             02
x            -35
eˣ            33
CHS          -22
GSB8       23 08
RCLD       36 14
x            -35
RCLE       36 15
RCLB       36 12
GSB5       23 05
→R            44
→P            34
P⇄S        16-51
GSB2       23 02
RCLC       36 13
RCLI       36 46
eˣ            33
1             01
RCLD       36 14
X²            53
-            -45
x            -35
RCLE       36 15
RCLB       36 12
GSB5       23 05
GSB3       23 03
P⇄S        16-51
RTN           24
*LBLa   21 16 11
STOI       35 46
RCLA       36 11
x            -35
3             03
0             00
.            -62
4             04
8             08
x            -35
F2?     16 23 02
GSB6       23 06
CHS          -22
STOC       35 13
R↓           -31
5             05
0             00
÷            -24
STOB       35 12
1             01
-            -45
RCLB       36 12
1             01
+            -55
÷            -24
STOD       35 14
0             00
F0?     16 23 00
RCL9       36 09
RCLI       36 46
x            -35
CHS          -22
STOI       35 46
RCLC       36 13
2             02
x            -35
X⇄Y          -41
CF0     16 22 00
RTN           24
*LBLB      21 12
GSBa    23 16 11
1             01
STOD       35 14
GTO1       22 01
*LBLC      21 13
GSBa    23 16 11
1             01
CHS          -22
STOD       35 14
*LBL1      21 01
RCLC       36 13
2             02
x            -35
STO0       35 00
RCLI       36 46
2             02
x            -35
eˣ            33
RCLD       36 14
x            -35
STO9       35 09
CHS          -22
GSB8       23 08
GSB9       23 09
STOD       35 14
R↓           -31
STOE       35 15
RCL0       36 00
RCL9       36 09
GSB8       23 08
RCLE       36 15
RCLD       36 14
GSB5       23 05
RCLB       36 12
x            -35
F1?     16 23 01
GSB9       23 09
GSB9       23 09
2             02
÷            -24
STOB       35 12
X⇄Y          -41
STOC       35 13
X⇄Y          -41
→R            44
1             01
+            -55
→P            34
1/X           52
X⇄Y          -41
CHS          -22
X⇄Y          -41
P⇄S        16-51
GSB3       23 03
F1?     16 23 01
GSB7       23 07
GSB7       23 07
CF1     16 22 01
RCL7       36 07
RCLC       36 13
+            -55
RCL3       36 03
RCLB       36 12
x            -35
→R            44
```

```
150      →P        34
151    GSB2     23 02
152     P⇄S     16-51
153     RTN        24
154  *LBLe   21 16 15
155     SF1  16 21 01
156     RTN        24
157  *LBLD      21 14
158    RCLA     36 11
159       ÷       -24
160      √X        54
161       x       -35
162       2        02
163       0        00
164       ÷       -24
165       1        01
166      eˣ        33
167     LOG     16 32
168       ÷       -24
169    STO9     35 09
170     SF0  16 21 00
171     RTN        24
172  *LBLE      21 15
173     SF2  16 21 02
174    STO0     35 00
175     RTN        24
176  *LBL2      21 02
177    STO1     35 01
178    STO4     35 04
179      R↓       -31
180    STO5     35 05
181    STO8     35 08
182     RTN        24
183  *LBL3      21 03
184    STO2     35 02
185    STO3     35 03
186      R↓       -31
187    STO6     35 06
188    STO7     35 07
189     RTN        24
190  *LBL5      21 05
191     X⇄Y       -41
192      R↓       -31
193       x       -35
194      R↓       -31
195       +       -55
196      R↑     16-31
197     RTN        24
198  *LBL6      21 06
199    RCL0     36 00
200      √X        54
201       x       -35
202     RTN        24
203  *LBL7      21 07
204    RCLB     36 12
205     CHS       -22
206    STOB     35 12
207     RTN        24
208  *LBL8      21 08
209      →R        44
210       1        01
211       +       -55
212      →P        34
213     RTN        24
214  *LBL9      21 09
215     1/X        52
216     X⇄Y       -41
217     CHS       -22
218     X⇄Y       -41
219     RTN        24
220     R/S        51
```

◁	EQU	CDVS a, Θ	VDVS μ, Θ	CDCS β, Θ	VDCS gm, Θ	MP-41
	TRF: Z_1, Z_2	TRF: N_1, N_2	SHO	OPE	IMG	

41.0

```
*LBLA    21 11
÷          -24
√X          54
GTO0     22 00
*LBLB    21 12
÷          -24
*LBL0    21 00
P⇄S      16-51
STO9     35 09
X²          53
1           01
-          -45
RCL9     36 09
X²          53
1           01
+          -55
÷          -24
STO1     35 01
CHS        -22
STO4     35 04
2           02
RCL9     36 09
x          -35
RCL9     36 09
X²          53
1           01
+          -55
÷          -24
STO2     35 02
STO3     35 03
0           00
STO5     35 05
STO6     35 06
STO7     35 07
STO8     35 08
GTO9     22 09
*LBLC    21 13
P⇄S      16-51
1           01
STO1     35 01
STO4     35 04
1           01
8           08
0           00
STO5     35 05
STO8     35 08
0           00
STO2     35 02
STO3     35 03
STO6     35 06
STO7     35 07
GTO9     22 09
*LBLD    21 14
P⇄S      16-51
1           01
STO1     35 01
STO4     35 04
0           00
STO2     35 02
STO3     35 03
STO5     35 05
STO6     35 06
STO7     35 07
STO8     35 08
GTO9     22 09
*LBLE    21 15
RCL6     36 06
RCL2     36 02
RCL5     36 05
RCL1     36 01
P⇄S      16-51
STO4     35 04
R↓         -31
STO8     35 08
R↓         -31
STO3     35 03
R↓         -31
STO7     35 07
P⇄S      16-51
RCL8     36 08
RCL4     36 04
RCL7     36 07
RCL3     36 03
P⇄S      16-51
STO2     35 02
R↓         -31
STO6     35 06
R↓         -31
STO1     35 01
R↓         -31
STO5     35 05
GTO9     22 09
*LBLa    21 16 11
RCL6     36 06
RCL2     36 02
RCL5     36 05
RCL1     36 01
P⇄S      16-51
STO1     35 01
R↓         -31
STO5     35 05
R↓         -31
STO2     35 02
R↓         -31
STO6     35 06
P⇄S      16-51
RCL8     36 08
RCL4     36 04
RCL7     36 07
RCL3     36 03
P⇄S      16-51
STO3     35 03
R↓         -31
STO7     35 07
R↓         -31
STO4     35 04
R↓         -31
STO8     35 08
GTO9     22 09
*LBLb    21 16 12
P⇄S      16-51
STO7     35 07
R↓         -31
2           02
x          -35
GSB1     23 01
÷          -24
STO3     35 03
0           00
STO2     35 02
STO6     35 06
1           01
STO1     35 01
STO4     35 04
1           01
8           08
0           00
STO5     35 05
STO8     35 08
GTO9     22 09
*LBLc    21 16 13
P⇄S      16-51
STO7     35 07
R↓         -31
2           02
x          -35
STO3     35 03
1           01
STO1     35 01
```

```
150   STO4      35 04
151   0            00
152   STO2      35 02
153   STO5      35 05
154   STO6      35 06
155   1            01
156   8            08
157   0            00
158   STO8      35 08
159   GTO9      22 09
160   *LBLd  21 16 14
161   P⇄S       16-51
162   1            01
163   8            08
164   0            00
165   STO5      35 05
166   +           -55
167   STO7      35 07
168   R↓          -31
169   2            02
170   x           -35
171   STO3      35 03
172   1            01
173   STO1      35 01
174   STO4      35 04
175   0            00
176   STO2      35 02
177   STO6      35 06
178   STO8      35 08
179   GTO9      22 09
180   *LBLe  21 16 15
181   P⇄S       16-51
182   1            01
183   8            08
184   0            00
185   +           -55
186   STO7      35 07
187   R↓          -31
188   2            02
189   x           -35
190   GSB1      23 01
191   x           -35
192   STO3      35 03
193   1            01
194   STO1      35 01
195   STO4      35 04
196   0            00
197   STO5      35 05
198   STO6      35 06
199   STO8      35 08
200   STO2      35 02
201   *LBL9     21 09
202   P⇄S       16-51
203   4            04
204   1            01
205   RTN          24
206   *LBL1     21 01
207   5            05
208   0            00
209   RTN          24
210   R/S          51
```

	Zoe, Zoo fo					MP-42.1
◁	CPL	DIR	ISO	SHO	OPE	

42.1

```
*LBLa   21 16 11
9             09
0             00
X⇄Y          -41
÷            -24
RCLA       36 11
×            -35
STOB       35 12
R↓           -31
STO0       35 00
X⇄Y          -41
STO9       35 09
X⇄Y          -41
P⇄S        16-51
÷            -24
STOI       35 46
1             01
-            -45
RCLI       36 46
1             01
+            -55
÷            -24
STOC       35 13
X²            53
CHS          -22
1             01
+            -55
STOD       35 14
√X            54
RCLB       36 12
COS           42
×            -35
RCLB       36 12
SIN           41
X⇄Y          -41
→P            34
R↓           -31
CHS          -22
STOE       35 15
RCLD       36 14
1             01
RCLB       36 12
COS           42
RCLC       36 13
×            -35
X²            53
-            -45
÷            -24
√X            54
STOB       35 12
X²            53
CHS          -22
1             01
+            -55
√X            54
STOC       35 13
RCLE       36 15
9             09
0             00
+            -55
STOD       35 14
RTN           24
*LBLA      21 11
SF1     16 21 01
*LBLB      21 12
GSBa    23 16 11
F1?     16 23 01
GTO0       22 00
RCLB       36 12
RCLC       36 13
STOB       35 12
R↓           -31
STOC       35 13
RCLD       36 14
RCLE       36 15
STOD       35 14
R↓           -31
STOE       35 15
*LBL0      21 00
1             01
STOI       35 46
2             02
F1?     16 23 01
3             03
R/S           51
*LBL1      21 01
STO1       35 01
R↓           -31
STO0       35 00
2             02
STOI       35 46
4             04
R/S           51
*LBL2      21 02
STO9       35 09
R↓           -31
STO8       35 08
CF1     16 22 01
RCLD       36 14
2             02
×            -35
RCLC       36 13
X²            53
RCL9       36 09
RCL8       36 08
GSB9       23 09
RCL1       36 01
RCL0       36 00
GSB9       23 09
CHS          -22
GSB7       23 07
1/X           52
STO3       35 03
R↓           -31
CHS          -22
STO7       35 07
RCLE       36 15
2             02
×            -35
RCLB       36 12
X²            53
RCL7       36 07
RCL3       36 03
GSB9       23 09
STOB       35 12
X⇄Y          -41
STOE       35 15
X⇄Y          -41
RCLD       36 14
RCLC       36 13
GSB9       23 09
RCL1       36 01
RCL0       36 00
GSB9       23 09
RCL9       36 09
RCL8       36 08
GSB9       23 09
RCLD       36 14
RCLC       36 13
GSB8       23 08
STO2       35 02
R↓           -31
STO6       35 06
RCLE       36 15
RCLB       36 12
RCL1       36 01
RCL0       36 00
GSB9       23 09
STO1       35 01
```

```
R↓        -31
STO5    35 05
RCL9    36 09
RCL8    36 08
RCLE    36 15
RCLB    36 12
GSB9    23 09
STO4    35 04
R↓        -31
STO8    35 08
RCL2    36 02
STO3    35 03
RCL6    36 06
STO7    35 07
*LBL5   21 05
PSE     16 51
RTN        24
*LBLD   21 14
1          01
ENT↑      -21
1          01
8          08
0          00
GTOi    22 45
*LBLE   21 15
1          01
ENT↑      -21
0          00
GTOi    22 45
*LBLC   21 13
GSBa 23 16 11
3          03
STOI    35 46
2          02
R/S        51
*LBL3   21 03
STO1    35 01
R↓        -31
STO0    35 00
4          04
STOI    35 46
3          03
R/S        51
*LBL4   21 04
STO9    35 09
R↓        -31
STO8    35 08
GTO5    22 05
*LBL7   21 07
→R         44
1          01
+         -55
→P         34
RTN        24
*LBL8   21 08
→R         44
R↓        -31
R↓        -31
→R         44
X⇄Y       -41
R↓        -31
+         -55
R↓        -31
+         -55
R↑     16-31
→P         34
RTN        24
*LBL9   21 09
X⇄Y       -41
R↓        -31
×         -35
R↓        -31
+         -55
R↑     16-31
RTN        24
```

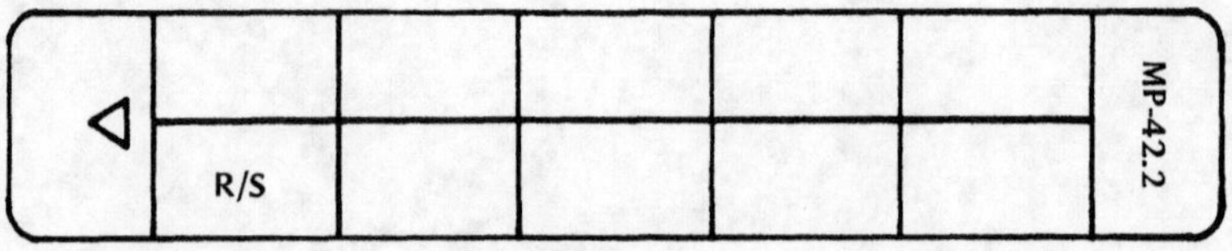

42.2

```
GTOi     22 45
*LBL4    21 04
RCL9     36 09
RCL8     36 08
RCL1     36 01
RCL0     36 00
GSB8     23 08
RCLE     36 15
RCLB     36 12
GSB9     23 09
RCLD     36 14
RCLC     36 13
GSB9     23 09
STO2     35 02
STO3     35 03
R↓         -31
STO6     35 06
STO7     35 07
RCLE     36 15
2           02
x          -35
STOE     35 15
RCLB     36 12
X²          53
STOB     35 12
RCL9     36 09
RCL8     36 08
GSB9     23 09
STO4     35 04
R↓         -31
STO5     35 05
RCLD     36 14
2           02
x          -35
STOD     35 14
RCLC     36 13
X²          53
STOC     35 13
RCL1     36 01
RCL0     36 00
GSB9     23 09
RCL5     36 05
RCL4     36 04
GSB8     23 08
STO4     35 04
R↓         -31
STO5     35 05
RCL1     36 01
RCL0     36 00
RCLE     36 15
RCLB     36 12
GSB9     23 09
STO0     35 00
R↓         -31
STO1     35 01
RCLD     36 14
RCLC     36 13
RCL9     36 09
RCL8     36 08
GSB9     23 09
RCL1     36 01
RCL0     36 00
GSB8     23 08
RCL5     36 05
RCL4     36 04
STO1     35 01
R↓         -31
STO5     35 05
R↓         -31
STO4     35 04
R↓         -31
STO8     35 08
*LBL2    21 02
P⇄S      16-51
1           01
RCL0     36 00
RCL9     36 09
x          -35
√X          54
5           05
0           00
÷          -24
STOI     35 46
-          -45
1           01
RCLI     36 46
+          -55
÷          -24
P⇄S      16-51
STOI     35 46
RCL6     36 06
RCL2     36 02
RCL7     36 07
RCL3     36 03
GSB9     23 09
RCLI     36 46
x          -35
STOD     35 14
X⇄Y        -41
STOE     35 15
RCL8     36 08
RCL4     36 04
RCLI     36 46
x          -35
CHS        -22
GSB7     23 07
STOB     35 12
R↓         -31
STOC     35 13
RCL5     36 05
RCL1     36 01
RCLI     36 46
x          -35
CHS        -22
GSB7     23 07
P⇄S      16-51
STO0     35 00
X⇄Y        -41
STO9     35 09
X⇄Y        -41
P⇄S      16-51
RCLC     36 13
RCLB     36 12
GSB9     23 09
RCLD     36 14
RCLI     36 46
x          -35
CHS        -22
RCLE     36 15
X⇄Y        -41
GSB8     23 08
1/X         52
STO0     35 00
R↓         -31
CHS        -22
STO9     35 09
0           00
RCLI     36 46
CHS        -22
RCL5     36 05
RCL1     36 01
GSB8     23 08
RCLC     36 13
RCLB     36 12
GSB9     23 09
RCLE     36 15
RCLD     36 14
GSB8     23 08
RCL9     36 09
```

```
150   RCL0      36 00
151   GSB9      23 09
152   STO1      35 01
153   R↓          -31
154   STO5      35 05
155   0            00
156   RCLI      36 46
157   CHS         -22
158   RCL8      36 08
159   RCL4      36 04
160   GSB8      23 08
161   P⇄S       16-51
162   RCL9      36 09
163   RCL0      36 00
164   P⇄S       16-51
165   GSB9      23 09
166   RCLE      36 15
167   RCLD      36 14
168   GSB8      23 08
169   RCL9      36 09
170   RCL0      36 00
171   GSB9      23 09
172   STO4      35 04
173   R↓          -31
174   STO8      35 08
175   1            01
176   RCLI      36 46
177   X²           53
178   -           -45
179   ST×0   35-35 00
180   RCL6      36 06
181   RCL2      36 02
182   RCL9      36 09
183   RCL0      36 00
184   GSB9      23 09
185   STO2      35 02
186   R↓          -31
187   STO6      35 06
188   RCL7      36 07
189   RCL3      36 03
190   RCL9      36 09
191   RCL0      36 00
192   GSB9      23 09
193   STO3      35 03
194   R↓          -31
195   STO7      35 07
196   P⇄S       16-51
197   R/S          51
198   *LBL7     21 07
199   →R           44
200   1            01
201   +           -55
202   →P           34
203   RTN          24
204   *LBL8     21 08
205   →R           44
206   R↓          -31
207   R↓          -31
208   →R           44
209   X⇄Y         -41
210   R↓          -31
211   +           -55
212   R↓          -31
213   +           -55
214   R↑        16-31
215   →P           34
216   RTN          24
217   *LBL9     21 09
218   X⇄Y         -41
219   R↓          -31
220   ×           -35
221   R↓          -31
222   +           -55
223   R↑        16-31
224   RTN          24
```

					MP-51.1
SERIES FB	SHUNT FB	CONFIG CONVER			

51.1

```
001  *LBLC     21 13
002  *LBLA     21 11
003   GSB4     23 04
004   GSB1     23 01
005   RCL5     36 05
006   RCL1     36 01
007   GSB7     23 07
008   STO1     35 01
009     R↓       -31
010   STO5     35 05
011   RCL6     36 06
012   RCL2     36 02
013   GSB7     23 07
014   STO2     35 02
015     R↓       -31
016   STO6     35 06
017   RCL7     36 07
018   RCL3     36 03
019   GSB7     23 07
020   STO3     35 03
021     R↓       -31
022   STO7     35 07
023   RCL8     36 08
024   RCL4     36 04
025   GSB7     23 07
026   STO4     35 04
027     R↓       -31
028   STO8     35 08
029   GSB1     23 01
030   GSB4     23 04
031   GTO0     22 00
032  *LBLB     21 12
033   GSB1     23 01
034   RCL5     36 05
035   RCL1     36 01
036   GSB7     23 07
037   STO1     35 01
038     R↓       -31
039   STO5     35 05
040   RCL6     36 06
041   RCL2     36 02
042   GSB6     23 06
043   STO2     35 02
044     R↓       -31
045   STO6     35 06
046   RCL7     36 07
047   RCL3     36 03
048   GSB6     23 06
049   STO3     35 03
050     R↓       -31
051   STO7     35 07
052   RCL8     36 08
053   RCL4     36 04
054   GSB7     23 07
055   STO4     35 04
056     R↓       -31
057   STO8     35 08
058   GSB1     23 01
059   GTO0     22 00
060  *LBL4     21 04
061      1        01
062      8        08
063      0        00
064   ST-5  35-45 05
065   ST-6  35-45 06
066   ST-7  35-45 07
067   ST-8  35-45 08
068    RTN        24
069  *LBL1     21 01
070   RCL7     36 07
071   RCL3     36 03
072   RCL6     36 06
073   RCL2     36 02
074   GSB9     23 09
075    CHS       -22
076   STO0     35 00
077     R↓       -31
078   STO9     35 09
079   RCL5     36 05
080   RCL1     36 01
081   GSB7     23 07
082   STOB     35 12
083     R↓       -31
084   STOC     35 13
085   RCL8     36 08
086   RCL4     36 04
087   GSB7     23 07
088   RCLC     36 13
089   RCLB     36 12
090   GSB9     23 09
091   RCL9     36 09
092   RCL0     36 00
093   GSB8     23 08
094    1/X        52
095      2        02
096      x       -35
097   STO0     35 00
098    X⇄Y       -41
099    CHS       -22
100      1        01
101      8        08
102      0        00
103   STOI     35 46
104      -       -45
105   STO9     35 09
106    X⇄Y       -41
107   RCL6     36 06
108   RCL2     36 02
109   GSB9     23 09
110   STO2     35 02
111     R↓       -31
112   STO6     35 06
113   RCL9     36 09
114   RCL0     36 00
115   RCL7     36 07
116   RCL3     36 03
117   GSB9     23 09
118   STO3     35 03
119     R↓       -31
120   STO7     35 07
121   RCL8     36 08
122   RCL4     36 04
123   GSB7     23 07
124   RCL9     36 09
125   RCL0     36 00
126   GSB9     23 09
127   GSB7     23 07
128   STOD     35 14
129     R↓       -31
130   STOE     35 15
131   RCL5     36 05
132   RCL1     36 01
133   GSB7     23 07
134   RCL9     36 09
135   RCL0     36 00
136   GSB9     23 09
137   GSB7     23 07
138   STO4     35 04
139     R↓       -31
140   STO8     35 08
141   RCLD     36 14
142   STO1     35 01
143   RCLE     36 15
144   STO5     35 05
145   RCLI     36 46
146   ST-5  35-45 05
147   ST-8  35-45 08
148    RTN        24
149  *LBL7     21 07
```

Step	Key	Code
150	→R	44
151	1	01
152	+	-55
153	→P	34
154	RTN	24
155	*LBL6	21 06
156	→R	44
157	1	01
158	-	-45
159	→P	34
160	RTN	24
161	*LBL8	21 08
162	→R	44
163	R↓	-31
164	R↓	-31
165	→R	44
166	X⇄Y	-41
167	R↓	-31
168	+	-55
169	R↓	-31
170	+	-55
171	R↑	16-31
172	→P	34
173	RTN	24
174	*LBL9	21 09
175	X⇄Y	-41
176	R↓	-31
177	×	-35
178	R↓	-31
179	+	-55
180	R↑	16-31
181	RTN	24
182	*LBL0	21 00
183	RCL1	36 01
184	STO0	35 00
185	RCL5	36 05
186	STO1	35 01
187	RCL7	36 07
188	RCL3	36 03
189	P⇄S	16-51
190	STO2	35 02
191	R↓	-31
192	STO3	35 03
193	P⇄S	16-51
194	RCL8	36 08
195	RCL4	36 04
196	P⇄S	16-51
197	STO6	35 06
198	R↓	-31
199	STO7	35 07
200	P⇄S	16-51
201	RCL6	36 06
202	RCL2	36 02
203	STO4	35 04
204	R↓	-31
205	STO5	35 05
206	5	05
207	1	01
208	.	-62
209	2	02
210	RTN	24
211	R/S	51

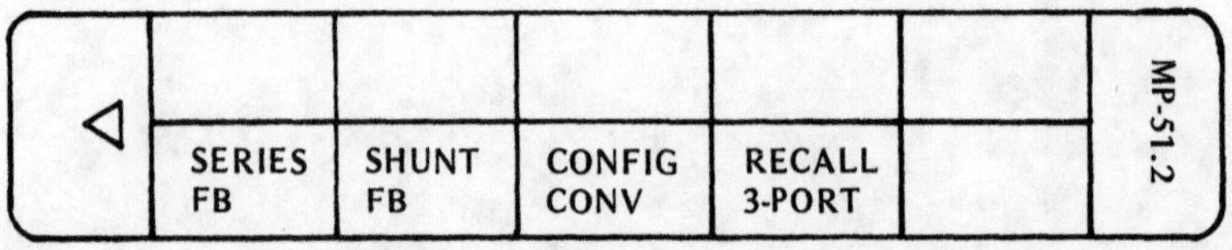

51.2

```
001  *LBLC     21 13
002  *LBLA     21 11
003   RCL5     36 05
004   RCL4     36 04
005    CHS       -22
006   RCL1     36 01
007   RCL0     36 00
008    CHS       -22
009   GSB8     23 08
010   GSB7     23 07
011   STO2     35 02
012     R↓       -31
013   STO3     35 03
014    P⇄S     16-51
015   RCL3     36 03
016   RCL2     36 02
017    CHS       -22
018    P⇄S     16-51
019   RCL1     36 01
020   RCL0     36 00
021    CHS       -22
022   GSB8     23 08
023   GSB7     23 07
024   STO6     35 06
025     R↓       -31
026   STO7     35 07
027    P⇄S     16-51
028   RCL7     36 07
029   RCL6     36 06
030    CHS       -22
031    P⇄S     16-51
032   RCL5     36 05
033   RCL4     36 04
034    CHS       -22
035   GSB8     23 08
036   GSB7     23 07
037    P⇄S     16-51
038   STO0     35 00
039     R↓       -31
040   STO1     35 01
041   RCL7     36 07
042   RCL6     36 06
043    CHS       -22
044   RCL3     36 03
045   RCL2     36 02
046    CHS       -22
047   GSB8     23 08
048   GSB7     23 07
049   STO4     35 04
050     R↓       -31
051   STO5     35 05
052   RCL1     36 01
053   RCL0     36 00
054    CHS       -22
055    P⇄S     16-51
056   RCL7     36 07
057   RCL6     36 06
058    CHS       -22
059   GSB8     23 08
060   GSB7     23 07
061   STO8     35 08
062     R↓       -31
063   STO9     35 09
064    RTN        24
065  *LBLB     21 12
066   RCL1     36 01
067   RCL0     36 00
068   RCL5     36 05
069   RCL4     36 04
070    CHS       -22
071   GSB8     23 08
072   GSB7     23 07
073   STO2     35 02
074     R↓       -31
075   STO3     35 03
076   RCL1     36 01
077   RCL0     36 00
078    P⇄S     16-51
079   RCL3     36 03
080   RCL2     36 02
081    CHS       -22
082   GSB8     23 08
083   GSB7     23 07
084    P⇄S     16-51
085   STO6     35 06
086     R↓       -31
087   STO7     35 07
088   RCL5     36 05
089   RCL4     36 04
090    P⇄S     16-51
091   RCL7     36 07
092   RCL6     36 06
093    CHS       -22
094   GSB8     23 08
095   GSB6     23 06
096   STO0     35 00
097     R↓       -31
098   STO1     35 01
099   RCL3     36 03
100   RCL2     36 02
101   RCL7     36 07
102   RCL6     36 06
103    CHS       -22
104   GSB8     23 08
105   GSB6     23 06
106   STO4     35 04
107     R↓       -31
108   STO5     35 05
109   RCL1     36 01
110   RCL0     36 00
111    CHS       -22
112    P⇄S     16-51
113   RCL7     36 07
114   RCL6     36 06
115   GSB8     23 08
116   GSB6     23 06
117   STO8     35 08
118     R↓       -31
119   STO9     35 09
120    RTN        24
121  *LBLD     21 14
122   RCL1     36 01
123   RCL0     36 00
124      1        01
125      1        01
126    R/S        51
127   RCL3     36 03
128   RCL2     36 02
129      1        01
130      3        03
131    R/S        51
132   RCL5     36 05
133   RCL4     36 04
134      1        01
135      2        02
136    R/S        51
137   RCL7     36 07
138   RCL6     36 06
139      3        03
140      1        01
141    R/S        51
142   RCL9     36 09
143   RCL8     36 08
144      3        03
145      3        03
146    R/S        51
147    P⇄S     16-51
148   RCL1     36 01
149   RCL0     36 00
```

```
P⇄S     16-51
3          03
2          02
R/S        51
P⇄S     16-51
RCL3    36 03
RCL2    36 02
P⇄S     16-51
2          02
1          01
R/S        51
P⇄S     16-51
RCL5    36 05
RCL4    36 04
P⇄S     16-51
2          02
3          03
R/S        51
P⇄S     16-51
RCL7    36 07
RCL6    36 06
P⇄S     16-51
2          02
2          02
RTN        24
0          00
1/X        52
*LBL7   21 07
→R         44
1          01
+         -55
→P         34
RTN        24
*LBL6   21 06
→R         44
1          01
-         -45
→P         34
RTN        24
*LBL8   21 08
→R         44
R↓        -31
R↓        -31
→R         44
X⇄Y       -41
R↓        -31
+         -55
R↓        -31
+         -55
R↑      16-31
→P         34
RTN        24
R/S        51
```

◁	FROM:	CE	CB	CC		MP-52.1
	TO:	CE	CB	CC		

52.1

```
*LBLb    21 16 12
RTN            24
*LBLc    21 16 13
SF0      16 21 00
RTN            24
*LBLd    21 16 14
SF1      16 21 01
RTN            24
*LBLB       21 12
F0?      16 23 00
GTO1        22 01
F1?      16 23 01
GTO2        22 02
GTO0        22 00
*LBLC       21 13
F1?      16 23 01
GTO4        22 04
F0?      16 23 00
GTO0        22 00
GTO1        22 01
*LBLD       21 14
F0?      16 23 00
GTO3        22 03
F1?      16 23 01
GTO0        22 00
GTO2        22 02
*LBL1       21 01
RCL1        36 01
RCL0        36 00
RCL9        36 09
RCL8        36 08
STO0        35 00
R↓            -31
STO1        35 01
R↓            -31
STO8        35 08
R↓            -31
STO9        35 09
RCL3        36 03
RCL2        36 02
RCL7        36 07
RCL6        36 06
STO2        35 02
R↓            -31
STO3        35 03
R↓            -31
STO6        35 06
R↓            -31
STO7        35 07
RCL5        36 05
RCL4        36 04
P⇄S         16-51
RCL1        36 01
RCL0        36 00
R↓            -31
R↓            -31
STO0        35 00
R↓            -31
STO1        35 01
P⇄S         16-51
R↓            -31
STO4        35 04
R↓            -31
STO5        35 05
P⇄S         16-51
RCL5        36 05
RCL4        36 04
RCL3        36 03
RCL2        36 02
STO4        35 04
R↓            -31
STO5        35 05
R↓            -31
STO2        35 02
R↓            -31
STO3        35 03
P⇄S         16-51
GTO0        22 00
*LBL2       21 02
RCL5        36 05
RCL4        36 04
RCL3        36 03
RCL2        36 02
STO4        35 04
R↓            -31
STO5        35 05
R↓            -31
STO2        35 02
R↓            -31
STO3        35 03
RCL7        36 07
RCL6        36 06
P⇄S         16-51
RCL3        36 03
RCL2        36 02
R↓            -31
R↓            -31
STO2        35 02
R↓            -31
STO3        35 03
P⇄S         16-51
R↓            -31
STO6        35 06
R↓            -31
STO7        35 07
RCL9        36 09
RCL8        36 08
P⇄S         16-51
RCL7        36 07
RCL6        36 06
R↓            -31
R↓            -31
STO6        35 06
R↓            -31
STO7        35 07
P⇄S         16-51
R↓            -31
STO8        35 08
R↓            -31
STO9        35 09
P⇄S         16-51
RCL1        36 01
RCL0        36 00
RCL5        36 05
RCL4        36 04
STO0        35 00
R↓            -31
STO1        35 01
R↓            -31
STO4        35 04
R↓            -31
STO5        35 05
P⇄S         16-51
GTO0        22 00
*LBL3       21 03
RCL0        36 00
STOB        35 12
RCL1        36 01
STOC        35 13
P⇄S         16-51
RCL6        36 06
STOD        35 14
RCL7        36 07
STOE        35 15
RCLB        36 12
STO6        35 06
RCLC        36 13
STO7        35 07
P⇄S         16-51
```

```
150  RCL8    36 08
151  ST00    35 00
152  RCL9    36 09
153  ST01    35 01
154   P⇄S    16-51
155  RCLE    36 15
156  RCLD    36 14
157   P⇄S    16-51
158  ST08    35 08
159    R↓      -31
160  ST09    35 09
161  RCL2    36 02
162  STOB    35 12
163  RCL3    36 03
164  STOC    35 13
165   P⇄S    16-51
166  RCL1    36 01
167  RCL0    36 00
168   P⇄S    16-51
169  ST02    35 02
170    R↓      -31
171  ST03    35 03
172   P⇄S    16-51
173  RCL2    36 02
174  ST00    35 00
175  RCL3    36 03
176  ST01    35 01
177  RCLB    36 12
178  ST02    35 02
179  RCLC    36 13
180  ST03    35 03
181  RCL4    36 04
182  STOB    35 12
183  RCL5    36 05
184  STOC    35 13
185   P⇄S    16-51
186  RCL5    36 05
187  RCL4    36 04
188   P⇄S    16-51
189  ST04    35 04
190    R↓      -31
191  ST05    35 05
192   P⇄S    16-51
193  RCL6    36 06
194  ST04    35 04
195  RCL7    36 07
196  ST05    35 05
197  RCLB    36 12
198  ST06    35 06
199  RCLC    36 13
200  ST07       35 07
201  *LBL0      21 00
202    CF0   16 22 00
203    CF1   16 22 01
204      5         05
205      2         02
206    RTN         24
207  *LBL4      21 04
208    CF0   16 22 00
209    CF1   16 22 01
210      5         05
211      2         02
212      .        -62
213      2         02
214    PSE      16 51
215    RTN         24
216    R/S         51
```

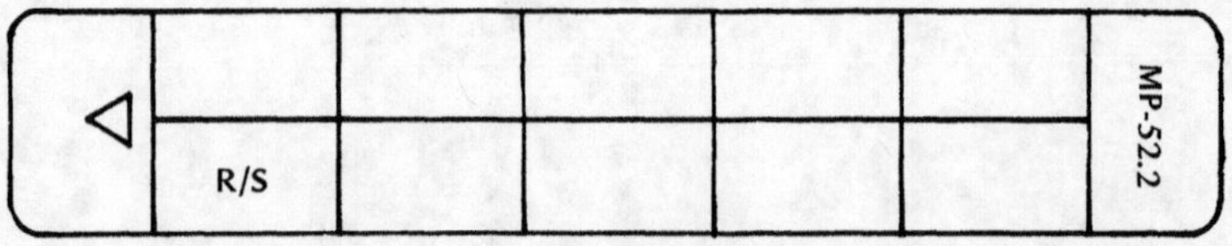

52.2

```
RCL8    36 08
STOB    35 12
RCL3    36 03
STOC    35 13
RCL0    36 00
STO8    35 08
RCL1    36 01
STO9    35 09
P⇄S     16-51
RCL7    36 07
RCL6    36 06
P⇄S     16-51
STO0    35 00
R↓        -31
STO1    35 01
P⇄S     16-51
RCLB    36 12
STO6    35 06
RCLC    36 13
STO7    35 07
RCL4    36 04
STOB    35 12
RCL5    36 05
STOC    35 13
P⇄S     16-51
RCL7    36 07
RCL6    36 06
P⇄S     16-51
STO4    35 04
R↓        -31
STO5    35 05
P⇄S     16-51
RCL4    36 04
STO6    35 06
RCL5    36 05
STO7    35 07
RCLB    36 12
STO4    35 04
RCLC    36 13
STO5    35 05
RCL2    36 02
STOB    35 12
RCL3    36 03
STOC    35 13
P⇄S     16-51
RCL3    36 03
RCL2    36 02
P⇄S     16-51
STO2    35 02
R↓        -31
STO3    35 03
P⇄S     16-51
RCL0    36 00
STO2    35 02
RCL1    36 01
STO3    35 03
RCLB    36 12
STO0    35 00
RCLC    36 13
STO1    35 01
5          05
2          02
P⇄S     16-51
R/S        51
```

◁	3-PORT ⇒ 2-PORT	RECALL 2-PORT				MP-53

53.0

```
001  *LBLA   21 11
002   RCL9   36 09
003   RCL8   36 08
004   GSB7   23 07
005    1/X      52
006   STOB   35 12
007    X⇄Y     -41
008    CHS     -22
009   STOC   35 13
010   RCL3   36 03
011   RCL2   36 02
012   RCL7   36 07
013   RCL6   36 06
014   GSB6   23 06
015   RCLC   36 13
016   RCLB   36 12
017    CHS     -22
018   GSB6   23 06
019   RCL1   36 01
020   RCL0   36 00
021   GSB8   23 08
022   STO0   35 00
023     R↓     -31
024   STO1   35 01
025   RCL3   36 03
026   RCL2   36 02
027    P⇄S   16-51
028   RCL1   36 01
029   RCL0   36 00
030    P⇄S   16-51
031   GSB6   23 06
032   RCLC   36 13
033   RCLB   36 12
034    CHS     -22
035   GSB6   23 06
036   RCL5   36 05
037   RCL4   36 04
038   GSB8   23 08
039   STO4   35 04
040     R↓     -31
041   STO5   35 05
042   RCL7   36 07
043   RCL6   36 06
044    P⇄S   16-51
045   RCL5   36 05
046   RCL4   36 04
047   GSB6   23 06
048   RCLC   36 13
049   RCLB   36 12
050    CHS     -22
051   GSB6   23 06
052   RCL3   36 03
053   RCL2   36 02
054   GSB8   23 08
055   STO2   35 02
056     R↓     -31
057   STO3   35 03
058   RCL5   36 05
059   RCL4   36 04
060   RCL1   36 01
061   RCL0   36 00
062   GSB6   23 06
063   RCLC   36 13
064   RCLB   36 12
065    CHS     -22
066   GSB6   23 06
067   RCL7   36 07
068   RCL6   36 06
069   GSB8   23 08
070   STO6   35 06
071     R↓     -31
072   STO7   35 07
073    P⇄S   16-51
074   RCL5   36 05
075   STO6   35 06
076   RCL1   36 01
077   STO5   35 05
078   RCL0   36 00
079   STO1   35 01
080   RCL4   36 04
081   STO2   35 02
082    P⇄S   16-51
083   RCL7   36 07
084   RCL6   36 06
085   RCL3   36 03
086   RCL2   36 02
087    P⇄S   16-51
088   STO3   35 03
089     R↓     -31
090   STO7   35 07
091     R↓     -31
092   STO4   35 04
093     R↓     -31
094   STO8   35 08
095      5      05
096      3      03
097    RTN      24
098  *LBLB   21 12
099   RCL5   36 05
100   RCL1   36 01
101      1      01
102      1      01
103    R/S      51
104   RCL6   36 06
105   RCL2   36 02
106      1      01
107      2      02
108    R/S      51
109   RCL7   36 07
110   RCL3   36 03
111      2      02
112      1      01
113    R/S      51
114   RCL8   36 08
115   RCL4   36 04
116      2      02
117      2      02
118    RTN      24
119  *LBL6   21 06
120    X⇄Y     -41
121     R↓     -31
122      ×     -35
123     R↓     -31
124      +     -55
125     R↑   16-31
126    RTN      24
127  *LBL7   21 07
128     →R      44
129      1      01
130      +     -55
131     →P      34
132    RTN      24
133  *LBL8   21 08
134     →R      44
135     R↓     -31
136     R↓     -31
137     →R      44
138    X⇄Y     -41
139     R↓     -31
140      +     -55
141     R↓     -31
142      +     -55
143     R↑   16-31
144     →P      34
145    RTN      24
146    R/S      51
```

▷	MAP:	CHART				MP-54.1
	MAP:	S_{11}	S_{12}	S_{21}	S_{22}	

54.1

```
001  *LBLB     21 12
002   GSBe  23 16 15
003   GTOc  22 16 13
004  *LBLC     21 13
005   GSB0     23 00
006   GSB4     23 04
007   GSBe  23 16 15
008   GSB0     23 00
009   GSB4     23 04
010   GTOc  22 16 13
011  *LBLD     21 14
012   GSB1     23 01
013   GSB3     23 03
014   GSBe  23 16 15
015   GSB1     23 01
016   GSB3     23 03
017   GTOc  22 16 13
018  *LBLE     21 15
019   GSB2     23 02
020   GSB3     23 03
021   GSB4     23 04
022   GSBe  23 16 15
023   GSB2     23 02
024   GSB3     23 03
025   GSB4     23 04
026  *LBLc  21 16 13
027      5        05
028      4        04
029      .       -62
030      2        02
031    RTN        24
032  *LBLe  21 16 15
033   RCL9     36 09
034   RCL8     36 08
035    CHS       -22
036   GSB7     23 07
037    1/X        52
038   STOB     35 12
039    X⇄Y       -41
040    CHS       -22
041   STOC     35 13
042   RCL9     36 09
043    CHS       -22
044   RCL8     36 08
045    CHS       -22
046   GSB7     23 07
047   RCLC     36 13
048   RCLB     36 12
049   GSB8     23 08
050   RCL7     36 07
051   RCL6     36 06
052   GSB8     23 08
053   RCL3     36 03
054   RCL2     36 02
055   GSB8     23 08
056      1        01
057   RCL8     36 08
058     X²        53
059      -       -45
060      ÷       -24
061     →R        44
062    P⇄S     16-51
063   STO8     35 08
064     R↓       -31
065   STO9     35 09
066    P⇄S     16-51
067   RCL9     36 09
068    CHS       -22
069   RCL8     36 08
070   RCL7     36 07
071   RCL6     36 06
072   GSB8     23 08
073   RCL3     36 03
074   RCL2     36 02
075   GSB8     23 08
076      1        01
077   RCL8     36 08
078     X²        53
079      -       -45
080      ÷       -24
081   RCL1     36 01
082   RCL0     36 00
083   GSB9     23 09
084     →R        44
085   STOD     35 14
086     R↓       -31
087   STOE     35 15
088   RCL9     36 09
089   RCL8     36 08
090   GSB7     23 07
091   RCLC     36 13
092   RCLB     36 12
093   GSB8     23 08
094     →R        44
095   STOB     35 12
096     R↓       -31
097   STOC     35 13
098    RTN        24
099  *LBL0     21 00
100   RCL0     36 00
101   RCL4     36 04
102   STO0     35 00
103     R↓       -31
104   STO4     35 04
105   RCL1     36 01
106   RCL5     36 05
107   STO1     35 01
108     R↓       -31
109   STO5     35 05
110    RTN        24
111  *LBL1     21 01
112    SF1  16 21 01
113  *LBL2     21 02
114   RCL1     36 01
115   RCL0     36 00
116    P⇄S     16-51
117    F1?  16 23 01
118   GTO5     22 05
119   RCL7     36 07
120   RCL6     36 06
121     R↓       -31
122     R↓       -31
123   STO6     35 06
124     R↓       -31
125   STO7     35 07
126   GTO6     22 06
127  *LBL5     21 05
128   RCL3     36 03
129   RCL2     36 02
130     R↓       -31
131     R↓       -31
132   STO2     35 02
133     R↓       -31
134   STO3     35 03
135  *LBL6     21 06
136     R↓       -31
137    P⇄S     16-51
138   STO0     35 00
139     R↓       -31
140   STO1     35 01
141    CF1  16 22 01
142    RTN        24
143  *LBL3     21 03
144   RCL3     36 03
145   RCL2     36 02
146    P⇄S     16-51
147   RCL5     36 05
148   RCL4     36 04
149     R↓       -31
```

```
150     R↓       -31
151   ST04     35 04
152     R↓       -31
153   ST05     35 05
154     R↓       -31
155    P⇄S     16-51
156   ST02     35 02
157     R↓       -31
158   ST03     35 03
159    RTN        24
160  *LBL4     21 04
161   RCL7     36 07
162   RCL6     36 06
163    P⇄S     16-51
164   RCL1     36 01
165   RCL0     36 00
166     R↓       -31
167     R↓       -31
168   ST00     35 00
169     R↓       -31
170   ST01     35 01
171     R↓       -31
172    P⇄S     16-51
173   ST06     35 06
174     R↓       -31
175   ST07     35 07
176    RTN        24
177  *LBL7     21 07
178     →R        44
179      1        01
180      +       -55
181     →P        34
182    RTN        24
183  *LBL8     21 08
184    X⇄Y       -41
185     R↓       -31
186      ×       -35
187     R↓       -31
188      +       -55
189     R↑     16-31
190    RTN        24
191  *LBL9     21 09
192     →R        44
193     R↓       -31
194     R↓       -31
195     →R        44
196    X⇄Y       -41
197     R↓       -31
198      +       -55
199     R↓       -31
200      +       -55
201     R↑     16-31
202     →P        34
203    RTN        24
204  *LBLb  21 16 12
205      0        00
206   STOD     35 14
207   STOE     35 15
208      1        01
209   STOB     35 12
210      0        00
211   STOC     35 13
212    P⇄S     16-51
213   STO9     35 09
214      1        01
215   STO8     35 08
216    P⇄S     16-51
217   GTOc  22 16 13
218    R/S        51
```

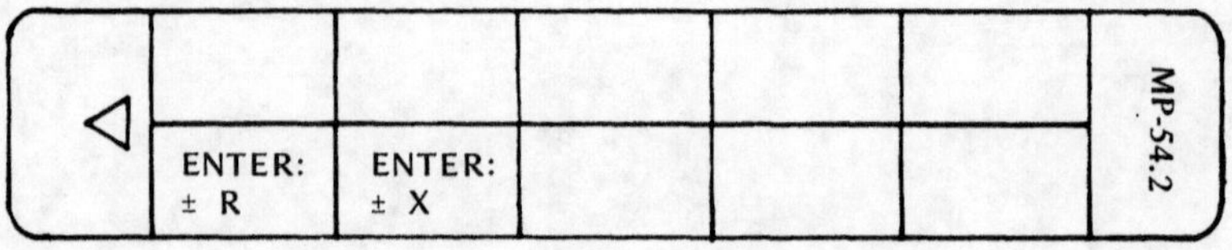

54.2

```
*LBLA     21 11
5            05
0            00
÷           -24
RCLB      36 12
÷           -24
STOI      35 46
P⇄S       16-51
RCL9      36 09
RCL8      36 08
→P           34
1            01
RCLI      36 46
+           -55
÷           -24
ABS       16 31
STOA      35 11
RCL9      36 09
RCL8      36 08
→P           34
P⇄S       16-51
RCLI      36 46
×           -35
1            01
RCLI      36 46
+           -55
÷           -24
RCLE      36 15
RCLD      36 14
→P           34
GSB9      23 09
RCLI      36 46
RCLB      36 12
×           -35
5            05
0            00
×           -35
RCLA      36 11
R↓          -31
RTN          24
*LBLB     21 12
5            05
0            00
÷           -24
RCLC      36 13
+           -55
1/X          52
RCLB      36 12
×           -35
STOI      35 46
P⇄S       16-51
RCL9      36 09
RCL8      36 08
→P           34
RCLI      36 46
×           -35
ABS       16 31
STOA      35 11
RCLI      36 46
1            01
→P           34
RCL9      36 09
RCL8      36 08
→P           34
GSB8      23 08
P⇄S       16-51
RCLE      36 15
RCLD      36 14
→P           34
GSB9      23 09
RCLI      36 46
RCLB      36 12
÷           -24
1/X          52
RCLC      36 13
-           -45
5            05
0            00
×           -35
RCLA      36 11
R↓          -31
RTN          24
*LBL8     21 08
X⇄Y         -41
R↓          -31
×           -35
R↓          -31
+           -55
R↑        16-31
RTN          24
*LBL9     21 09
→R           44
R↓          -31
R↓          -31
→R           44
X⇄Y         -41
R↓          -31
+           -55
R↓          -31
+           -55
R↑        16-31
→P           34
RTN          24
R/S          51
```

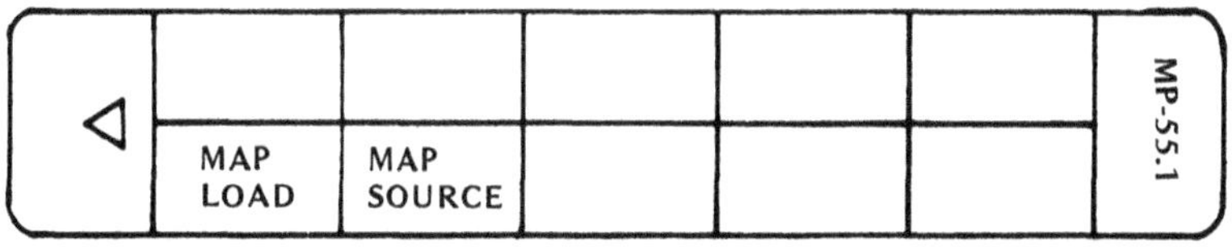


```
55.1

001  *LBLB       21 12
002    SF1    16 21 01
003  *LBLA       21 11
004    F1?    16 23 01
005   GSB2       23 02
006   RCL5       36 05
007   RCL1       36 01
008   GSB4       23 04
009   RCL7       36 07
010   RCL3       36 03
011   GSB5       23 05
012   RCL6       36 06
013   RCL2       36 02
014   GSB6       23 06
015   RCL8       36 08
016   RCL4       36 04
017   GSB3       23 03
018  *LBL0       21 00
019    F1?    16 23 01
020   GSB2       23 02
021    CF1    16 22 01
022    P⇄S       16-51
023   RCL5       36 05
024   RCL4       36 04
025   RCL3       36 03
026   RCL2       36 02
027   GSB9       23 09
028   STO2       35 02
029    X⇄Y         -41
030   STO3       35 03
031   RCL7       36 07
032   RCL6       36 06
033    CHS         -22
034   GSB7       23 07
035    1/X          52
036   STOA       35 11
037     R↓         -31
038    CHS         -22
039   STOB       35 12
040   RCL7       36 07
041    CHS         -22
042   RCL6       36 06
043    CHS         -22
044   GSB7       23 07
045   RCLB       36 12
046   RCLA       36 11
047   GSB9       23 09
048   RCL3       36 03
049   RCL2       36 02
050   GSB9       23 09
051      1          01
052   RCL6       36 06
053     X²          53
054      -         -45
055   STOI       35 46
056      ÷         -24
057     →R          44
058   STO8       35 08
059     R↓         -31
060   STO9       35 09
061   RCL7       36 07
062    CHS         -22
063   RCL6       36 06
064   RCL3       36 03
065   RCL2       36 02
066   GSB9       23 09
067   RCLI       36 46
068      ÷         -24
069   RCL1       36 01
070   RCL0       36 00
071   GSB8       23 08
072     →R          44
073   STOD       35 14
074     R↓         -31
075   STOE       35 15
076   RCL7       36 07
077   RCL6       36 06
078   GSB7       23 07
079   RCLB       36 12
080   RCLA       36 11
081   GSB9       23 09
082     →R          44
083   STOB       35 12
084     R↓         -31
085   STOC       35 13
086    P⇄S       16-51
087      5          05
088      5          05
089      .         -62
090      2          02
091    RTN          24
092  *LBL2       21 02
093   RCL1       36 01
094   RCL4       36 04
095   STO1       35 01
096     R↓         -31
097   STO4       35 04
098   RCL2       36 02
099   RCL3       36 03
100   STO2       35 02
101     R↓         -31
102   STO3       35 03
103   RCL5       36 05
104   RCL8       36 08
105   STO5       35 05
106     R↓         -31
107   STO8       35 08
108   RCL6       36 06
109   RCL7       36 07
110   STO6       35 06
111     R↓         -31
112   STO7       35 07
113    RTN          24
114  *LBL3       21 03
115    P⇄S       16-51
116   STO6       35 06
117     R↓         -31
118   STO7       35 07
119     R↑       16-31
120    P⇄S       16-51
121    RTN          24
122  *LBL4       21 04
123    P⇄S       16-51
124   STO0       35 00
125     R↓         -31
126   STO1       35 01
127     R↑       16-31
128    P⇄S       16-51
129    RTN          24
130  *LBL5       21 05
131    P⇄S       16-51
132   STO2       35 02
133     R↓         -31
134   STO3       35 03
135     R↑       16-31
136    P⇄S       16-51
137    RTN          24
138  *LBL6       21 06
139    P⇄S       16-51
140   STO4       35 04
141     R↓         -31
142   STO5       35 05
143     R↑       16-31
144    P⇄S       16-51
145    RTN          24
146  *LBL7       21 07
147     →R          44
148      1          01
149      +         -55
```

```
150     ÷P       [illegible]
151    RTN       [illegible]
152  *LBL8       21 08
153     ÷R          44
154     R↓         -31
155     R↓         -31
156     ÷P          44
157    X⇄Y         -41
158     R↓         -31
159      +         -55
160     R↓         -31
161      +         -55
162     R↑      16-31
163     ÷P          34
164    RTN          24
165  *LBL9       21 09
166    X⇄Y         -41
167     R↓         -31
168      X         -35
169     R↓         -31
170      +         -55
171     R↑      16-31
172    RTN          24
173    R/S          51
```

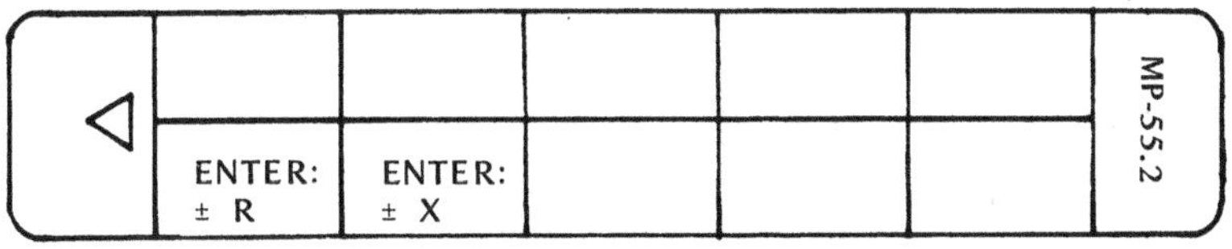


```
*LBLA     21 11
5            05
0            00
÷           -24
RCLB      36 12
÷           -24
STOI      35 46
P⇄S       16-51
RCL9      36 09
RCL8      36 08
→P           34
1            01
RCLI      36 46
+           -55
÷           -24
ABS       16 31
STOA      35 11
RCL9      36 09
RCL8      36 08
→P           34
P⇄S       16-51
RCLI      36 46
×           -35
1            01
RCLI      36 46
+           -55
÷           -24
RCLE      36 15
RCLD      36 14
→P           34
GSB9      23 09
RCLI      36 46
RCLB      36 12
×           -35
5            05
0            00
×           -35
RCLA      36 11
R↓          -31
RTN          24
*LBLB     21 12
5            05
0            00
÷           -24
RCLC      36 13
+           -55
1/X          52
RCLB      36 12
×           -35
STOI      35 46
P⇄S       16-51
RCL9      36 09
RCL8      36 08
→P           34
RCLI      36 46
×           -35
ABS       16 31
STOA      35 11
RCLI      36 46
1            01
→P           34
RCL9      36 09
RCL8      36 08
→P           34
GSB8      23 08
P⇄S       16-51
RCLE      36 15
RCLD      36 14
→P           34
GSB9      23 09
RCLI      36 46
RCLB      36 12
÷           -24
1/X          52
RCLC      36 13
-           -45
5            05
0            00
×           -35
RCLA      36 11
R↓          -31
RTN          24
*LBL8     21 08
X⇄Y         -41
R↓          -31
×           -35
R↓          -31
+           -55
R↑        16-31
RTN          24
*LBL9     21 09
→R           44
R↓          -31
R↓          -31
→R           44
X⇄Y         -41
R↓          -31
+           -55
R↓          -31
+           -55
R↑        16-31
→P           34
RTN          24
R/S          51
```

◁	TRL	$Z \Rightarrow \Gamma$ Θ, Z	USE $Z\angle\Theta$ Z, Θ		USE PA	MP-61
	SE R	SE SLC	SE PLC	SE SST	SE OST	

```
001  *LBL1     21 01
002   GSBb  23 16 12
003   RCLA     36 11
004    R/S        51
005  *LBL2     21 02
006   RCLB     36 12
007   RCLC     36 13
008   RCLA     36 11
009      +       -55
010   STOA     35 11
011   X≤Y?     16-35
012   GTO1     22 01
013    R/S        51
014  *LBLA     21 11
015      0        00
016   GTO6     22 06
017  *LBLC     21 13
018    SF1  16 21 01
019  *LBLB     21 12
020    EEX       -23
021      3        03
022    X⇄Y       -41
023      ÷       -24
024   RCLA     36 11
025      2        02
026      ×       -35
027     Pi     16-24
028      ×       -35
029   STOI     35 46
030      ÷       -24
031    F1?  16 23 01
032    1/X        52
033    X⇄Y       -41
034   RCLI     36 46
035      ×       -35
036    F1?  16 23 01
037    1/X        52
038    X⇄Y       -41
039      -       -45
040    F1?  16 23 01
041    1/X        52
042   GTO5     22 05
043  *LBLE     21 15
044    CHS       -22
045   GSB7     23 07
046      ÷       -24
047   GTO5     22 05
048  *LBLD     21 14
049   GSB7     23 07
050      ×       -35
051  *LBL5     21 05
052      9        09
053      0        00
054  *LBL6     21 06
055    X⇄Y       -41
056    F0?  16 23 00
057   GSB8     23 08
058   RCL2     36 02
059   RCL1     36 01
060    F0?  16 23 00
061   GSB8     23 08
062   GSB9     23 09
063    F0?  16 23 00
064   GSB8     23 08
065    CF0  16 22 00
066    CF1  16 22 01
067   STO1     35 01
068    X⇄Y       -41
069   STO2     35 02
070    X⇄Y       -41
071    RTN        24
072  *LBL7     21 07
073      3        03
074      0        00
075      .       -62
076      4        04
077      8        08
078      ×       -35
079   RCLA     36 11
080      ×       -35
081    TAN        43
082    RTN        24
083  *LBL8     21 08
084    1/X        52
085    X⇄Y       -41
086    CHS       -22
087    X⇄Y       -41
088    RTN        24
089  *LBL9     21 09
090     →R        44
091     R↓       -31
092     R↓       -31
093     →R        44
094    X⇄Y       -41
095     R↓       -31
096      +       -55
097     R↓       -31
098      +       -55
099     R↑     16-31
100     →P        34
101    RTN        24
102  *LBLa  21 16 11
103   GSB7     23 07
104   STOI     35 46
105   RCL2     36 02
106   RCL1     36 01
107     →R        44
108   STO1     35 01
109    X⇄Y       -41
110   STO2     35 02
111     R↓       -31
112      ×       -35
113     R↑     16-31
114   RCLI     36 46
115      ×       -35
116    CHS       -22
117     R↑     16-31
118      +       -55
119     →P        34
120    1/X        52
121     R↑     16-31
122      ×       -35
123     R↑     16-31
124   RCLI     36 46
125      ×       -35
126   RCL2     36 02
127      +       -55
128   RCL1     36 01
129     →P        34
130    X⇄Y       -41
131     R↑     16-31
132      -       -45
133   STO2     35 02
134     R↓       -31
135      ×       -35
136   STO1     35 01
137     R↑     16-31
138    X⇄Y       -41
139    RTN        24
140  *LBLc  21 16 13
141   GTO6     22 06
142  *LBL0     21 00
143     R↓       -31
144    CHS       -22
145    R/S        51
146    X⇄Y       -41
147   GSB9     23 09
148     X²        53
149   ST+0  35-55 00
```

```
150     RTN         24
151   *LBLb   21 16 12
152       5         05
153       0         00
154       ÷        -24
155      →R         44
156       1         01
157       -        -45
158      →P         34
159    GSB8      23 08
160       2         02
161       ×        -35
162      →R         44
163       1         01
164       +        -55
165      →P         34
166    GSB8      23 08
167     RTN         24
168   *LBLe   21 16 15
169     SF0   16 21 00
170     RTN         24
171     R/S         51
```

NOTES

NOTES

NOTES

NOTES

NOTES

NOTES

NOTES

www.ingramcontent.com/pod-product-compliance
Lightning Source LLC
LaVergne TN
LVHW080030210926
848481LV00011B/3151